# Reducing Poverty, Protecting Livelihoods, and Building Assets in a Changing Climate

Analyzing the social impacts of climate change in Latin America and the Caribbean, this book presents evidence that we must improve our efforts on resilience and adaptation measures to counter the consequences of climate change on the most vulnerable population groups. The model developed in the book illustrates the necessity of addressing climate change in its proper developmental context. This model may well be used on other continents; in particular, it may be relevant in Africa.
—**Søren Pind,** *Minister for Development Cooperation, Denmark*

This book brings an important contribution for our understanding of how climate change and variability impact society and livelihoods and in particular on how the poor are affected. It should be read by everyone who has a responsibility in development policies and in capacity building in LAC.
—**Antonio Magalhães,** *Former Secretary of Planning, Brazil*

This immensely important book provides one of the earliest evidence-based assessments of climate threats to development. Its arguments and organization will be a model for future efforts to understand and address climate impacts on development outcomes.
—**Arun Agrawal,** *Professor and Associated Dean University of Michigan, USA*

Democratization and institutional reform and strengthening was, during the last three decades, the Latin American way to respond to the challenges of coping with severe macroeconomic imbalances, recovering growth, and reducing poverty. This important book shows that this process has to be deepened to deal with the social consequences of climate change.
—**Andre Urani,** *President of Instituto Natura, Brazil*

It is right to focus on the impact of climatic change on poor people. We all need to be cognizant of the fact that those who have contributed the least to the depletion of natural resources and to the changes in climate patterns are those who will suffer the most from unsustainable human ecosystems.
—**Lauritz B. Holm-Nielsen,** *Rektor at University of Aarhus, Denmark*

# Reducing Poverty, Protecting Livelihoods, and Building Assets in a Changing Climate

*Social Implications of Climate Change in Latin America and the Caribbean*

Dorte Verner, Editor

**THE WORLD BANK**
**Washington, DC**

© 2010 The International Bank for Reconstruction and Development / The World Bank
1818 H Street NW
Washington DC 20433
Telephone: 202-473-1000
Internet: www.worldbank.org
E-mail: feedback@worldbank.org

This volume is a product of the staff of the International Bank for Reconstruction and Development / The World Bank. The findings, interpretations, and conclusions expressed in this volume do not necessarily reflect the views of the Executive Directors of The World Bank or the governments they represent. The World Bank does not guarantee the accuracy of the data included in this work.

The maps in this book were produced by the Map Design Unit of The World Bank. The boundaries, colors, denominations, and any other information shown on these maps do not imply, on the part of The World Bank Group, any judgement on the legal status of any territory, or any endorsement or acceptance of such boundaries.

ISBN: 978-0-8213-8238-7
eISBN: 978-0-8213-8378-0
DOI: 10.1596/978-0-8213-8238-7

Cover photos: Dorte Verner
Cover design: Naylor Design, Inc.

**Library of Congress Cataloging-in-Publication Data has been requested.**

# Contents

**Boxes**

**Figures**

**Tables**

# Foreword

This book provides a much needed look at the impact of climate change on the poor. It convincingly demonstrates that issues of poverty and livelihoods must be integrated into climate change policies to help achieve sustainable development gains.

The high incidence of natural disasters, growing urbanization, and increased water scarcity—combined with the acute impact of these phenomena on the poor and vulnerable—complicates the already enormous challenge of reducing poverty and inequality in Latin America and the Caribbean. This publication lays bare the social implications of climate change and equips the reader with a framework for understanding how climate change and climate variability affect livelihoods, poverty, income, health, and migration.

Based on a study carried out by a multidisciplinary team in the Sustainable Development department, this book's timely analysis complements the regional flagship report *Low Carbon, High Growth*. It describes much-needed policy options on climate change adaptation.

Without appropriate adaptive responses, the impacts of climate change on the vulnerable will be severe in the region. Key findings show that changes in water flows brought about by climate change could place 70 percent of the Andean population at risk of fresh water scarcity by 2020. Furthermore, the prevalence of some vector- and waterborne diseases could grow two- to five-fold in parts of South America, while the residents of some of the poorest areas in the region could experience a reduction in income of more than 20 percent by 2050.

These scenarios call for greater efforts to incorporate poverty, livelihood, and social considerations into climate change adaptation and mitigation policies. Purposefully targeted policies and investments can support economic growth and poverty reduction efforts and help achieve sustainable development goals. In other words, good climate change adaptation policies can also be good development policies.

This book will change the way we think about the relationship between poverty, social development, and climate change. It provides climate-smart policy options to help reduce vulnerability, protect livelihoods, and build communities that are resilient to changing climate conditions. And it gives us hope that if we act now, act together, and act differently, we can successfully meet the defining environmental challenge of our generation while sustaining development for all.

**Pamela Cox**
Vice President
Latin America and Caribbean Region, World Bank

# Acknowledgments

The report and team were developed and managed by Dorte Verner (Team Leader). The team members included Lykke Andersen, Inger Brisson, Jens Hesselbjerg Christensen, John Bjerg Geary, Jakob Kronik, Lotte Lund, Sara Trab Nielsen, Jørgen Eivind Olesen, Benjamin Orlove, Claus Pörtner, Tine Rossing, and Olivier Rubin, and Sanne Agnete Tikjøb.

Special thanks to the team of advisers: Jocelyne Albert, Shelton Davis, Estanislao Gacitua-Mario, Andrew Norton, Walter Vergara, and Alonso Zarzar.

The team is grateful to peer reviewers Kirk Hamilton and Andrea Liverani for comments and suggestions. It also gratefully acknowledges helpful comments and suggestions from Anjali Acharya, Arun Agrawal, Maximilian Shen Ashwill, James Azueta, Herman Belmar, McDonald Benjamin, Winston Bennett, Carter Brandon, Rita Cestti, Alejandro Deeb, Pablo Fajnzylber, Dennis Garbutt, Maninder Gill, Jose Gutierrez, Gillette Hall, Marea Hatziolos, Rasmus Heltberg, Willem Janssen, Ottis Joslyn, Michel Kerf, Pilar Larreamendy, Celia Mahung, Alexandre Marc, Melanie McField, Robin Mearns, Augusta Molnar, Edmundo Murrugarra, John Nash, Frode Neergaard, Nicolas Perrin, Howie Prince, Gustavo Santiel, Emmanuel Skoufias, and Senator Eddie Webster, of Stann Creek District,

Belize. The team is grateful to the Latin America and Caribbean Region management team and especially to Pamela Cox, Laura Tuck, McDonal Benjamin, and Maninder Gill.

The team would also like to thank Ramon Anria and Jorge Hunt for managing all the paperwork very effectively and Rachel Weaving for editing the study. Moreover, the team would like to thank Pat Katayama, Rick Ludwick, and Denise Bergeron, of the World Bank Office of the Publisher, and Jeff Lecksell, of the Bank's General Services Division, Cartographic Services. Finally, the World Bank is grateful for financial support from DANIDA, Denmark.

# Abbreviations

| | |
|---|---|
| 4th AR | Fourth Assessment Report |
| AdapCC | Adaptation for Smallholders to Climate Change |
| ALA | Alaska |
| AMZ | Amazonia |
| ANT | Antarctic |
| AOGCM | *Atmosphere-Ocean Coupled General Circulation Model* |
| BAMS | Bulletin of American Meteorological Society |
| CAM | Central America |
| CAN | Central North America |
| CAS | Central Asia |
| CC | climate change |
| CCER | Civic Coalition for Emergency and Reconstruction |
| CDM | Clean Development Mechanism |
| CARICOM | Caribbean Community |
| CCER | Civic Coalition for Emergency and Reconstruction |
| CHA | Caribbean Hotel Association |
| CNI | Congreso Nacional Indígena |
| $CO_2$ (1) | carbon dioxide |
| COGERH | Companhia de Gestão dos Recursos Hídricos |
| CONAIG | Interinstitucional del Agua |

| | |
|---|---|
| CONAGUA | The National Water Commission of Mexico |
| COTAS | Technical Committees for Groundwater |
| CSUTCB | United Union Confederation of Campesino Workers of Bolivia |
| CTO | Caribbean Tourism Organization |
| DALY | disability-adjusted life years |
| DEMO | District Emergency Management Organization |
| DFID | Department For International Development |
| DHS | demographic health survey |
| DJF | December, January, February |
| DMI (2) | Danish Meteorological Institute |
| EAF | East Africa |
| EAS | Eastern Asia |
| EIA | environmental impact assessment |
| ENA | Eastern North America |
| ENSO | El Niño/Southern Oscillation |
| EMIC | Earth models of Intermediate Complexity |
| EZLN | Ejército Zapatista de Liberación Nacional |
| FAO | Food and Agricultural Organization |
| FSLN | Frente Sandinista de Liberación Nacional |
| GCM | General Circulation Model |
| GDP (1) | gross domestic product |
| GHG (1) | greenhouse gas |
| GRL | Greenland |
| IADB | Inter-American Development Bank |
| IAS | Intra-Americas Seas |
| IBGE | Instituto Brasileiro de Geografia e Estatística |
| ICZM | Integrated Coastal Zone Management |
| IDEAM | Instituto de Hidrologia, Meteorologia y Estudios Ambientales |
| IO | international organization |
| IPCC | Intergovernmental Panel on Climate Change |
| ISDR | International Strategy for Disaster Reduction |
| ITCZ | Inter-Tropical Convergence Zone |
| IWRM | integrated water resource management |
| JJA | June, July, August |
| LAC (1) | Latin America and the Caribbean (geographical location) |
| LCR | The Latin American and Caribbean Region (unit in World Bank) |

| | |
|---|---|
| LECZ | Low Elevation Coastal Zones |
| LLJ | low level jet |
| MAS | Movement toward Socialism |
| MASL | meters above sea level |
| MCDW | Monthly Climatic Data for the World |
| MED | Mediterranean |
| Mercosur | Mercado Comun del Sur |
| MIC | middle-income countries |
| MoSSaic | Management of Slope Stability in Communities |
| MPA | marine protected area |
| MSD | mid-summer drought |
| NARCCAP | North American Regional Climate Change Assessment Program |
| NAS | Northern Asia |
| NAU | Northern Australia |
| NAFTA | North American Free Trade Agreement |
| NAPA | National Adaptation Programmes of Action |
| NCC | no climate change |
| NCDC | National Climatic Data Center |
| NEMO | National Emergency Management Organization |
| NEU | Northern Europe |
| NGO | nongovernmental organization |
| NMACC | National Mechanism of Adaptation to Climatic Change |
| NOAA | National Oceanic and Atmospheric Administration |
| OECD | Organization for Economic Cooperation and Development |
| PHMR | Port Honduras Marine Reserve |
| PLC | Partido Liberal Constituyente |
| PRUDENCE | Prediction of Regional scenarios and Uncertainties for Defining European Climate change risks and Effects |
| PSIA | Poverty and Social Impact Analysis |
| RAAN | North Atlantic Autonomous Region |
| RAAS | South Atlantic Autonomous Region |
| RCM | regional climate model |
| RMA | risk management approach |
| SACZ | South Atlantic Convergence Zone |
| SAF | Southern Africa |
| SAH | Sahara |
| SAM | Southern Annual Mode |

| | |
|---|---|
| SAMS | South American Monsoon System |
| SAS | Southern Asia |
| SAU | Southern Australia |
| SEA | South East Asia |
| SEAR | Sistema Educativo Autonómico Regional |
| SENAMHI | Peru's National Meteorology and Hydrology Service |
| SINCHI | Instituto de Investigaciones de la Amazonía |
| SLF | Sustainable Livelihoods Framework |
| SLR | sea level rise |
| SRES | Special Report on Emissions Scenarios |
| SSA | Southern South America |
| SST | sea surface temperature |
| TIB | Tibet |
| TIDE | Toledo Institute for Environment and Development |
| UNWTO | United Nations World Tourism Organization |
| YLD | years of life with disability |
| YLL | years of life lost |
| VIDECICODI | Vice-Minister of Civil Defense and Cooperation of Integrated Development |
| WAF | West Africa |
| WHO | World Health Organization |
| WNA | Western North America |
| WMO | World Meteorological Organization |
| WWF | World Wildlife Fund |

# CHAPTER 1

# Introduction

## Dorte Verner

*"Why are the mountains crying?"*

~Aymara women, Bolivia

Climate change is the defining development challenge of our time. More than a global environmental issue, climate change is also a threat to poverty reduction and economic growth and may unravel many of the development gains made in recent decades. Both now and over the long run, climate change and variability threaten human and social development by restricting the fulfillment of human potential and disempowering people and communities, constraining their ability to protect and enrich their livelihoods.[1]

Latin America and the Caribbean account for a relatively modest 12 percent of the world's greenhouse gas (GHG) emissions,[2] but communities across the region are already suffering adverse consequences from climate change and variability (De la Torre, Fajnzylber, and Nash 2009). Looking ahead, climate change is likely to have unprecedented social, economic, environmental, and political repercussions. Already precipitation has increased in the southeastern part of South America and now often comes in the form of sudden deluges, leading to flooding and soil erosion, that endanger people's lives and livelihoods. Southwestern parts of

South America and western Central America are seeing decreases in precipitation and an increase in droughts. The Andean intertropical glaciers are shrinking and are expected to disappear altogether within the next 20 to 40 years. Although this process will lead to increased water flow in the short term and indeed pose a risk of glacier lake outbursts and flooding, in the longer term the glaciers' disappearance will severely curtail the supply of water for drinking, irrigation, and power generation (box 1.1). For Latin America and the Caribbean (LAC) as a whole, estimates of the cost of damage due to climate change and climatic variability vary from 1.3 percent to 7 percent of gross domestic product (GDP)

**Box 1.1**

## The Melting Glaciers in Bolivia: A Threat to People's Water Supply

Chacaltaya was famous for being the world's highest ski run, at 5,300 meters. In the last two decades it has lost 80 percent of its area. The Tuni Condoriri Mountains' 15 original glaciers have been reduced by more than 30 percent since 1983. Five may already have disappeared.

The Tuni Condoriri, roughly an hour away from La Paz, hold a reservoir that provides about 80 percent of the drinking water of El Alto and large parts of La Paz, home to nearly two million people. If the glaciers were stable, they would not contribute to the net amount of water available; they would absorb as much water in the form of ice and snow as they would release. They would, however, play an important role in regulating the flow of water and evenly distributing the supply over the year. But in the short run, as the glaciers melt it can be expected that the water supply will increase. It is estimated that the glaciers will completely disappear at some point between 2025 and 2050.

Bolivia has a pressing need for more water. Around 12,000 migrants from rural areas arrive in El Alto every year in search of better economic opportunities or in pursuit of education, raising the demand for water. The water supply has been able to keep up with demand over recent decades only because of the boost in supply from the melting of the glaciers. But even with the current increased supply, Bolivia's leading hydrologist, Edson Ramirez, predicted that as early as 2009 demand would exceed the supply of water available in the reservoirs. Once the glaciers are gone, the boost to the water supply will also disappear, and water shortages can be expected because precipitation is insufficient to meet current demand. With the glaciers' loss

*(continued)*

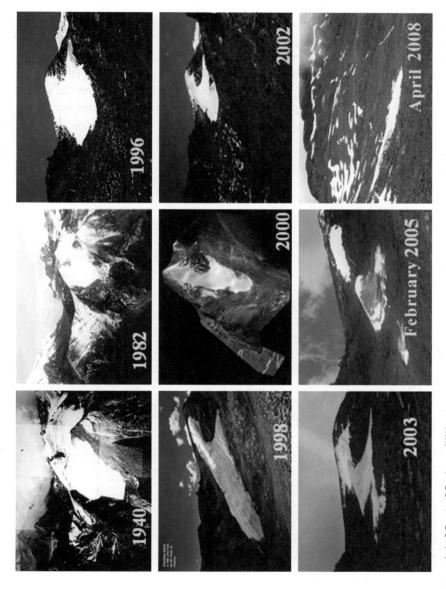

*Source:* Photographs by B. Francou, E. Ramirez, and W. Vergara.

**Box 1.1** *(continued)*

the regulatory function they currently perform of evening out the flow of water across the year will also cease. Indeed, unless measures are taken to capture runoff, additional water may be lost, as the glaciers will no longer be there to absorb precipitation as it falls and release it slowly. Water will run off the mountains into the ocean at greater speed than it does now.

Water quality is also an issue. El Alto has no sewerage treatment system, so a smaller water supply means a higher concentration of waste. The resulting threat to health will intensify. Diarrhea is already a serious problem as a result of contaminated water and food (for example, meat and vegetables grown outside La Paz that have been irrigated by contaminated river water).

The melting of the glaciers also threatens agriculture and the energy supply from hydroelectric plants. The implications are dire, not just for La Paz, but also for Quito and Bogota, two other large cities that depend on glacial water. More than 11 million people now live in these three Andean cities, and the population of El Alto alone is growing by 3 percent to 5 percent annually.

*Source:* Information gathered by the author during fieldwork in Bolivia in March 2008.

by 2050, in the absence of adaptation.[3] For the period 2000–05, the annual average cost of climate-related damage may be 0.7 percent to 0.8 percent of GDP (Nagy and others 2006). Although the overall effect on GDP is significant, of greater concern is that such damage disproportionately affects the poor, who number around 100 million people in the region (World Bank 2008).

This book examines the social implications of climate change and climatic variability in the LAC region and the options for improving resilience and adaptability to these phenomena. By "social implications" we mean direct and indirect effects in the broad sense of the word *social*, including factors contributing to human well-being, health, livelihoods, human agency, social organization, and social justice. Until recently, the growing literature on climate change has focused on the biophysical sciences and macroeconomic dimensions.[4] Providing a complementary focus, this book uses innovative analyses and develops a framework to address the social implications of climate change and variability, drawing on consultations with experts within and outside the World Bank (see also Mearns and Norton 2009). Its goal is to generate, consolidate, and share knowledge and information on the social dimensions of climate change in the region and the policy options for addressing them.

Much of this book relies on new empirical work. The authors undertook quantitative analyses of relationships in LAC countries between climate change and (a) health indicators (life expectancy and child mortality), (b) income or consumption and inequality, and (c) migration, using municipal-level data. Research included extensive field surveys in indigenous communities across the region to investigate how individuals and communities perceive and respond to climate change and variability (complete analyses can be found in Kronik and Verner 2010).

The book focuses on climate change's social impact on the most vulnerable population groups in the LAC region and on necessary elements of adaptation measures that can reduce the negative effects that they will otherwise face. The book does not address mitigation or any negative social impacts potentially associated with it, despite their obvious importance.[5] Neither does the book address issues of energy, forests, or infrastructure, since those areas are the subjects of other ongoing work.[6]

The book's focus is on the most vulnerable people, who tend to be the poorest.[7] The poor are more vulnerable than the nonpoor because they depend more heavily on natural resources for their livelihoods and well-being and have limited capacity to cope with unpredictable and extreme weather. Traditional knowledge about their environment—a key asset of indigenous peoples and poor farmers, in particular—may no longer be reliable in the face of climatic changes. In addition, the poor tend to live in areas at greater risk from extreme weather events (for example, on flood plains, unstable slopes, or unproductive land) and to lack the assets necessary to adapt to rapidly changing circumstances. Some poor and vulnerable people share a strong sense of injustice, believing that state policies have not benefited them. A large proportion of poor and vulnerable people identify insecurity, income inequality, and lack of political representation as critical factors reinforcing the vicious cycle of deprivation and destitution that they find themselves in. Others, including many indigenous people, blame the effects of climate change on their own failings.[8]

The progression of cause and effect underlying this book is sketched in figure 1.1, showing how greenhouse gas emissions resulting from human activity are linked to their environmental impacts (step 1). This environmental degradation affects the availability and quality of water for human consumption (including domestic, agricultural, and industrial

**Figure 1.1    Climate Change and Its Social Implications**

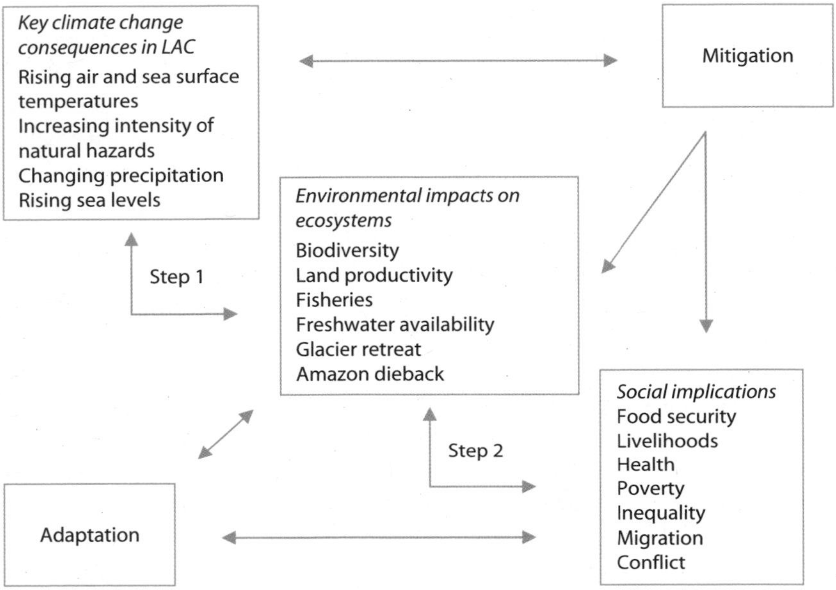

| | |
|---|---|
| *Key climate change consequences in LAC*<br>Rising air and sea surface temperatures<br>Increasing intensity of natural hazards<br>Changing precipitation<br>Rising sea levels | Mitigation |

Step 1

*Environmental impacts on ecosystems*
Biodiversity
Land productivity
Fisheries
Freshwater availability
Glacier retreat
Amazon dieback

Step 2

*Social implications*
Food security
Livelihoods
Health
Poverty
Inequality
Migration
Conflict

Adaptation

*Source:* Author's elaboration.

use and power generation), as well as terrestrial and marine flora and fauna ecosystems. The environmental impacts have social implications, as presented in step 2, affecting people's livelihoods, food security, and health. Excessive stress on those determinants of human well-being will increase poverty and income inequality. It may also cause migration to swell and has the potential to heighten the risk of conflict.

Although many of the effects of climate change and variability are already unavoidable, much scope remains for human agency and ingenuity in crafting strategies to mitigate them by addressing the causes of climate change itself and for adaptation to address the consequences.[9] An optimal national strategy would employ both mitigation and adaptation efforts and should embody good governance, providing for the exercise of voice, representation, and social accountability.

It is critical that policy makers in Latin America and the Caribbean address the social issues related to climate change. Even if global mitigation efforts improve, the climate trends that are already under way have considerable momentum and will dramatically affect economic, human, and social development for years to come. Thus, as the *Stern Review on the*

*Economics of Climate Change* argues, it is paramount that climate change and variability become fully integrated into development policy (Stern 2007). Social development is key to efforts to reduce loss of livelihood systems, forced migration, and potential conflicts. Indeed, an overriding message that emerges from the study that this book describes is that almost all of the policies, investments, and institutional reforms advocated here are good development policy. The realities of climate change may give them added value, but in most cases they would have significant benefits even in its absence. Thus, rather than devising adaptation policies per se, more appropriate will be to ensure that development policies are "climate-proofed" in the sense that they enhance resilience and enable adaptation. For example, an urgent need exists to improve physical infrastructure, both to improve people's living conditions and economic opportunities and to enable people to cope with climate change. At-risk communities need special attention because risky climate situations are costly not just to the households affected but to society at large. Although this book focuses on the external aspects of household and community transitions, climate change is also affecting the psychological transition— a period of breaking free of traditional patterns of thinking and resolving problems, of increased competition over resources, and of emphasis on the present over the future. The poor and vulnerable who live a day-to-day existence and lack the assets to allow long-term planning have little experience addressing these issues; they typically behave in ways that are rational for their objectives and perceptions of risk, which by necessity employ a very short-term perspective. Planning for a future with a changing climate is extremely difficult when an individual's or a household's asset base is barely sufficient to survive from day to day, especially given frustrations that arise from economic turbulence, blocked participation, and marginalization in the community.

Although fraught with risks, climate change and variability also present opportunities for households, communities, local and national governments, society, and the economy. Decisions about adaptation strategies, developing skills, and engaging with the broader civic community will determine the quality of life of the next generation. With more knowledge about how climate change affects the poor and vulnerable, governments may be better able to understand and serve this group. If policy makers do not invest now in mitigation and adaptation, they will miss a unique opportunity to equip this group with the tools to break the downward spiral of poverty and inequality and become drivers of growth and sustainable development.

## Assumptions and Analytical Framework Used in the Book

To understand the social implications of climate change and variability it is useful to identify risk factors and protective factors. Although not necessarily causal, these factors can be important predictors, and an understanding of them can help to shape policies and programs to strengthen people's resilience and capacity to adapt. Risk or vulnerability factors increase the likelihood that a person or community will experience negative outcomes: experience shows that the risks associated with climate change increase when combined with poverty, poor governance, and poorly maintained infrastructure. How vulnerable people are to these risks often depends on local social, political, and economic realities and government policies.

Protective factors increase the likelihood that a person or community will make a successful transition. Important factors that protect against the negative impacts of climate change exist at the household, community, and societal levels. They include good public policies, such as provision of public health services, education, social protection schemes, and the like; social connectedness, whether to relatives, neighbors, civil society organizations, or government agencies; solid and well-maintained infrastructure; good governance; and healthy public finances.

Climate change compounds existing vulnerabilities by eroding livelihood assets. For the poor in particular, the detrimental effects of climate change on the environment erode a broad set of livelihood assets—natural, physical, financial, human, social, and cultural. The resilience of the poor to disaster is already low. Many depend directly on fragile natural resources for their livelihoods and well-being, and many live in environmentally fragile areas that are especially prone to natural hazards such as drought, floods, rising sea levels, and landslides. When circumstances change for the worse, the poor are hard put to adapt. For many, the effects of climate change are compounded by other pressures, including a growing scarcity of land viable for agriculture, joblessness, difficulty obtaining enough food, poor health, lack of education, social marginalization, and lack of access to credit and insurance. Although livelihoods have constantly adapted to change throughout history, it is very likely that the impacts of climate change and variability will push poor people beyond their capacity to cope.

The next section outlines the book's assumptions about climate change and variability in the LAC region.[10] The following sections explain the analytical framework that is used to assess the vulnerability of the poor to the changes and to identify ways to help them increase their resilience and adaptive capacity.[11]

### Climate Change and Climatic Variability in Latin America and the Caribbean

Precise projections about climate change and variability in the LAC region cannot be made. Too little detailed historical information is available on the region's weather conditions, sea levels, and extreme events to allow robust regional climate models to be developed, and global climate studies yield relatively few robust statements and projections for the region (IPCC 2007a).

Nonetheless, the available evidence shows clearly that climate change is taking place and gathering speed in the region (a more detailed scientific description and explanation can be found in appendix A, and summaries of likely effects for individual countries in appendix B, both by Jens Hesselberg Christensen):

- *Projected temperature changes.* Overall, like the world as a whole, the LAC region is projected to warm. Most of South America is forecast to warm more than the global average, the exception being the Southern Cone. This implies that, over most of the region, temperatures in all seasons will continue to rise during the 21st century. Heat waves are likely to be more frequent and more intense, and the higher temperature level in general will tend to favor a longer warm season, with possible related extreme events such as hurricanes. In the high Andes, the temperature rise is projected to be greater than the mean values for the region. This means that less water will be stored as snow and ice and glaciers will continue melting. In the Amazon region, the expected higher temperatures are likely to worsen the destructive effects of deforestation and increase the risk of wildfires.

- *Projected precipitation.* The patterns of probable change show dry areas becoming dryer and wet areas becoming wetter.[12] Mean annual precipitation is projected to decrease over northern South America near the Caribbean coasts, as well as over large parts of northern Brazil, Chile, and Patagonia, and to increase in Colombia, Ecuador, and Peru, around the equator, and in southeastern South America. The midcontinental areas, such as the inner Amazonian and northern Mexico, are projected to become dryer during the summer months, with increased risk of droughts and forest fires. Annual precipitation is likely to decrease in the southern Andes, with relative changes being largest in summer. How annual and seasonal mean rainfall will change over northern South America, including the Amazon forest, is uncertain (figure 1.2).

*Extreme events.* Most areas of Latin America and the Caribbean have experienced several instances of severe weather in recent decades, with torrential rain and hurricanes causing thousands of deaths and damaging properties, infrastructure, and natural resources. These events are widely interpreted as reflecting climate change and increased variability, but no formal scientific detection of such a relationship at the regional level has been made.[13] Neither is there evidence that the extreme events recently seen are less severe, or more so, than those that may be experienced in the future. This said, the available models clearly suggest that changes will

**Figure 1.2    Changes in Water Runoff for the LAC Region by 2050 Due to Climate Change**

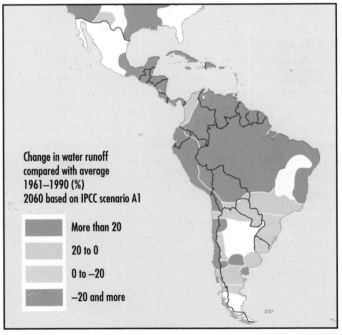

Change in water runoff
compared with average
1961–1990 (%)
2060 based on IPCC scenario A1

More than 20

20 to 0

0 to −20

−20 and more

IBRD 37780
MAY 2010

*Source:* Author's extraction of the LAC region from global map (Arnell 2004).

take place over this century that will generally be in the direction of more of the extremes, that is, more intensive precipitation, longer dry spells and warm spells, heat waves with higher temperatures than generally experienced up to now, and more numerous severe hurricanes. According to the Intergovernmental Panel on Climate Change (IPCC 2007a), further increases are expected in floods and droughts and in the intensity of tropical cyclones. Regional hot spots are shown in figure 1.3.

The countries in the Caribbean and the Gulf of Mexico, which are often assailed by intense hurricanes, can expect these storms to become even fiercer as a result of climate change. Other important issues are the destruction of coral reefs and a growing threat to southeast Pacific fish stocks due to increasing sea surface temperatures. A rise in sea level would likely bring flooding to low-lying regions such as the coasts of El Salvador and Guyana and could exacerbate social and political tensions in the region.

**Figure 1.3    Key Climate Change Hot Spots for Latin America**

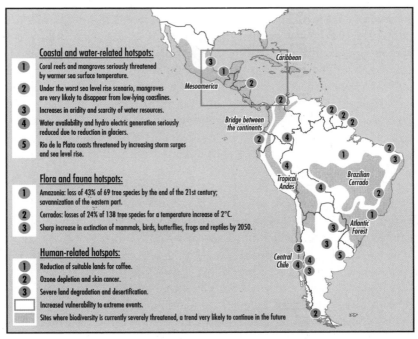

Coastal and water-related hotspots:
1  Coral reefs and mangroves seriously threatened by warmer sea surface temperature.
2  Under the worst sea level rise scenario, mangroves are very likely to disappear from low-lying coastlines.
3  Increases in aridity and scarcity of water resources.
4  Water availability and hydro electric generation seriously reduced due to reduction in glaciers.
5  Rio de la Plata coasts threatened by increasing storm surges and sea level rise.

Flora and fauna hotspots:
1  Amazonia: loss of 43% of 69 tree species by the end of the 21st century; savannization of the eastern part.
2  Cerrados: losses of 24% of 138 tree species for a temperature increase of 2°C.
3  Sharp increase in extinction of mammals, birds, butterflies, frogs and reptiles by 2050.

Human-related hotspots:
1  Reduction of suitable lands for coffee.
2  Ozone depletion and skin cancer.
3  Severe land degradation and desertification.
☐  Increased vulnerability to extreme events.
▨  Sites where biodiversity is currently severely threatened, a trend very likely to continue in the future

IBRD 37781
MAY 2010

*Source:* Adapted from IPCC 2007b.

For the Andean countries, the most momentous climate effects include major warming, changes in rainfall pattern, rapid tropical glacier retreat (box 1.1), and impact on mountain wetlands. These factors will combine with increasing precipitation variability to significantly affect water availability (IPCC 2007c). These factors may lead to greater migration and risk of conflicts.

In the Amazon region, the most pressing issue is the risk that the forest will die back—that rising temperatures and decreases in soil moisture in the eastern Amazon region will lead to the replacement of tropical forest by savannah. When that danger is combined with the deforestation caused by human activities, the outlook is dire. If present deforestation trends continue, 30 percent of the Amazonian forest will have disappeared by 2050. That, together with the transformation of tropical rainforest into dry grassland savannah, would lead to the extinction of a great number of plant and animal species unique to the area.

The overall effects of climate change will be negative in the LAC region. Exceptions will be seen, however, particularly with respect to increasing temperatures. In the southernmost part of South America and in the Andes, rising temperatures will expand the range of some crops, and the higher concentration of $CO_2$ in the atmosphere will increase yields. But elsewhere, crops such as coffee and maize are already being grown in close-to-optimum temperatures, so temperature increases, especially if coupled with declining rainfall, will lead to significant yield reductions. Higher temperatures not only will affect plant growth but also will cause more heat- and disease-related stress and mortality for livestock and humans alike. The remainder of this book focuses on these effects associated with climate change and variability.

## Analytical Framework

To permit a systematic analysis of the effects of climate change on the poor and vulnerable, a slightly adapted version of the United Kingdom Department for International Development's sustainable livelihoods framework (SLF), described in box 1.2, is used (DFID 2001). This framework makes it possible to see how different aspects of climate change and climatic variability affect people's assets, their livelihood strategies and livelihood outcomes, and hence their well-being. Thus it also sheds light on possible entry points for efforts to raise income, increase well-being, reduce vulnerability, improve food security, or achieve more sustainable use of resources.

Box 1.2

## The Sustainable Livelihoods Framework (SLF)

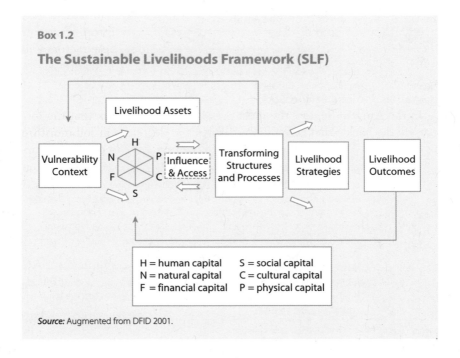

H = human capital    S = social capital
N = natural capital    C = cultural capital
F = financial capital    P = physical capital

*Source:* Augmented from DFID 2001.

To understand people's vulnerability, the SLF focuses on five aspects to assess their livelihood outcomes:

- *Vulnerability context.* This refers to the environment in which people live. People's livelihoods and the wider availability of assets are affected by all types of external trends and shocks, including seasonality, and climatic variability. Gradual climate change and global warming may produce a wide range of shocks, such as flooding, hurricanes, tornadoes, and droughts, that in turn may cause negative outcomes, such as changes in economic status and health, that may lead to conflict, migration, and so on. Climate change may also affect people's vulnerability by altering seasonality (for example, of rainfall) and increasing climatic variability.

- *Livelihood assets of individuals, households, or communities.* These include physical, financial, human, social, natural, and cultural capital (table1.1). The more assets a person has available, the less vulnerable he or she is. Access to livelihood assets determines a person's level of resilience and capacity to adapt to climate change. The reverse relationship is just as important; climate change and variability can affect

**Table 1.1    Definition of Livelihood Assets**

| Capital | Assets |
|---------|--------|
| Physical | The stock of plants, equipment, infrastructure, and other productive resources owned by individuals, the business sector or the country itself. |
| Financial | The financial resources available to people (savings, supplies of credit). |
| Human | Investments in education, health, and the nutrition of individuals. Labor is linked to investments in human capital; health status determines people's capacity to work, and skill and education determine the returns from their labor. |
| Social | An intangible asset, defined as the rules, norms, obligations, reciprocity, and trust embedded in social relations, social structures, and societies' institutional arrangement. It is embedded at the microinstitutional level (communities and households) as well as in the rules and regulations governing formalized institutions in the marketplace, political system, and civil society. |
| Cultural | The knowledge, experience and connections people have had throughout their lives, which enable them to succeed better than someone without such a background. |
| Natural | The stock of environmentally provided assets such as soil, atmosphere, forests, minerals, water, and wetlands. In rural communities land is a critical productive asset for the poor; while in urban areas, land for shelter is also a critical productive asset. |

*Source:* Augmented version of DFID 2001; see appendix C.

access to assets. For this book, the authors have added cultural capital (Bourdieu 1973, 1986; Bourdieu, de Saint Martin, and Clough 1996) to the five livelihood assets considered in DFID's original SLF. Field research revealed that, particularly for indigenous people, the cultural dimension of livelihood strategies and social institutions is important for understanding the impacts of climate change and climatic variability.

- *Transformational structures and processes at play within the community.* These are the institutions, organizations, policies, and legislation that shape livelihoods. They operate at all levels, from the household, community, and municipality to the national and international levels, and in all spheres from the most private to the most public. They effectively determine access (to various types of assets, to livelihood strategies, and to decision-making bodies and sources of influence); the terms of exchange between different types of assets; and the returns (economic and otherwise) on any given livelihood strategy. They also directly affect whether people achieve a feeling of inclusion and well-being and account for otherwise unexplained differences in the way things are done in different societies. These structures and processes can amplify vulnerabilities or be harnessed to enhance adaptive capacity and resilience.

- *Livelihood strategies.* The SLF provides insight into how vulnerability, assets, structures, and processes influence livelihood strategies, and how these strategies may be improved. The expansion of choice and value is important because it provides people with opportunities for self-determination and the flexibility to adapt over time. It is most likely to be achieved by working to improve poor people's access to assets—the building blocks for livelihood strategies—and to make the structures and processes that transform these into livelihood outcomes more responsive to their needs.

- *Livelihood outcomes.* The SLF examines how livelihood strategies, given the other factors, result in different livelihood outcomes, and how the livelihood outcomes feed back into available assets, creating either a virtuous or a vicious circle.

The SLF is a tool for assessing the vulnerability of the poor and their capacity to cope with shocks and adapt to changing trends—all highly important from a climate change perspective. Vulnerability, and the added vulnerability resulting from climate change and climatic variability, can be assessed for different population groups, as can people's ability to adapt to climate change within their specific environmental context.

## Outline of the Book

This book is in 11 chapters and follows the steps outlined in figure 1.1. The key climate change and variability consequences have been described briefly in this introductory chapter. Step 1, which traces the environmental results of climate change and variability, is addressed in chapter 2, which deals with water scarcity, and chapter 3, which discusses climate change, vulnerability hotspots, and asset erosion. Step 2, which focuses on the social implications of environmental impacts, is addressed in chapters 4 and 5, which analyze the effects on the livelihoods of some of the largest groups among the poor—agrarian, coastal, and urban poor; chapter 6, which addresses effects on health and life expectancy; chapter 7, which discusses migration induced by climate change; chapter 8, which explores links between climate change and conflict; and chapter 9, which analyzes the correlations between climate change and poverty and inequality. The book then explores possible means of countering the effects of climate change and variability: chapter 10 argues the importance of building social capital and livelihood resilience and offers asset-based adaptation strategies as tentative areas for further discussion. Chapter 11 concludes and proposes policy directions.

## Notes

1. "Climate change" refers to long-term change, that is, slow changes in, for example, mean annual temperature and precipitation levels, leading to sea level rise, melting of glaciers, and so on. "Change in climatic variability" refers to increased unpredictability with respect to seasons, rainfall, hurricanes, and the like, and is largely related to the El Niño-Southern Oscillation (ENSO).

2. WRI 2005. The figure falls to 6 percent of global emissions if emissions from energy use only are taken into account. When land use changes are included, the proportion rises to 12 percent of the world's green house gas (GHG) emissions, mainly as a result of the large-scale deforestation taking place in the region. Yet, despite the region's small contribution to global warming, its people still find their well-being, homes, and livelihoods threatened by climate change and climatic variability. The inverse relationship between responsibility for global warming and vulnerability to its effects is often ignored (UNDP 2007).

3. These estimates are higher than earlier ones that only included agricultural losses for the region, ranging from US$ 35 billion per year, equivalent to 0.23 percent of gross domestic product (GDP) (Mendelsohn and Williams 2004), to US$ 120 billion per year, equivalent to 0.56 percent of GDP (Tol 2002), by 2100. A more recent study projects total losses for the region of about US$ 91 billion, equivalent to around 1 percent of GDP, by 2050, if warming reaches 1.79°C relative to 1900 (Medvedev and Van Der Mensbrugghe 2008). This estimate is also lower than the range estimated by Nagy and others (2006), but it excludes noneconomic sectors. It also does not take account of the effects of possibly increased frequency and severity of natural disasters as a result of climate change or of potentially catastrophic climate change effects, such as collapsing ice sheets or melting permafrost.

4. The Intergovernmental Panel on Climate Change's landmark *Fourth Assessment Report*, released in 2007, provides scientific evidence that emissions from human activity, particularly burning fossil fuels for energy, are causing changes to the Earth's climate (IPCC 2007c). *The Stern Review* examined the economics of climate change (Stern 2007). Other authoritative sources include the *Human Development Report 2007/2008* (UNDP 2007); Mathur, Burton, and van Aalst 2004; and De la Torre, Fajnzylber, and Nash 2009.

5. Reducing emissions from deforestation and degradation (REDD) and other mitigation measures, energy policies, and infrastructure can all have positive or negative social impacts, depending on their design. While much effort is invested into finding viable ways to mitigate greenhouse gases, only recently have the social effects of mitigation measures surfaced as an area requiring attention. A few examples illustrate these complex relations. Switching to renewable energy sources, such as hydropower or biofuels, can have negative social implications unless attention is paid to how the switch takes place. Thus, building dams for hydropower often displaces large numbers of people, and

reducing the social costs requires close consultation and cooperation with and among the affected people to ensure that the relocation takes into account their wishes and needs, including for livelihood. Similarly, large-scale cultivation of biofuels may displace subsistence farmers from their land, forcing them either to move to less-productive land or to clear forest for new land. At the same time, a switch to growing biofuels may displace food crops, resulting in food price hikes that threaten the food security of the poor. "Hard" adaptation measures, such as sea defenses and other infrastructure, can also affect vulnerable segments of the population in positive or negative ways, depending on their configuration.

6. Given the scale of these areas, the World Bank is undertaking work that covers the forest sector and REDD exclusively and similarly covers the energy and infrastructure sectors extensively. It would seem a natural follow-up to this volume to initiate a similar study of the social impacts associated with mitigation. A number of countries within LAC may introduce REDD programs in an effort to reduce greenhouse gas emissions and receive credits for it. However, any effort to nationalize forests must be considerate of indigenous peoples, who may fear the impact on their livelihoods. Forming REDD partnerships recognizing these peoples' autonomy and authority can be a strategy for effective and sustainable forest management.

7. The effects of climate change have very unequal distributional implications and as such are socially differentiated. Hence some have called climate change the "greatest social injustice of our time" (as stated by Mary Robinson, International Institute for Environment and Development, on December 11, 2006).

8. Interviews of indigenous people for this book; see also Kronik and Verner 2010.

9. The essence of mitigation policies should be to price carbon and carbon equivalent gases so as to reflect their true costs—including social costs. This book does not address these policies in any detail.

10. Climate change assumptions are detailed in appendixes A and B, which summarize existing knowledge about recent and most likely future climate change and variability in the LAC region, building on the recent Intergovernmental Panel on Climate Change (IPCC) *Fourth Assessment Reports* (2007a; 2007b; 2007c) and recent literature.

11. Appendix C provides further details of the analytical framework and methodology underlying the book.

12. There will be local exceptions to these broad tendencies, but existing models do not give robust results for every part of this large region of interest. In particular, there are still many unresolved issues related to changes in the Amazonas, as important aspects of the interaction between vegetation and climate are still not well understood.

13. To formally attribute change or occurrence of particular events to a cause, a statistically sound number of events must normally be considered. By nature, extreme events are rare at a given location. Events that occur over a large geographical region, such as the LAC, cannot simply be lumped together for study because the causal chains leading to them most likely differ from event to event and from location to location. That precludes a simple statistical treatment of the data. Therefore, the recent IPCC reports have very little to say about recent changes at a regional level, and even less about the national or provincial level, and their possible links with global climate change and variability.

## References

Agrawal, Arun. 2008. "The Role of Local Institutions in Adaptation to Climate Change." Paper presented at World Bank Workshop on Social Dimensions of Climate Change, Social Development Department, World Bank, Washington, DC.

Arnell, N. W. 2004. "Climate Change and Global Water Resources: SRES Scenarios and Socio-economic Scenarios." *Global Environmental Change* 14: 31–52.

Bourdieu, Pierre. 1973. "Cultural Reproduction and Social Reproduction." In *Knowledge, Education, and Cultural Change*, ed. Richard Brown. London: Willmer Brothers Ltd.

———. 1983, 1986. "The Forms of Capital." Translated into English by Richard Nice. In *Handbook of Theory and Research for the Sociology of Education*, ed. J. G. Richardson, 241–58. New York: Greenwood Press.

Bourdieu, Pierre, Monique de Saint Martin, and Laurette C. Clough. 1996. *The State Nobility*. Stanford, CA: Stanford University Press.

Buhaug, H., N. Gleditsch, and O. Theisen. 2008. "Implications of Climate Change for Armed Conflict." Paper presented at World Bank Workshop on Social Dimensions of Climate Change, Social Development Department, World Bank, Washington, DC.

De la Torre, A., P. Fajnzylber, and J. Nash. 2009. *Low Carbon High Growth: Latin American Responses to Climate Change. An Overview*. World Bank Latin American and Caribbean Studies. Washington, DC: World Bank.

DFID (Department for International Development, UK). 2001. *Sustainable Livelihoods Guidance Sheet*. http://www.nssd.net/pdf/sectiont.pdf.

IPCC (Intergovernmental Panel on Climate Change). 2007a. *Climate Change 2007: The Physical Science Basis*. Contribution of Working Group I to the Fourth Assessment Report of the IPCC. Geneva: IPCC.

———. 2007b. *Climate Change 2007: Impacts, Adaptation, and Vulnerability*. Contribution of Working Group II to the Fourth Assessment Report of the IPCC. Geneva: IPCC.

———. 2007c. Climate Change 2007: *Synthesis Report*. Contribution of Working Groups I, II, and III to the Fourth Assessment Report of the Intergovernmental Panel on Climate Change. Geneva: IPCC.

Kronik, Jakob, and Dorte Verner. 2010. *Indigenous Peoples and Climate Change in Latin America and the Caribbean*. Washington, DC: World Bank.

Mathur, A., I. Burton, and M. van Aalst, eds. 2004. "An Adaptation Mosaic: A Sample of the Emerging World Bank Work in Climate Change Adaptation." Global Climate Change Team, World Bank, Washington, DC.

Mearns, R., and A. Norton, eds. 2009. *The Social Dimensions of Climate Change: Equity and Vulnerability in a Warming World*. Washington, DC: World Bank.

Medvedev, D., and D. Van Der Mensbrugghe. 2008. *Climate Change in Latin America: Impact and Mitigation Policy Options*. Washington, DC: World Bank.

Mendelsohn, R., and L. Williams. 2004. "Comparing Forecasts of the Global Impacts of Climate Change." *Mitigation and Adaptation Strategies for Global Change* 9: 315–33.

Nagy, G. J., R. M. Caffera, M. Aparicio, P. Barrenechea, M. Bidegain, J. C. Jiménez, E. Lentini, G. Magrin, and co-authors. 2006. "Understanding the Potential Impact of Climate Change and Variability in Latin America and the Caribbean." Report prepared for the *Stern Review on the Economics of Climate Change*. http://www.hm-treasury.gov.uk/media/6/7/Nagy.pdf.

Stern, Nicholas. 2007. *Stern Review on the Economics of Climate Change*. Cambridge, U.K.: Cambridge University Press

Tol, R. S. J. 2002. "Estimates of the Damage Costs of Climate Change." *Environmental and Resource Economics* 21: 47–73.

UNDP (United Nations Development Program). 2007. *Fighting Climate Change: Human Solidarity in a Divided World, Human Development Report 2007/08*. New York: Oxford University Press.

World Bank. 2006. "Project Document for Regional Implementation of Adaptation Measures in Coastal Zones (SPACC)." World Bank Latin America and Caribbean Region/Global Environment Facility, World Bank, Washington, DC.

———. 2008. "Poverty Data: A Supplement to *World Development Indicators 2008*." Washington, DC: World Bank.

WRI (World Resources Institute). 2005. "Climate and Atmosphere—$CO_2$ Emissions: Cumulative $CO_2$ Emissions, 1990–2002." *Earthtrends*. http://earthtrends.wri.org/text/climate-atmosphere/variable-779.html.

CHAPTER 2

# Water Scarcity, Climate Change, and the Poor

## Tine Rossing

Few resources have more critical bearing on human livelihoods and well-being than water. Water is essential for all socioeconomic development and for maintaining healthy ecosystems. Throughout history, the success or failure of societies in harnessing the productive potential of water, while limiting its destructive potential, has determined human progress. This chapter examines how climate change and variability will affect water scarcity and thereby human livelihoods in Latin America and the Caribbean (LAC)—a region in which about 60 percent of the population is concentrated in 20 percent of the land area, largely in arid and semiarid coastal and mountainous zones that hold only 5 percent of the region's water resources (WMO/IADB 1996).

Water scarcity is relative. In general, it is defined as "the point at which the aggregate impact of all users impinges on the supply or quality of water under prevailing institutional arrangements to the extent that the demand by all sectors, including the environment, cannot be satisfied fully" (UN Water 2007). As might be expected, water scarcity is often rooted in water shortage; it tends to be most acute in arid or semiarid regions that are affected by droughts and climatic variability, combined with rapid population growth and economic development. However, water scarcity can also be a social construct, reflected in, for example, imbalances between water availability and demand, the degradation of groundwater and surface water, and intersectoral competition. Water scarcity can also be the consequence of altered supply patterns, such as less rain or increased glacial runoff, resulting from climate change and variability.

The first section of this chapter shows that, despite an overall abundance of water in the LAC region, water scarcity is a growing problem at subregional and local levels, mostly as the result of population growth, urbanization, and economic development. Particularly vulnerable to water scarcity are (a) arid or semiarid rural subregions affected by drought and wide climatic variability, and (b) large urban centers, where high population densities and economic growth coincide with limited availability of freshwater. Both types of areas tend to have high concentrations of poor people, who tend to be those most affected by water scarcity. The second section of the chapter looks at the impact of climate change on freshwater resources. It highlights the ways in which climate change and climatic variability are already altering the volume, timing, and quality of both surface water and groundwater. Looking ahead to the medium and longer term, water flows for millions of people in Latin America and the Caribbean will be subject to mounting uncertainty and unpredictability. The third section of the chapter provides case studies of water scarcity and decentralized governance in Bolivia, Brazil, and Mexico. These are used to support the argument that because both water scarcity and the impacts of climate change and variability on water resources tend to be local, integrated water resource management needs to be improved by tailoring it to specific local circumstances, mainly through the formulation of location-specific policies and institutional arrangements in which water users participate in decision making. The chapter concludes with recommendations for policy and research.

## Current Water Scarcity

Compared to other regions, the LAC region overall has abundant renewable freshwater resources. The Falkenmark Water Stress Index used in table 2.1 (data column 6) describes the amount of long-term, mean renewable water resources available to a population.[1] If all the freshwater in the LAC region were divided equally within the three subregional populations, on average, 89,000 cubic meters ($m^3$) of water would be available annually per person in South America, 25,802 $m^3$ per capita per year in Central America, and 1,843 $m^3$ per capita per year in the Caribbean. When compared to the Falkenmark threshold of 1,700 $m^3$ per capita per

**Table 2.1    Supply and Demand for Renewable Freshwater Resources in the LAC Region  (2005 or Most Recent Data)**

| LAC Sub-Region / Country | Population | Freshwater supply (Actual renewable water resources)[1] | | | | | Freshwater demand (Total annual water withdrawals)[1] | | | | | | Access to improved water source[2] | | Access to improved sanitation[2] | |
| | Total population | Average precipitation in volume | Surface water: total renewable (actual) | Groundwater: total renewable (actual) | Water resources: total renewable (actual) | Falkenmark indicator: water resources: total renewable per capita (actual)[3] | Total water withdrawal (summed by sector) | Per capita | Agriculture | Industry | Domestic | Total water withdrawal as % of total renewable water resource (actual)[3] | Rural population with access | Urban population with access | Rural population with access | Urban population with access |
| | (1,000 inhab) | (1,000 m³/yr) | (1,000 m³/yr) | (1,000 m³/yr) | (1,000 m³/yr) | (m³/ inhab/yr) | (1,000 m³/yr) | (1,000 m³/yr) | (%) | (%) | (%) | (%) | (%) | (%) | (%) | (%) |
|---|---|---|---|---|---|---|---|---|---|---|---|---|---|---|---|---|
| **Central** | | | | | | | | | | | | | | | | |
| America | 143,775 | 2,650 | 1,036 | 335 | 1,165 | 25,802 | 87.3 | 378 | 60.2 | 18 | 21.9 | 3.7 | 80 | 96 | 60 | 81 |
| Belize | 276 | 39 | 75 | 37 | 19 | 68,722 | 0.2 | 593 | 20 | 73.3 | 6.7 | 0.8 | 82 | 100 | 25 | 71 |
| Costa Rica | 4,327 | 150 | 75 | 6 | 112 | 25,976 | 2.7 | 654 | 53.4 | 17.2 | 29.5 | 2.4 | 96 | 99 | 95 | 96 |
| El Salvador | 6,668 | 36 | 25 | 6 | 25 | 3,667 | 1.3 | 196 | 59.4 | 15.6 | 25.0 | 5.1 | 68 | 94 | 80 | 90 |
| Guatemala | 12,710 | 217 | 103 | 34 | 111 | 8,832 | 2.0 | 172 | 80.1 | 13.4 | 6.5 | 1.8 | 94 | 99 | 79 | 90 |
| Honduras | 6,834 | 221 | 87 | 39 | 96 | 13,314 | 0.9 | 128 | 80.2 | 11.6 | 8.1 | 0.9 | 74 | 95 | 55 | 78 |
| Mexico | 104,266 | 1,472 | 409 | 139 | 457 | 4,272 | 78.2 | 760 | 77.1 | 5.48 | 17.4 | 17.0 | 85 | 98 | 48 | 91 |
| Nicaragua | 5,463 | 311 | 193 | 59 | 197 | 35,847 | 1.3 | 252 | 83.1 | 2.31 | 14.6 | 0.7 | 63 | 90 | 34 | 57 |
| Panama | 3,232 | 203 | 145 | 21 | 148 | 45,786 | 0.8 | 268 | 28 | 4.88 | 67.1 | 0.6 | 81 | 96 | 63 | 78 |

(continued)

**Table 2.1 Supply and Demand for Renewable Freshwater Resources in the LAC Region (2005 or Most Recent Data)** (continued)

| LAC Sub-Region / Country | Population Total population (1,000 inhab) | Freshwater supply (Actual renewable water resources)[1] Average precipitation in volume (1,000 m³/yr) | Surface water: total renewable (actual) (1,000 m³/yr) | Groundwater: total renewable (actual) (1,000 m³/yr) | Water resources: total renewable (actual) (1,000 m³/yr) | Falkenmark indicator: water resources: total renewable per capita (actual) (m³/inhab/yr) | Freshwater demand (Total annual water withdrawals)[1] Total water withdrawal (summed by sector) (1,000 m³/yr) | Per capita (1,000 m³/yr) | Withdrawals by sector Agriculture (%) | Industry (%) | Domestic (%) | Total water withdrawal as % of total renewable water resource (actual)[3] (%) | Access to improved water source[2] Rural population with access (%) | Urban population with access (%) | Access to improved sanitation[2] Rural population with access (%) | Urban population with access (%) |
|---|---|---|---|---|---|---|---|---|---|---|---|---|---|---|---|---|
| **South America** | 373,679 | 28,375 | 17,084 | 3693 | 17,132 | 89,185 | 164.7 | 789 | 74.5 | 8.2 | 17.3 | 1.3 | 75 | 97 | 60 | 87 |
| Argentina | 38,747 | 1,642 | 814 | 128 | 814 | 21,008 | 29.2 | 775 | 73.7 | 9.46 | 16.8 | 3.6 | 80 | 98 | 83 | 92 |
| Bolivia | 9,182 | 1,259 | 596 | 130 | 623 | 67,799 | 1.4 | 166 | 80.6 | 6.88 | 12.5 | 0.2 | 69 | 96 | 22 | 54 |
| Brazil | 186,831 | 15,236 | 8,233 | 1,874 | 8,233 | 44,167 | 59.3 | 331 | 61.8 | 18 | 20.3 | 0.7 | 58 | 97 | 37 | 84 |
| Chile | 16,295 | 1,152 | 922 | 140 | 922 | 56,582 | 12.6 | 796 | 63.5 | 25.2 | 11.3 | 1.4 | 72 | 98 | 74 | 97 |
| Colombia | 44,946 | 2,975 | 2,132 | 510 | 2,132 | 46,754 | 10.7 | 246 | 45.9 | 3.73 | 50.3 | 0.5 | 77 | 99 | 58 | 85 |
| Ecuador | 13,061 | 592 | 424 | 134 | 424 | 32,083 | 17.0 | 1,340 | 82.2 | 5.3 | 12.5 | 4.0 | 91 | 98 | 72 | 91 |
| Guyana | 739 | 513 | 241 | 103 | 241 | 320,905 | 1.6 | 2,195 | 97.6 | 0.61 | 1.8 | 0.7 | 91 | 98 | 80 | 85 |
| Paraguay | 5,904 | 460 | 336 | 41 | 336 | 54,563 | 0.5 | 85 | 71.4 | 8.16 | 20.4 | 0.1 | 52 | 94 | 42 | 89 |

| | | | | | | | | | | | | | | | | |
|---|---|---|---|---|---|---|---|---|---|---|---|---|---|---|---|---|
| Peru | 27,274 | 2,234 | 1,913 | 303 | 1,913 | 68,400 | 20.1 | 752 | 81.6 | 10.1 | 8.4 | 1.1 | 63 | 92 | 36 | 85 |
| Suriname | 452 | 381 | 122 | 80 | 122 | 271,715 | 0.7 | 1,519 | 92.5 | 2.99 | 4.5 | 0.5 | 79 | 97 | 60 | 89 |
| Uruguay | 3,326 | 223 | 139 | 23 | 139 | 40,139 | 3.2 | 929 | 96.2 | 1.27 | 2.5 | 2.3 | 100 | 100 | 99 | 100 |
| Venezuela | 26,726 | 1,710 | 1,211 | 227 | 1,233 | 46,102 | 8.4 | 330 | 47.4 | 7.05 | 45.5 | 0.7 | 72 | 93 | 51 | 91 |
| Caribbean | 40,525 | 328 | 70 | 24 | 94 | 1,843 | 13.4 | 331 | 51 | 17 | 32.0 | 28.4 | 86 | 94 | 82 | 85 |
| Bahamas | 323 | 18 | ~ | ~ | 0 | 62 | ~ | ~ | ~ | ~ | ~ | ~ | 100 | 98 | 100 | 100 |
| Barbados | 292 | 1 | 0 | 0 | 0 | 296 | 0.1 | 336 | 22.2 | 44.4 | 33.3 | 113.0 | 86 | 100 | 100 | 99 |
| Cuba | 11,260 | 148 | 32 | 6 | 38 | 3,383 | 8.2 | 733 | 68.8 | 12.2 | 19.0 | 21.5 | 78 | 95 | 95 | 99 |
| Dominican Republic | 9,470 | 69 | 21 | 12 | 21 | 2,360 | 3.4 | 398 | 66.1 | 1.77 | 32.2 | 16.1 | 91 | 97 | 74 | 81 |
| Haiti | 9,296 | 40 | 12 | 2 | 14 | 1,645 | 1.0 | 121 | 93.9 | 1.01 | 5.1 | 7.1 | 51 | 70 | 12 | 29 |
| Jamaica | 2,682 | 23 | 6 | 4 | 9 | 3,547 | 0.4 | 157 | 48.8 | 17.1 | 34.1 | 4.4 | 88 | 97 | 84 | 82 |
| Puerto Rico | 3,947 | 18 | ~ | ~ | 7 | 1,795 | ~ | ~ | ~ | ~ | ~ | ~ | ~ | ~ | ~ | ~ |
| S. Kitts and Nevis | 49 | 1 | 0 | 0 | 0 | 558 | ~ | ~ | ~ | ~ | ~ | ~ | 99 | 99 | 96 | 96 |
| Trinidad and Tobago | 1,324 | 11 | ~ | ~ | 4 | 2,943 | 0.31 | 240 | 6.45 | 25.8 | 67.7 | 8.07 | 93 | 97 | 92 | 92 |
| Total LAC | 557,97 | 31,353 | 18,190 | 4,052 | 18,391 | 38,943 | 265.4 | 499 | 62 | 14.4 | 23.6 | 11 | 86 | 94 | 82 | 85 |

**Sources:** UN Population Division 2006; World Bank 2005; Quijandria, Monares, and de Peña Montenegro 2001; FAO Aquastat Water Data 2008; and World Bank LAC Water and Sanitation, Data and Statistics Website. The most recent data are used and stem from 2005–07, but others are long-term averages from multiple sources and years.

**Notes:** 1) FAO (2008); 2) World Bank LAC Water – Data and Statistics website. The most recent data are used and most stem from 2005–2007, but other are long-term averages from multiple sources and years.

year, these figures give the impression of abundant supply throughout Latin America and the Caribbean. At the national level, the only countries that fall under this threshold are in the Caribbean: Haiti (1,645 m³), St. Kitts and Nevis (558 m³), Barbados (296 m³), and Bahamas (62 m³) (table 2.1 and box 2.1).

Water availability, however, is not a proxy for well-being; it is simply a measure of the natural endowment of a given area and says little about how people in that area mobilize or use the water available. Table 2.1 (data column 12) therefore includes another commonly used water stress index, which combines water availability and use, showing

### Box 2.1

## Water Scarcity and Climate Change Impacts in The Bahamas

Freshwater in The Bahamas is limited, vulnerable, and scarce. It comes only from groundwater, which stems solely from rainfall accumulating over years. The country has no bodies of surface water. Freshwater resources are therefore limited to very fragile freshwater "lenses" lying atop shallow saline water, in shallow karstic limestone aquifers less than five feet below ground. Precipitation has been decreasing from north to south through the archipelago, so that the more southerly islands have less freshwater. Inagua, the southernmost island, is practically a desert.

The various islands obtain water in several ways:

- Groundwater provided via water authority on a large scale
- Private water wells
- Groundwater barged from one island to another
- Fresh groundwater blended with brackish groundwater
- Groundwater piped from one island to another by underwater lines
- Desalination (usually reverse osmosis)
- Water trucking from one part of an island to another
- Bottled water for drinking and cooking

Water losses are great. For New Providence, water loss is estimated at 53 percent, roughly equivalent to the amount barged in from Andros. In addition, as a result of overextraction of the limited freshwater reserves, saltwater intrusion is already occurring on New Providence, which has the country's greatest demand for water. The nature of the geology and the lack of proper sewage collection and treatment systems also contribute to the contamination of groundwater.

*(continued)*

**Box 2.1** *(continued)*

The country's freshwater supplies are highly vulnerable to the effects of climate change. Being very shallow, the aquifers are at great risk of becoming inundated with saline water if even a small rise in sea level occurs. Less precipitation over the years in some islands as a result of climate change is reducing freshwater availability. Natural disasters and severe weather, such as hurricanes, are probably the greatest threat to the health of the freshwater reserves. Once polluted, groundwater is very expensive to clean up. Protecting it from contamination is preferable and more cost-effective than remediation. Little, if anything, can be done to protect the groundwater from natural disasters, however. As a result of these and other factors, reverse osmosis (RO) is a key to future water supply for New Providence and many other islands in The Bahamas, particularly the central and southernmost islands. At present, Grand Bahama, Abaco, and Andros have enough fresh groundwater reserves to meet their demands.

Since water is a limiting factor for economic and social development, The Bahamas should treat groundwater as a strategic national resource. Regulating and protecting its water resources is essential, and the use of integrated groundwater management for this purpose is recommended. Current laws and regulations, particularly those governing land use, planning, and water management, are unclear and inadequate. Ignoring the overexploitation of water resources will have severe repercussions, including health risks from waterborne diseases and much greater water costs. Water will become much more costly if it has to be treated because of groundwater contamination, if RO needs to be used, or if more water has to be barged to meet demand. Proper land use planning and regulations, currently lacking, must play an important role in protecting this resource. The Ministry of Health and the Environment has proposed creating a new department of environmental planning and protection to regulate groundwater extraction and control pollution.

*Sources:* U.S. Army Corps of Engineers 2004; Bahamas Water and Sewerage Corporation 2008, http://www.wsc.com.bs/waterandsewerage.asp.

the total water withdrawal as a percentage of total renewable water resources at the country level. Generally, if more than 40 percent of the renewable water resources (supply) in a given location is withdrawn by people (demand), the area is deemed to be suffering from severe water stress. Table 2.2 shows the three water stress levels that this water scarcity index generally uses to characterize the degree of water scarcity in a given location. Column 12 in table 2.1 shows that

**Table 2.2    Water Scarcity Levels**

| Water scarcity indicator | Water scarcity level |
|---|---|
| > 40 percent withdrawal | Severe water stress |
| 20–40 percent | Medium water stress |
| 0–20 percent | Low water stress |

*Source:* WBGU 2008.

according to this index, Barbados (113 percent withdrawal) is the only country in the LAC region that suffers from severe water stress. Cuba (22 percent) appears to face medium water stress, and Mexico (17 percent) and the Dominican Republic (16 percent) have some degree of water shortage.[2]

Although both indexes of water well-being are in broad use, neither fully reflects the complexity of water scarcity, which encompasses not only water supply and use but also water quality and environmental demand. Moreover, although both indexes provide a picture of water resources at the national level, neither depicts the local contexts that affect levels of water stress and vulnerability (recommendations for local-level, alternative water indexes are included in the Policy Implications section at the end of this chapter).

Although the LAC region enjoys abundant water resources when viewed from a regional or national perspective, it suffers from serious water scarcity in numerous places, particularly in its arid and semiarid rural zones and in some large urban centers. In Mexico, 80 percent of the water is concentrated in the south and southeast, but the main consumption centers are located in the arid central and northern parts of the country.[3] In Brazil, 50 percent of the renewable water supply for the entire LAC region is concentrated in the Amazon basin, yet most of Brazil's population lives elsewhere, including in the 900,000-square-kilometer "drought polygon" in the Northeast (Ringler, Rosegrant, and Paisner 2000; UNDP 2006). Areas at particular risk of water scarcity are northeast Brazil; central and northern Mexico; northeast Venezuela; the Andean regions of Bolivia, Peru, Ecuador, and Colombia; and the Pacific coastal regions of Nicaragua and Honduras. Large cities in the region that increasingly face serious water shortages include Fortaleza, Recife, and Rio de Janeiro (Brazil); Mexico City (Mexico); La Paz and El Alto (Bolivia); Quito (Ecuador); Lima (Peru); Cali, Medellin, and Baranquilla (Colombia); Maracaibo (Venezuela); and Managua (Nicaragua) (OECD 2007).

*Drivers of Water Scarcity in Latin America and the Caribbean*

The main causes of water scarcity in Latin America and the Caribbean are population growth, urbanization, and economic development. Over the last century the region's population has tripled, and domestic water use has grown more than twice as fast as population. In 1995, according to the Falkenmark indicator described above, about 22.2 million people in the region lived in watersheds where water was already scarce (less than 1,000 m$^3$ per capita per year) (IPCC 2008). With the present population of 558 million expected to increase to 688 million by 2025 (see appendix C), significantly more people will suffer either severe or medium water stress.

Another key driver of water scarcity is urbanization, caused mainly by migration from rural areas and villages. The LAC region is now the most urbanized in the developing world, with 80 percent of its population in urban settings (UN Population Division 2006). From 2000 to 2005, the average urban growth rate was about 1.9 percent, compared to a population decline of 0.6 percent in rural areas. In cities in the arid and semiarid zones, water stress is compounded by rapid growth of domestic and industrial uses. The region's cities already accommodate roughly 430 million people but are expected to grow to more than 574 million by 2025—an 83.5 percent urbanization rate.

Though urban water scarcity stems mostly from growth in domestic and industrial uses, another important contributing factor is pollution. Table 2.3 highlights large cities that already face major water quality problems. As much as 80 percent of industrial effluent is discharged untreated directly into surface waters, contaminating freshwater supplies with heavy metals, chemicals, and sediments. In addition, although roughly 50 percent to 80 percent of the urban population has access to improved sanitation, 90 percent to 95 percent of domestic sewage is still discharged untreated directly into surface waters (UNEP FI 2004).[4] For example, the Tiete and La Paz Rivers, which flow through São Paulo, Brazil, and La Paz, Bolivia, respectively, are chronically polluted with untreated effluent and high concentrations of lead, cadmium, and other heavy metals (Ohno and others 1997). It is not uncommon for these contaminated waters to be reused in downstream rural communities for domestic purposes and irrigation. In Mexico City County, for example, highly contaminated water from Mexico City irrigates about 90,000 hectares of fields in the Tula Valley (table 2.3).

A third driver of water scarcity is economic development (IADB 1999). Accounting for 7 percent of regional gross domestic product (GDP),

**Table 2.3    Climate Change Impact on Selected Major Cities in Latin America and the Caribbean with Severe Existing Water Problems**

| Main threats from climate change | City | Country | Population, millions (year) | Share of urban HH connected to water (%) | Share of urban HH connected to sewage (%) | Water demand (m³/sec) | Primary water source – Surface water | Primary water source – Groundwater | Primary water source – Surface + groundwater | Existing water quality problems |
|---|---|---|---|---|---|---|---|---|---|---|
| Glacier melt | Bogota | Colombia | 5.1 (1995) | ? | ? | 17 | X | | | Bogota River highly polluted by draining urban wastewaters. |
| | El Alto | Bolivia | 0.7 (1998) | 33 | 20 | ? | X | | | La Paz River highly polluted from draining urban wastewaters. |
| | La Paz | Bolivia | 1.2 (1998) | 55 | 58 | ? | X | | | La Paz River highly polluted from draining urban wastewaters. |
| Glacier melt, coastal inundation, sea level rise | Lima | Peru | 7.1 (1995) | 70 | 69 | 25 | | | X | Overexploitation of aquifers leading to saline intrusion; 2,000 hectares of vegetable crops irrigated with urban wastewaters. |
| Coastal inundation; sea level rise | Buenos Aires | Argentina | 12.6 (1995) | ? | ? | 85 | | | X | Highly polluted surface water from urban wastewaters draining into the Riachuelo. |
| | Montevideo | Uruguay | 1.6 (1995) | ? | ? | 5 | X | | | Pollution; limited resources. |

| | | | | | | Drought | |
|---|---|---|---|---|---|---|---|
| Guatemala City | Guatemala | 1.1 (1995) | ~ | ~ | 5 | X | Location on a divide; lack of water. |
| Mexico City | Mexico | 22.8 (1995) | ~ | ~ | 50 | X | Severe groundwater overuse and contamination; about 90,000 hectares of agricultural land in the Tula Valley irrigated with wastewater from the city. |
| Sao Paulo | Brazil | 16.8 (1995) | ~ | ~ | 50–55 | X | Contaminated waters from the Tiete River used to irrigate vegetable gardens downstream from the urban core. |
| Santiago | Chile | 4.8 (1995) | 98 | 92 | 20 | X | Mapocho River highly polluted from draining urban wastewaters; contaminated wastewaters used for irrigation. |

*Sources:* Author's preparation based on data from Alfaro and Marin 1994; Anton 1993; Fernandez and Graham 1999; Ringler, Rosegrant, and Paisner 2000; and WRI 1998.

*Note:* HH = household; m³/sec = cubic meters per second; ~ = no data available.

agriculture is still the mainstay of many rural communities in Latin America and the Caribbean, and over the last decade, agricultural production grew at an average of 3 percent a year. Irrigated agriculture and livestock production, the engines of growth, have been the main activities behind increasing water scarcity, especially in areas already short of water. Agriculture accounts for 62 percent of the freshwater consumed in the region as a whole, although its share ranges from as high as 97.6 percent in Guyana to as low as 7 percent in Trinidad and Tobago (table 2.1). The region's 18.4 million hectares of irrigated land represent 14 percent of the world's total cultivated land.[5] Irrigated agriculture not only consumes large volumes of water, but it is also a major source of damage to water quality through sedimentation and through pollution from chemicals. In arid areas, the return flow from agriculture itself and multiple reuses of water progressively degrade water quality (UN Water 2006).

Industrial development also contributes to water scarcity in Latin America and the Caribbean. Industry accounted for 14.4 percent of the freshwater resources withdrawn in 2005 (table 2.1), but its share is expected to double by 2025 (World Bank 2008a). As mentioned above, industrial pollution has a large negative impact on water quality.

### Impact of Water Scarcity on the Rural and Urban Poor

Water scarcity in Latin America and the Caribbean is a local and subregional phenomenon and is a problem particularly for the poor. This section looks first at water as an element of production for poor farmers and then at issues in water supply and sanitation for poor users.

### Water for Nonirrigated Agriculture

In many of the countries with high rural poverty rates the water scarcity-poverty nexus is particularly observable in arid and semiarid regions. Overall, 32 percent of the LAC region's rural poor live in the water-stressed

**Table 2.4    Rural Poverty in Countries with Arid and Semiarid Regions**

| Country | Total rural population 2005 (1,000) | Rural population below the national poverty line (%) |
|---|---|---|
| Brazil | 29,462 | 41.0 |
| Honduras | 3,657 | 70.4 |
| Mexico | 24,702 | 27.9 |
| Nicaragua | 2,407 | 64.3 |
| Peru | 7,881 | 72.1 |
| Venezuela | 2,051 | n.a. |

*Source:* UN Population Division. 2006; World Bank 2008b.
*Note:* n.a. indicates not available.

regions of six countries: Brazil, Honduras, Mexico, Nicaragua, Peru, and Venezuela (Quijandria, Monares, and de Peña Montenegro 2001). Many of these are indigenous peoples (Kronik and Verner 2010).

Although irrigation is widespread and sustains rural economies in large parts of Latin America and the Caribbean, most of the rural poor in the region are smallholders who depend mainly on rainfall for their production. The poor cannot afford to invest in irrigation and lack the power to pressure their governments to undertake the necessary investments. Small-scale family farms make up 63 percent of the total farmland in the region on average. The proportion is much higher in some countries—91 percent in Ecuador and 80 percent in Peru, for example (Estrada 2006). Most of these farmers are small-scale livestock breeders, small-scale arable farmers, or people living from a mix of livestock and crops at subsistence levels (chapter 4).

Their livelihood conditions, coupled with their dependence on local precipitation, make them very vulnerable to climatic variability, such as the unpredictability of rain, and disasters such as droughts and floods. Just to produce their annual food supply takes 800 to 1,000 m$^3$ of water per person per year, water that must be available locally. It is this dependence on local rainfall for food production that makes the impacts of drought so dramatic and widespread. Moreover, any reduction in rainfall will lead to further degradation of the agricultural lands in arid and semiarid regions, many of which are subject to overpopulation and poor agricultural management. More water shortages will therefore result not only in increasing variation in crop yields, but also in long-term loss of crop productivity in the affected areas. Most of the people in these drylands are directly or indirectly dependent on agriculture for their livelihood, and climate change will thus further reduce the assets on which many households depend. Unfortunately, rainwater is rarely integrated into water management strategies, which usually focus exclusively on surface and groundwater (UN Water 2007) result.

## Household Water Supply and Sanitation

About 15 percent of the population of Latin America and the Caribbean— roughly 76 million people—lack access to safe water, and 116 million people lack access to improved sanitation. On average, the urban poor have much better access than the general rural population, both poor and nonpoor, to these basic services. For example, whereas 58.7 percent of the urban poor have piped water on their premises, only 31.4 percent of the entire rural population has this kind of access to water. In addition, about a third (33.6 percent) of the urban poor have access to a flush toilet, but only 12.6 percent of the rural population do (National Research Council 2003).[6] Notably, access is much more limited in smaller cities than in larger ones. The greater ease of providing services to clustered rather than scattered populations helps explain some of the service differences between the urban and rural poor (Fay 2005).

Despite the better services in urban areas, the basic service needs of the urban poor are seldom fully met, for several reasons. First, coverage remains far from universal. Even in a relatively wealthy country such as Argentina, 47 percent of the urban poor lacked adequate sanitation in 1998 (World Bank 2000). In Cali, Colombia, a fifth of the poorest did not have access to a private toilet in 2000 (World Bank 2002). And half of the favela dwellers in Rio de Janeiro, Brazil, lived without a sewerage connection in 2000 (World Bank 2001). Second, quality and reliability tend to be inadequate, and services are often unaffordable. In Guayaquil, Ecuador, about 40 percent of the city's population has to make do with only 3 percent of the piped water, and some 800,000 people living in low-income or informal settlements depend on water vendors (UNDP 2006). Without access to safe water, the poor must often rely on bottled water, paying 5 to 10 times more than households that have piped water. Third, access to services also reflects disparities between indigenous and nonindigenous people. For example, in Bolivia the average rate of access to piped water is 49 percent for indigenous people, compared to 80 percent for nonindigenous people (UNDP 2006).

Because of crowding, enormous repercussions for public health—often leading to decreased economic opportunities—may follow if even a fraction of a neighborhood lacks access to sanitation. People who cannot afford clean water use contaminated water, which, together with poor hygiene practices and poor sanitation, greatly increases their exposure to waterborne diseases (chapter 6). Often people are forced by poverty to settle on any available land they can find. They live in great numbers along water drains, river banks, and the like, using them for their sanitary needs. They use the water that flows in them for washing and laundry, and that both adds to the pollution and exposes them to it. In heavy rains these water drains and rivers can turn into virtual death traps, as people's belongings and litter cause blockages that result in flash floods.

As water becomes scarcer, more of it will be withdrawn from low-quality sources. More frequent heavy rainfall in some areas will increase the pollution and sediment loads carried by streams and rivers through higher runoff and infiltration. The associated risks include the malfunctioning of water infrastructure during floods. Drainage systems are often built with insufficient capacity to cope with the sudden, large volumes of water that have been seen in recent years, and which may recur with greater frequency, and that may allow contamination of drinking water and overloading of water and wastewater treatment plants during extreme rainfall. The problem of drinking water contamination is particularly severe in cities where storm and sanitary sewers are not separate. These causal links help to explain why infant and child mortality are higher among the urban than the rural poor in a number of countries, despite better urban access to both infrastructure and health services (Fay 2005).

Although urban water demands are still several times smaller than the demands of irrigated agriculture, growing water scarcity in urban areas is a consequence of the spatial concentration of urban demand. The population of the LAC region continues to urbanize at some of the fastest rates in the world. Urbanization in the region is expected to increase from its current rate of 79 percent to 81 percent by 2015 (World Bank 2008a; Ravallion, Chen, and Sangraula 2008). With this trend comes increased pressure on urban water supply and sanitation systems. Unless poverty rates decrease, population growth and continued urbanization imply an additional 22 million urban poor in the LAC region by 2015, 9 million of whom will be living in extreme poverty (Fay 2005). Notably, the urbanization is likely to be fastest in the least urbanized countries in the region (the Central American countries, Bolivia, Ecuador, and Paraguay). Many of them, such as Bolivia, are already struggling with water scarcity challenges (UN Water 2006).

## Climate Change and Freshwater Resources

This section assesses the implications of climate change and variability for freshwater resources and water scarcity in Latin America and the Caribbean. While the drivers of water demand described above will continue to make water scarcer in parts of the region, climate change and climatic variability will also aggravate existing water scarcity by increasingly causing changes in freshwater quantity and quality, some of which have potentially disastrous implications.

### Impact of Climate Change on Water Resources: Trends and Projections

Projections by the Intergovernmental Panel on Climate Change (IPCC) suggest that climate change and variability will make water supplies more variable and uncertain in the LAC region, compounding the difficulties facing people already affected by water scarcity, especially the poor.

Climate change and variability have been affecting the region's water resources for some time. A key factor has been an increase in climate-related disasters such as floods, droughts, and landslides. La Niña–related droughts have severely restricted water supplies for irrigation in central western Argentina and in central Chile, and El Niño–related droughts have reduced the flow of the Cauca River in Colombia. Over the past three decades, some areas in the LAC region have seen an increase in rainfall. In northeast Argentina (the pampas), parts of Bolivia, southern Brazil, Ecuador, northwest Mexico, Paraguay, northwest Peru, and Uruguay, the increases have caused more frequent flooding (10 percent more frequent in the Amazon River) and more instances of high stream flow in rivers (50 percent greater in the Uruguay, Parana, and Paraguay Rivers). The LAC region has also seen an increase in intense rainfall events and consecutive dry days. Less rain than before has been observed in southwest Argentina, northeast Brazil, Chile, southern Peru, and western Central America (including Nicaragua) (IPCC 2007b).

Both sea level rise and glacial melt have already affected water resources in the region. During the last 10 to 20 years, the sea level has been rising at 2 to 3 millimeters a year along the southeastern part of South America (IPCC 2007b). Generally, sea-level rise poses a risk to coastal groundwater aquifers. With excessive pumping in some areas, groundwater tables have fallen so far that pumping is now difficult or too costly (World Bank 2008a). In coastal aquifers, falling groundwater tables increase vulnerability to saline intrusion, which will get worse in depleted aquifers as sea levels rise. In Lima, Peru, for example, the combination of sea level rise and overexploitation of aquifers has caused saline intrusion into groundwater resources that are vital for the city's water supply (table 2.3 above).

Glaciers in the tropical Andes region are shrinking fast, having lost 20 percent of their volume, on average, since 1970.[7] The smallest glaciers are the most affected; over the past century many have been reduced to mere vestiges of their former expanses.[8] Modeling work and projections indicate that many of the lower-altitude glaciers could completely disappear in the next 10 to 20 years (Vergara 2008).

Weather-related natural hazards pose increasing threats to both water quantity and quality in many parts of Latin America and the Caribbean. In Mexico, for example, the Balsas, South Pacific, and southern frontier basins, along with the Yucatan, are very vulnerable to hurricanes and flooding. The North Gulf, the South Frontier, and the Center Gulf are very vulnerable to seawater intrusion (which could be exacerbated by intensified storms) into river mouths and adjacent shorelines. With regard to droughts, the Rio Bravo and the Central North basins are very vulnerable, and among the existing high water-demand "hot spots," the Rio Bravo, Central North, Balsas, Northeast, and Baja California basins are of particular concern.[9] Notably, the Valley of Mexico and the Lerma-Santiago basin already face significant water deficits due to overuse of groundwater and pollution of surface water.

Modeling exercises that look at future climate change suggest a complex range of possible outcomes but three main trends, as outlined in chapter 1. First, dry areas will get drier and wet areas will get wetter. For example, rainfall is likely to increase over parts of southeastern South America and western-to-central Amazonia, while northern Brazil will receive less rain. That is accompanied by a shift toward more heavy-rain events, except perhaps in easternmost Brazil, where rainfall may become less severe.[10] Second, dry spells are expected to be longer, reducing water availability during the critical low-flow season. Finally, water flows will become more unpredictable, possibly linked to more frequent and extreme weather events, including those associated with the periodic El Niño–Southern Oscillation (ENSO).[11]

Climate change and variability will compound existing water quality problems.[12] Changes in intense rainfall events will affect the rate at which materials are flushed into rivers and groundwater, and changes in flow volumes will affect the dilution of loads. Water temperature changes will affect the autopurification rate at which streams degrade organic loads. The overall result will be a climate change–induced decline in water quality, which will lead to increasing water withdrawals from low-quality sources and greater pollution loads from diffuse sources due to heavy precipitation (via more runoff and infiltration). The associated risks include malfunctioning of water infrastructure during floods and overloading of water and wastewater treatment plants during extreme rainfall.

## Implications for Water Scarcity

The consequences of climate change for water scarcity in Latin America and the Caribbean will vary within subregions and individual countries. Any given country may experience numerous shifts in hydrological cycles linked to microclimates.

According to IPCC projections of water availability for 2050, about half of the LAC region will experience a significant decrease in water runoff from rainfall by 2050. Figure 1.2 in chapter 1 highlights the way in which, according to IPCC scenario A1, climate change and variability will cause a decline of at least 20 percent in water runoff from rainfall in large areas in South America.[13] Much of Brazil—particularly the semiarid regions of the Northeast—along with parts of Bolivia, Chile, Colombia, Guyana, Suriname, and Venezuela will be affected, as will all of Central America, most of the Caribbean, and the southeastern part of Mexico. These projections probably underestimate the decreases in water availability, which will also be influenced by changes in temperature and the timing of flows. For example, whereas a given area could experience more water runoff, actual water availability could diminish as a result of increased evaporation.

The consequences of climate change and variability for water scarcity will occur mainly through four processes—glacier melt, reductions in rainfall, rising sea levels, and extreme hydrological climate events—now discussed in turn.

First, the shrinking of tropical glaciers in the Andes has potentially disastrous implications for economic growth and human development in the region. The situation is so grave that over the next 50 years, glacial melt could emerge as one of the most severe threats to human progress and food security in the Andean region (UNDP 2006). One immediate danger is that melt water will greatly enlarge glacial lakes, leading to increased risk of flooding, avalanches, mudslides, and dam bursts. The warning signs are already evident. For instance, the surface area of Lake Safuna Alta, in the Cordillera Blanca in Peru, has increased by a factor of five since 1975 (Maskrey and others 2007). Many glacier-fed basins have also seen an increase in runoff in recent years.

After the initial melt-off that will occur by 2050, models predict a rapid decline in stream flows, especially in the dry season. As glaciers retreat, water stocks will be depleted on a large scale. By 2020, deglaciation in the Andes could put close to 40 million people (70 percent of the Andean population) at risk of reduced water supply for drinking, hydropower generation,[14] and farming, particularly in and around Quito, Lima, and La Paz. Starting in the not-too-distant future, water managers can expect a rapid diminution of flows into reservoirs and irrigation systems during the non-rainy season. Urban consumers will face increased costs to finance new reservoirs. Longer-term effects will include a reduced flow of water for agriculture during the dry season. Even worse, if the temperature rises by more than 2°C, the Amazon rainforest could begin to collapse, as glacial retreat jeopardizes its supply of water (Amat y Leon and others 2008).

Second, reductions in rainfall could create severe water shortages in arid and semiarid regions of Argentina, northeast Brazil, Chile, and northern

Mexico. Some models also predict a drastic reduction in rainfall in the western Amazon (up to 90 percent) by the end of the century (Cox and others 2004). The Amazonian rainforest plays a crucial role in the climate system. It helps to drive atmospheric circulation in the tropics by absorbing energy and by recycling about half of the rainfall that falls on it (World Bank 2008c). The moisture that the Amazon ecosystem injects into the atmosphere is also critical to the precipitation patterns in the region. Disruptions in this process could trigger desertification over vast areas of Latin America and the Caribbean, and even in North America (Avissar and Werth 2005).

Third, both rising sea levels and coastal flooding could dramatically reduce the availability of freshwater for many countries as saltwater intrudes into aquifers (Oki and Kanae 2006). Among future vulnerability "hot spots," therefore, will be areas with large coastal populations. A substantial proportion of the population of Latin America and the Caribbean lives in coastal zones: 49 percent of the population in Brazil, 45 percent in Argentina, 82 percent in Chile, 30 percent in Colombia, 24 percent in Mexico, and virtually 100 percent in the Caribbean Community (CARICOM) countries (Miller 2006; see also chapter 5). In the Caribbean in particular, surface water is scarce, and groundwater depletion will become of increasing concern as well (box 2.2). Havana (Cuba) depends totally on groundwater, and Kingston and Montego Bay (Jamaica), San Juan (Puerto Rico), Port-au-Prince (Haiti), Nassau (Bahamas), and Bridgetown (Barbados) rely heavily on it (Ringler, Rosegrant, and Paisner 2000).

Fourth, water scarcity in Latin America and the Caribbean could be worsened by extreme events. Based on IPCC projections, an increasing share of annual rainfall will be concentrated in intense precipitation events—enormous rainfall concentrated in a few days—and longer periods without rainfall will occur. These trends increase both the risk of floods and the frequency and intensity of droughts (IPCC 2008). The region can expect an increase in the intensity of rainfall and more frequent El Niño–like conditions, leading to floods on the west coasts of Central and South America and droughts in some parts of the region.

Without doubt, the projected decrease in water availability will significantly increase the number of people facing water scarcity. As mentioned earlier, about 22 million people were already challenged by water scarcity in the LAC region in 1995. According to various projections based on the IPCC Special Report on Emissions Scenarios (SRES), the number of people living in water-stressed areas is likely to reach between 79 million and 178 million in the 2050s (IPCC 2001; 2008).[15]

**Box 2.2**

## Water Scarcity, Climate Change, and Small Island Dynamics in St. Vincent and Bequia

In response to the threats of climate change and variability, many Caribbean governments are establishing national adaptation and management frameworks, based on an integrated ecosystems approach, to promote resilience at the local level. One of the first Caribbean countries to adopt such a framework is St. Vincent and the Grenadines. Although there is still no specific agency or institution to manage water, the islands have already made basic decisions on integrating climate change and variability concerns in the broader development process. In addition, the government acknowledges the importance of empowering communities. Given the size of the islands, all adaptation measures will inevitably involve implementation at the community level.

An important element of the climate adaptation process is to identify how best to address the very different conditions found on each of the islands. St. Vincent is relatively water abundant, having both surface water and groundwater resources, but Bequia (the largest island of the Grenadines) has no surface water, no potable water system, and only very limited groundwater resources. Bequia is both a major tourism site and an established fishing community, with a population of about 6,000 people. Some of its water is imported, and the remainder is obtained from desalination plants.

Rainwater harvesting and storage are necessities but are also a significant challenge on Bequia because of limited public storage infrastructure and widespread poverty. Rainfall tends to be intense, lasting a few days, and to be followed by dry periods. Rain normally stops in December, and the dry season lasts from January until July. The island's two small reservoirs can only supply public buildings in Port Elisabeth, the main population center. All households are, therefore, responsible for catching and storing their own water supply, which they do with a rooftop and storage tank system in each home. Outside Port Elisabeth, few households can afford storage tanks or the pump and diesel fuel that are needed to pump water up and into homes.

Paget Farm, one of the poorest communities on the island, consisting mainly of poor fisherfolk, is a case in point. At present the people there have no rain harvesting or storage facility. In the dry season they rely on obtaining rainwater from neighboring villages and managing it very carefully. People who can afford it travel the 5 kilometers to Port Elisabeth, where they pay US$150 for a weekly supply of water.

*(continued)*

**Box 2.2** *(continued)*

Bequians receive no government support for their water supply, and the state-owned Central Water and Sewerage Authority (in St. Vincent) does not even operate on the island. Consequently, even though the Bequian population is poorer, they pay considerably more for their water than people in St. Vincent. The cost of water shipped from St. Vincent during droughts and the dry seasons is borne entirely by Bequians.

The prospect of climate change and variability is troublesome for Bequia since, as the IPCC projects, summer rainfall in the Caribbean, especially in the vicinity of the Greater Antilles, will likely decline. It is also unlikely that increased rainfall during the winter months will compensate for the decline, inasmuch as storage facilities in Bequia are modest at best. Add to this the likelihood of increased demand due to population growth, and the risks of sea level rise and coastal flooding, and the future looks extremely challenging for the people of Bequia.

*Sources:* Based largely on interviews and information gathered during a field trip for this study to St. Vincent and Bequia. Other sources include World Bank 2006; National Environmental Advisory Board/Ministry of Health and the Environment 2000; IPCC 2008; Oki and Kanae 2006.

## Case Studies: Water Scarcity and Decentralized Governance

Current management practices and existing water infrastructure are very unlikely to be able to offset the negative impacts of climate change on the reliability of water supply (IPCC 2008). Poor rural and urban communities will typically be the most affected and the least equipped to cope with the impact of water scarcity, especially with the added climate change challenge. Moreover, they are often marginalized by laws and institutions.

In the face of water scarcity and climate change and variability, improved water governance through integrated water resources management (IWRM) is widely recognized as a critical need throughout the developing world. IWRM is based on a multistakeholder, multilevel approach to sustainable water resource development and management. It basically advocates cross-sectoral coordination and a sound governance framework at all levels (especially the river basin)[16] and aims at balancing "water for livelihoods" and "water as a resource." Notably, IWRM was recently proposed by the Global Water Partnership to be the appropriate response framework for climate change adaptation, indicating that the way people use and manage water resources must become the focus of adaptation (DANIDA 2008).

Given that both water scarcity and the effects of climate change and variability on water resources tend to be local, IWRM must be tailored to specific local circumstances, mainly through the formulation of location-specific policies and institutional arrangements. Solutions will also need to consider the local dynamics of both water supply and demand, along with the expected consequences of climate change and variability. For water managers, it is crucial both to reduce the level of demand and to distribute supplies efficiently, while resolving conflicts of interest cooperatively and allocating rights of access and use equitably. Moreover, because both water scarcity and climate impacts are location specific, work is needed to enhance the capacity of local authorities and communities to initiate and implement adaptive measures and to develop lasting mechanisms that enable them to work in partnership with one another. Such a locally focused approach will help strengthen the sustainability of adaptation initiatives, while avoiding exclusion of the needs and priorities of marginal groups, particularly the poor.

Latin America and the Caribbean are increasingly joining the global trend toward IWRM and the delegation of water management responsibility to local authorities and water user groups, thereby promoting the principle of subsidiarity. Over the past 10 to 15 years, many river basin management organizations have been created with the goal of facilitating stakeholder involvement to achieve more integrated water management. But while appropriate legal and institutional frameworks have been created in a variety of places, the implementation of IWRM has met a number of important obstacles. Initial conditions and different stakeholder capacities and interests can have an important influence on the success of decentralized water resources management.

The case studies below provide examples in which responsibility for water management and adaptation to climate change and variability is being devolved or decentralized from national governments and agencies to regional or local government authorities. The cases are drawn from locations that have experienced—and will likely continue to experience—water scarcity compounded by adverse climate-related impacts: northeast Brazil, Mexico, and Bolivia.

### Water Scarcity, Rainfall Reduction, and River Basin Management in Northeast Brazil

The experience of Jaguaribe River basin, in Ceará, northeast Brazil, provides an important example of a successful water resource decentralization process that started under relatively unfavorable conditions.

Situated in Brazil's "drought polygon," which regularly experiences severe and progressively longer water shortages (UNDP 2006), Ceará has Brazil's highest rural poverty rate, at 76 percent, and the largest concentration of rural poor in the LAC region (World Bank 2008a). The state has five large river basins but no perennial rivers, making its water supply vulnerable to climate change and climatic variability. IPCC projections show that the area will become even shorter of water over the next 50 years.

In 1992, in response to a long series of droughts, Ceará embarked on a significant reform of its water resources management. When the process started, stakeholder participation was at odds with local tradition: water was controlled by either the government or the owner of the land where the water could be found. A small elite dominated the local political culture. But conflicts over the distribution of water resources during periods of drought had provided an incentive for better coordination of water resources management functions. A new law (Law 11.996) created several levels of water management, including watershed users commissions, watershed committees, and a state-level Water Resources Council (Lemos and Oliveira 2004). The law defined the watershed as the planning unit of action.

The relatively unfavorable initial conditions were largely overcome through continued political commitment by the state government to reduce clientelism and promote the principle of participatory, decentralized, water resource management.[17] In 1997, a new National Water Act revolutionized water management in Brazil. New legislation was prepared based on five years of structured national dialogue that included numerous public hearings (UNDP 2006). Decentralized water management emerged as a critical policy objective, with river basins as the unit for the devolved authority. In addition, new institutions were created at all levels of governance in which representatives of all ministries with water functions, state representatives, water users, and nongovernmental organizations (NGOs) were brought together in a strong technical advisory body.

While federal institutions continued to support specific drought relief programs, a self-financed, state-level river basin management company— COGERH (Companhia de Gestão dos Recursos Hídricos)—was created to carry out management, monitoring, and enforcement and to control state infrastructure. The creation of COGERH marked the start of a new, more participatory approach to water resources management. One of the most innovative aspects of water reform in the state was the creation of an interdisciplinary users commission within COGERH, charged with determining how best to implement all aspects of the reform. The commission, which represented industry, commercial farmers, rural labor unions, and cooperatives, developed an operational plan for managing water use in the river basin, with technical advice from COGERH hydrologists. Implementation has been overseen by a committee of representatives elected by the commission.

Water management practices have been transformed. Water users have become active participants in water management. Among other things, the government and local stakeholders have cooperated to develop an emergency operation plan to better manage water allocations in drought periods. Success was made possible by high levels of user participation and public debate within the users commission, which helped to institutionalize the rules for managing competition. A strong technical advisory body, perceived as competent and independent of individual user interest groups, contributed to the success of the users commission.

The organization of stakeholder councils and an effort to use technological knowledge to support the decision-making process may have both strengthened the capacity of Ceará's river basin systems to adapt to climate variability and improve the sustainability of local water resources (Formiga-Johnsson and Kemper 2005; Engle 2007). Participation can empower and legitimize water reform institutions. It provides a forum for a diverse group of stakeholders to deliberate about water use and availability, while also improving accountability and providing a way for reform-oriented policy networks to push for further reform. Water rationing, which was commonplace before the changes were introduced, no longer poses a major problem. A tangible result of promoting the participation of interested parties was that in 1997—the year of the new National Water Act—80 percent of the water-related conflicts were resolved by the communities rather than by the courts (Porto 1998).

Even so, effective participation does not necessarily equal equitable participation. A survey of Lower Jaguaribe River Basin Committee members revealed that many still perceived water management as an exclusionary process. Survey respondents pointed out that the main constraint on the democratization of decision making within the committee was the disparate levels of knowledge between tecnicos and general members (Taddei 2005). Knowledge can enhance effectiveness and democratize decision making, given that better-informed participants can make sounder decisions.[18] However, unequal access to knowledge may skew decision making and benefit some participants over others (Lemos 2007). Moreover, improved governance at the local level is necessary, but far from sufficient, to build resilience to water scarcity and climate change for the most vulnerable consumers.

## Water Scarcity, Interannual Precipitation Variability, and Local Governance in Mexico

Mexico, like Brazil, faces serious challenges in the water sector. Both countries are also in the process of transforming a centralized, top-down system into a more decentralized one in which subnational structures have greater roles. Thus far they are the only two countries in the LAC region where river basin organizations are legally mandated (Tortajada 2006). This case study highlights how the effects of climate change and climatic variability are likely to further complicate Mexico's challenging transformation from a highly centralized water sector into a decentralized one that is better equipped to make real decisions at the local river basin level.

As is widely recognized, the unsustainable use of water in Mexico is a threat to development and particularly affects the poor (World Bank 2008d). An underlying reason is that both economic growth and population growth have been concentrated in precisely the areas where water availability is lowest. As mentioned earlier, those areas include the arid and semiarid north and central parts of the country, which are plagued by droughts and ongoing desertification (Reyes Castañeda 1981). Meanwhile, southern Mexico, with 68 percent of the country's water, has about 23 percent of the national population and contributes less than 15 percent of its GDP (World Bank 2008d).

In some of the arid and semiarid regions, groundwater has been a major supplier of water to meet household and irrigation demands. Overexploitation of groundwater has increased steadily over the last decades, partly because of a heavy subsidy to electricity used to pump water. By 2006, the aquifers that provided almost 60 percent of the groundwater extracted were overexploited (meaning that the annual extraction rate exceeded their natural recharge). Of Mexico's 653 aquifers, 104 are considered overexploited (World Bank 2008d). As a result of overexploitation, water tables in these aquifers have been dropping at a rate of 1 to 5 meters per year, placing increasing constraints on economic activity and aggravating existing water scarcity.

Climate change and variability will very likely increase water scarcity in Mexico. With the increased risk of drought and the persistent vulnerability of the poor, a serious impact may be expected not only on water supply, but also subsequently on health, food security, and poverty reduction, particularly in the coastal areas and the arid interior regions (VARG 2006).

In response to increasing water scarcity and water contamination, the Mexican government has sought for more than two decades to promote decentralization in the water sector. The National Water Law of 1992 provides the overall policy framework for water resources. CONAGUA—the National Water Commission of Mexico—was established in 1989 to modernize and decentralize the management of the nation's water resources (OECD 2003). According to the National Water Law (art. 13), coordination of the country's federal, state, and local water management bodies is to take place at the river basin level, through river basin councils to be established by CONAGUA. The law charges these councils with formulating and implementing programs and actions to improve water administration, developing water infrastructure and the associated services, preserving basin resources, and approving river basin plans. The river basin plans must then be integrated within a national water master plan, at which point they become mandatory for the federal government and indicative for the local and the state governments and water users (Tortajada 2000b).

Seventeen years after the law was passed 25 out of the planned 26 river basin councils had been established, but so far only one is fully functional. The councils do not yet have any say in the planning, design, or operation of water infrastructure in the country.[19]

The institutional setup of the councils is problematic for several reasons. First, the National Water Law does not establish an effective legal framework for the decentralization of water management at the river basin level. Tortajada (2006) stresses that the federal government has still not formulated, let alone implemented, strategies to decentralize and transfer the functions, responsibilities, and funds from CONAGUA to the basin councils or to the appropriate authorities at the regional, state, or municipal level. The law itself gives the councils very little real decision-making power; CONAGUA retains the absolute unilateral and discretionary power to accept or disregard any decision taken or agreement reached within the councils (Wester, Merrey, and de Lange 2003). Whether the traditional centralized decision making will continue, or the central authorities will allow the councils to use their decision-making powers on matters of importance, remains to be seen (Tortajada 2006). Neither do the river basin councils have any real implementation role, given that they are only coordinating and consensus-building agencies between government and local groups.

Another shortcoming is that water users are poorly represented. Even though representatives from different sectors attend council meetings, regrettably they do not necessarily represent the views of the majority of the stakeholders in their respective sectors, largely because only people or agencies who have titles to withdraw and use water may attend. Citizens and civil society organizations that do not have such titles may not participate unless invited by CONAGUA (Tortajada 2000a). Also, not all members have the right to vote. According to Guerrero and Garcia-Leon (2003), only the representatives of the users, the state governments, and the president of the council have voting rights.

In summary, despite decentralization efforts, water planning and management in Mexico are still very much vested in a single institution, CONAGUA, at the central level. A key reason is that CONAGUA has been unwilling to decentralize decision-making powers, investment funds, or technical and managerial expertise. As a result, it has been unable to respond adequately to the escalating needs of the water sector in various parts of the country (Tortajada 2006). Progress has also been hampered by the limited capacity of the municipal governments to handle new management and technical responsibilities. For example, the river basin councils lack experience in formulating water policies. Other obstacles are the authorities' lack of appreciation of the importance of stakeholders' participation and the absence of economic instruments such as water pricing and demand management.

The potential of those economic instruments is illustrated in a simulation exercise, reported in De la Torre, Fajnzylber, and Nash (2009), that estimated the economic costs of water shortages forecast for the Rio Brava basin by 2100.[20] Two scenarios were compared: first, a maladaptation scenario, in which the response to shortages is proportional reductions in all types of uses (agricultural, industrial, residential); and second, a scenario in which water is allocated to the highest-value uses in accordance with efficient pricing. The results indicate that economic costs would be 100 times greater under the first scenario than under the second, illustrating the large potential of efficient adaptation policy to reduce the costs of climate change. From a social protection point of view, however, an important caveat is that measures must be taken to ensure that basic water needs are met. Such a measure could be in the form of a lifeline tariff that ensures that people pay only a minimal rate for the quantity of water essential for basic needs, or it could be differentiated tariffs for different neighborhoods, with poor neighborhoods either exempt from the tariffs or paying only very low rates.

The main institutional challenge for the future is how best to transform the basin councils into bodies that not only advise, but govern, plan, organize, run, control, and supervise water management at the river basin level.

## Water Scarcity, Glacial Melt, and Local Governance Responses in Bolivia

By 2020 climate change is likely to affect the water supply and overall livelihood of up to 77 million people who live in river basins supplied by glacial melt or snowmelt in the Andes region (IPCC 2007b). The projected disappearance of tropical glaciers will cause severe water shortages in the arid and semiarid Andes regions of Bolivia, Colombia, Ecuador, and Peru (Kaser 1999). We use the case of glacial melt in Bolivia to highlight some of the resulting water stress dynamics. Bolivia has one of the largest concentrations of poverty and social and economic inequality in Latin America, and glacial melt threatens not just to diminish water availability but to exacerbate existing inequalities in the country. A promising move is the current decentralization of water resources management, under the policy framework known as Agua para Todos (Water for Everybody).

Glacial melt has been occurring in the Andes for centuries, but so slowly that the retreat of the glaciers was not widely noted. Within the last century, however, melting has accelerated and has been documented with photographs showing glacier coverage declining year after year. This acceleration in the disappearance of glaciers can be linked to higher summer temperatures and declining precipitation. If the glaciers were stable, that is, if they released no more water than they received in the form of precipitation, their net overall effect on the annual water supply would be zero. Their effect would be restricted to flow regulation: during the summer, which is the wet season, they would absorb precipitation thereby reducing runoff and stream flows. During the winter, which is the dry season, that accumulation would gradually melt, contributing to stream flow, which would otherwise be low during the dry season. In other words, a stable glacier would help even out the flow of water over the year but would not affect the total volume of water available.

A melting glacier, on the other hand, does increase water supply while it is melting, but of course, once it has disappeared completely it will no longer add to the flow or affect the timing of water flow. The current situation, with melting glaciers, has helped accommodate increasing water demand and has ensured a continuing flow of water during the dry season. However, the melting has accelerated, causing the volume of glacial lakes to increase to levels at which glacial lake outbursts probably now present a real threat to people's lives and property. When the glaciers have disappeared completely, not only will the people of the Andes no longer receive the boost to water supply they have come to rely on, but they will also lose the regulatory function that the glaciers perform. Thus water availability will decrease during the winter. Summer precipitation will no longer be held back (stored) by the glaciers and will run off into the ocean much faster than previously, leaving less water for consumption unless measures are taken to capture it. In summary, average annual stream flows will increase in the short term, as glaciers melt, but will decline sharply once the glaciers have disappeared. The changes in stream flow are likely to have a large impact on water availability (Vergara and others 2007; Coudrain, Francou, and Kundzewicz 2005).

As detailed in box 1.1 in chapter 1, currently the retreat of the Tuni and Condoriri glaciers gives cause for serious concern, as it threatens the future water supply of two of the largest cities in Bolivia: La Paz and its twin city, El Alto, which are home to around 1.8 million people (Bolivian National Statistical Institute 2009). These glaciers have lost 43 percent and 56 percent of their surface area, respectively, in the last 50 years and are projected to disappear by 2035, if not earlier (World Bank 2008e). Compounding the problem, the population of the twin cities is expected increase by 30 percent by 2020. As changing weather patterns increasingly cause crops to fail on the Andes high plain, subsistence farmers, mostly indigenous, migrate especially to El Alto in search of work. This is a disaster in the making. Without the glaciers, Bolivia's highlands may have to depend on just 400 millimeters of rain annually, unless an alternative is found. Beyond water quantity, water quality is becoming an acute problem in La Paz and El Alto. The area has no sewerage treatment system, and with a reduced water supply the concentration of waste, including solid waste and contamination from mining, will increase (Carvajal 2007), intensifying the danger to health.

The overall implications of glacial melting for Bolivia are dire. Compensating for the loss of glacial flows in the medium term would require billions of dollars of investment.[21] Adaptation measures could include conservation of water supplies in urban areas, a shift to less water-intensive agriculture, both facilitated by water pricing,[22] or the creation of new highland reservoirs to stabilize the cycle of seasonal runoff (Bradley and others 2006). Building massive rain catchment dams in this earthquake-prone zone would be very complex and expensive, however, and to pump up water from Bolivia's distant but water-abundant Amazon basin is not economically viable. Shifting from hydropower to thermal power generation would also be costly.

Added to these challenges, access to water and the price of water for low-income groups have already become highly politicized issues in Bolivia (see chapter 9). That fact, coupled with less water availability in the dry season, will force both national and municipal authorities to pay considerable attention to including stakeholders in discussions of water-related issues, to avoid serious unrest.

Some promising steps have already been taken.[23] In 2002, the Interinstitucional del Agua (CONAIG) was created (through Decree no. 26599) to provide a forum in which government, social, and economic organizations could discuss legal, institutional, and technical aspects of water resources management. The cornerstones of the new Agua para Todos framework are

- establishing sustainable, participatory, and integral water resource management through the Water Resources Management National Strategy, River Basin National Plan, and Water Sector Information Program;
- protecting ecosystems through a desertification prevention program; and
- increasing civil society collaboration through a coordination mechanisms strengthening plan.

This framework is aimed at increasing participation, especially for rural and indigenous communities, while renouncing previous privatization policies within the sector. In addition, the Water Ministry (the national water authority created in 2006) is drafting a water law that integrates the demands for water from multiple users. The new law includes provisions for protecting traditional water-sharing arrangements through a system of official licenses, and it increases the participation of local communities in decision-making processes such as the establishment of water tariffs.

Although these governance-related moves will not solve Bolivia's water scarcity and climate challenges, they will enable the much-needed increased involvement of those most affected at the local level, including organizations of water users and farmers, campesino communities, indigenous peoples, public and private business enterprises, NGOs, universities, and public local entities.

To sum up, all three country cases show that momentum exists in Latin America and the Caribbean toward integrated river basin management. Authorities in the region increasingly accept that institutional frameworks for water management should respond to local needs, taking account of hydrological, social, political, and cultural specifics. Those frameworks should also be set up to conserve water and create incentives to ensure that it is allocated to its highest-valued uses. However, the cases also clearly illustrate that achieving integrated river basin management is very complex. They highlight that integrating the impacts of climate change into integrated river basin management is still in its infancy. So far the greatest progress has been made in Brazil, especially in the northeastern state of Ceará, where decentralized legal, institutional, technical, and social frameworks are showing positive results. That case also illustrates, however, that even with participatory institutional reform, unequal access to knowledge may skew decision making and benefit some participants over others. Despite Mexico's decentralization efforts, the real authority to plan and manage water resources is still vested in a single institution, CONAGUA, at the central level. River basin councils have been established but without a legal framework to grant them real autonomy, funds, and responsibilities. In Bolivia, the decentralization process is off to a promising start, but its impact cannot yet be assessed.

## Policy and Research Perspectives

This chapter has stressed that to meet the increasing challenge of water scarcity in Latin America and the Caribbean will require integrated management of water use at the river basin level, to achieve better allocation across sectors and greater efficiency in the use of water in irrigation systems. The following section provides some suggestions for policy implications.

### Policy Implications

The compounding impact of climate change on water resources increases the urgency of using integrated planning approaches, as is already evident, especially in arid regions. Both water managers and decision makers should be encouraged to engage in greater dialogue with climate and development specialists to better understand the climate-related challenges and how to deal with them. A suggested first step in reducing vulnerability to climate change would be a "no-regrets" approach of improving or encouraging adaptation to existing climate variability while also improving the governance and management of water resources.

To function well, decentralized institutions need solid management and technical expertise, as well as timely access to necessary funds. Although it is crucial to establish the proper legal and institutional frameworks, human and financial resources must also be considered in the long-term planning and management of water resources at the local level.

It is important, too, to recognize that not all solutions to local-level water scarcity and climatic impacts will be found at the local level. National government and institutions need to support local-level adaptation efforts because the combined challenge of water scarcity and climate-related risks can overwhelm local adaptive capacity and institutions. For example, although decentralized, "soft" adaptation measures, such as capacity building and promotion of stakeholder participation, are crucial, they will need to be complemented by more costly, "hard" adaptation measures, such as building infrastructure and providing technology and communication facilities, projects that often require action by the national government. In addition, the cost of such measures can be so overwhelming that the international donor community needs to step in, and that again necessitates significant national government involvement. For instance, the Bolivia case stresses the need for hard adaptation measures, such as extensive rain catchment systems. In addition, while reforming governance is necessary, it is far from the only step needed to build resilience to water scarcity and climatic changes, particularly for the most vulnerable. Often the causes of vulnerability or barriers to coping are societal, depending on social and economic realities or government policies. Increased need therefore exists for enhanced national-level social policies that are responsive to climate risks, such as indexed insurance and social safety nets. The importance of the asset base in coping with natural disasters is discussed in chapter 3.

### Topics for Future Research
An important point highlighted by this chapter is that most of the conventional water well-being indexes, such as the Falkenmark Index, fail to fully measure the complexity of water scarcity, which encompasses not only water availability and use, but also water quality and environmental demand. In addition, these indexes are generally only applied at the national level. They therefore do not capture water stress and vulnerabilities at the subregional and local levels. They also do not consider the climate change factor at all.

Data problems and constraints tend to be the most important limitation to the development of the more comprehensive and useful indexes that are needed (Gleick and others 2002). Particularly with respect to climate change, scaling remains a severe challenge to converting data from global climate models and scenario results into information for the operational local level. Despite these challenges, promising new initiatives are aiming at the development of more sophisticated water indicators. By taking advantage of improving data sets, they seek to address some of the limitations of single-factor indicators. Future fieldwork would therefore benefit from applying the new indexes to better assess the effects of climate change for water resources and water systems.

A promising example is the Vulnerability of Water Systems index developed by Peter Gleick.[24] This index measures five aspects of vulnerability at the watershed or basin level: (a) storage volume relative to total renewable water resources; (b) consumptive use relative to total renewable water resources; (c) proportion of hydroelectricity relative to total electricity; (d) groundwater overdraft relative to total groundwater withdrawals; and (e) variability of flow. The combination of these five aspects makes it possible to measure both water problems and the effects of climate change at the local level. For example, the first measure is an indicator of the ability of an area to withstand prolonged droughts or severe flooding. With large storage compared to supply, short-term droughts are less likely to cause major water shortages. In contrast, when this ratio is small, changes in the intensity of floods and droughts may be more strongly felt.

Another example is the Climate Vulnerability Index (CVI) being developed by the Oxford University Centre for the Environment. It provides a powerful tool to express systematically the vulnerability of human communities in relation to water resources, using a holistic approach that integrates the physical, social, economic, and environmental factors. Notably, the CVI is suitable for examining vulnerability to present levels of climate variability. It can also be used to examine the impacts of climate change, combining climate scenarios with expected changes in technological, social, economic, and environmental conditions. Another promising feature is that the index's outputs can be linked directly to impacts on people through application of the sustainable livelihoods framework (SLF), which allows the poverty implications to be examined. Tentative fieldwork from locations in South Africa, Sri Lanka, and Tanzania confirms that the poor are the most vulnerable to climate impacts. Finally, the identification of zones of vulnerability provides a systematic rationale for determining priority areas where proactive measures to protect populations should be taken.[25]

Finally, to develop a set of best practices of use to the LAC region, it would be interesting to focus future case studies on how IWRM can be tailored to local circumstances and how scientific knowledge about climate change and IWRM principles is being applied on the ground.[26]

## Notes

1. For the past two decades, the Falkenmark Water Stress Index has been the most influential and powerful water measure. Generally, 1,700 cubic meters ($m^3$) per person per year is considered the level above which water shortages are rare and localized. Between 1,700 and 1,000 $m^3$ per person per year is considered regular water *stress*. Between 1,000 and 500 $m^3$ per person per year is considered chronic water *scarcity*, which begins to hamper health, economic development, and well-being, and below 500 $m^3$ per person per year is considered *absolute/severe* water scarcity, where water availability is a primary constraint on life. For more elaboration on the Falkenmark Water Stress Index, see Gleick and others 2002.

2. Notably, a figure for The Bahamas is not available.

3. The arid northwest and central regions contain 77 percent of Mexico's population and generate 85 percent of GDP (*Wikipedia.org*, "Mexico").

4. See also World Bank Water and Sanitation Web site for info on the LAC region, http://web.worldbank.org/WBSITE/EXTERNAL/TOPICS/EXTWAT/0,,contentMDK:21706928~menuPK:4602430~pagePK:148956~piPK:216618~theSitePK:4602123,00.html. Surface water and, to a lesser extent, groundwater in these cities has become contaminated with varying degrees of detergents (soaps and solvents), pesticides, petroleum and its derivatives, toxic metals (such as lead and mercury), fertilizers and other plant nutrients, oxygen-depleting compounds (for example, waste from canneries, meat-processing plants, slaughterhouses, and paper and pulp-processing operations), and disease-causing agents responsible for hepatitis and infections of the intestinal tract such as typhoid fever, cholera, and dysentery (Anton 1993).

5. World Bank Water and Sanitation Web site, http://web.worldbank.org/WBSITE/EXTERNAL/TOPICS/EXTWAT/0,,contentMDK:22005375~pagePK:148956~piPK:216618~theSitePK:4602123,00.html.

6. These figures are obtained from a probit analysis. Poverty is identified as the lowest quartile of a composite assets-and-durables index, as there are no monetary values in the demographic and health surveys on which the analysis is based. Data are from various years.

7. Peru's 18 glaciers have lost 22 percent of their surface over the past 27 to 35 years, according to a report by Peru's National Meteorology and Hydrology Service (SENAMHI); see Vergara 2008.

8. The number of glaciers that have entirely melted is small; many are reduced to small remnants. However, these small remnants can persist, especially those at higher elevations that face away from the sun (southern exposures in the Southern Hemisphere) (Ben Orlove, pers. comm.; IPCC 2007b).

9. Referring back to the rural poverty discussion, it should be noted that there are very few small-scale poor farmers in northern Mexico. They are mostly located in the southern part of the country.

10. Information provided by Jens Hesselberg Christensen. More details on these projections are provided in IPCC 2007a (Working Group I Table 11.1, 11.6); and IPCC 2007b (Working Group II 13.3.1).

11. Precipitation—the principal input to freshwater systems—is not adequately simulated in present climate models. Quantitative projections of changes in river flow at the basin scale—of prime importance for water management— remain largely uncertain (Milly, Dunne, and Vecchia 2005; Nohara and others 2006). However, it can safely be asserted that the much greater warming predicted for the 21st century will produce significant changes in evaporation and precipitation, allied with a more unpredictable hydrological cycle (IPCC 2007b).

12. Quantifying this impact poses a challenge. To date only a few studies have analyzed the potential impact of climate change and variability on water quality (Jiménez 2003). Moreover, most of those focus on developed countries and do not address notable, distinct differences in water quality problems between developed and developing countries.

13. The A1 scenario is one of the IPCC's constructed story lines of plausible human, demographic, economic, political, and technological futures related to future emissions levels. The A1 scenario describes a future world of very rapid economic growth, global population that peaks in midcentury and declines thereafter, and rapid introduction of new technologies. For more information, see IPCC 2001.

14. Hydropower supplies more than 70 percent of total energy needs in Ecuador and 76 percent in Peru (World Bank 2008b).

15. These estimates are higher by between 6 million and 20 million additional people than earlier estimates of populations living in water-stressed areas by 2055 (Arnell 2004).

16. The principles and practices of IWRM are largely based on four principles of the 1992 Dublin Declaration. In particular, the second principle states that "water development and management should be based on a participatory approach, involving users, planners and policy makers at all levels" (IRC International Water and Sanitation Centre 2006).

17. Tendler (1997) provides a comprehensive account of the process.

18. Smit and others (2000) note that the ability to transfer knowledge and adopt innovation is an essential factor in building adaptive capacity to climate variability and change.

19. The most progress has been made in the basin with the only functional council, the Lema-Chapala basin—or rather in the states of Guanajuato and Queretaro, which are two of the five states in which the basin lies. In these two states, technical committees for groundwater (COTAS) have been established with the purpose of reducing serious overexploitation of the aquifers (Tortajada 2006). With the participation of the different water users, new

criteria for water allocation have been developed. But as Tortajada points out, this innovative initiative is not enough to affect water management in the basin overall. The COTAS are not managed in a uniform manner, and they exist in only two of the basin's five states.

20. This exercise was undertaken by Mendelsohn (2008) in a background paper for De la Torre, Fajnzylber, and Nash (2009).

21. According to Amat y Leon and his coauthors (2008), climate change and variability losses could amount to US$30 billion annually in the four Andean countries by 2025.

22. Efficient water pricing would allocate water to its highest-valued uses. It should reduce wasteful use, which is a big problem particularly affecting agriculture, the biggest consumer of water. However, in the absence of programs ensuring the poor a basic income through a cash safety net, which they could spend as they wish, paying the same price for each good as everyone else, it may be necessary to introduce a measure to protect the poor, who otherwise might not be able to afford water for their basic needs. For example, a lifeline tariff would price a quantity of water sufficient for basic needs very low, with a higher price applying to quantities over the basic needs threshold. The drawback of this option is that it requires metering, which rarely is present. Another option could be a system of fixed charges based on neighborhood, with a very low (or no) charge in poorer neighborhoods.

23. This and the following paragraph are based on information from *Wikipedia.org,* "Water Resources Management in Bolivia."

24. More details on this index, along with additional indexes, are provided in Gleick and others 2002.

25. More information can be found at http://ocwr.ouce.ox.ac.uk/research/wmpg/cvi.

26. Thanks to R. Cessti for bringing our attention to this recommendation.

## References

Alfaro, J. F., and J. Marin. 1994. "On-farm Water and Energy Use for Irrigation in Latin America." In H. Garduño and F. Arreguín-Cortés (eds.), *Efficient Water Use.* Montevideo, Uruguay: UNESCO–ROSTLAC.

Amat y Leon, Carlos, B. Seminario, M. P. Cigaran, S. Bambaren, L. Macera, M. T. Cigaran, and D. Vasquez. 2008. *El Cambio Climatico no Tiene Fronteras— Impacto del Cambio Cliamtico en la Comunidad Andina.* Secretaria General De la Comunidad Andina. Lima, Peru.

Anton, D. J. 1993. *Thirsty Cities: Urban Environments and Water Supply in Latin America.* Ottawa, Canada: International Development Research Center.

Arnell, N. W. 2004. "Climate Change and Global Water Resources: SRES Scenarios and Socio-economic Scenarios," *Global Environmental Change* 14: 31–52.

Avissar, R., and D. Werth. 2005. "Global Hydroclimatological Teleconnections Resulting from Tropical Deforestation," *Journal of Hydrometeorology* 6: 134–45.

Bolivian National Statistical Institute. 2009. http://www.ine.gov.bo (accessed May 21, 2009).

Bradley, R., M. Vuille, H. F. Diaz, and W. Vergara. 2006. "Threats to Water Supplies in the Tropical Andes," *Science* 312 (23), June.

Carvajal, Liliana. 2007. *Impacts of Climate Change on Human Development*. Background paper to *Fighting Climate Change: Human Solidarity in a Divided World, Human Development Report 2007/08* (UNDP 2007/08). New York: United Nations Development Program.

Coudrain, Anne, Bernard Francou, and Zbigniew Kundzewicz. 2005. "Glacial Shrinkage in the Andes and Consequences for Water Resources—Editorial." *Hydrological Sciences–Journal des Sciences Hydrologiques* 50 (6): 925–32.

Cox, P. M., R. A. Betts, M. Collins, P. P. Harris, C. Huntingford, and C. D. Jones. 2004. "Amazonian Forest Dieback under Climate Carbon Cycle Projections for the 21st Century." *Theoretical and Applied Climatology* 78: 137–56.

DANIDA (Danish International Development Agency). 2008. "Dialogue on Climate Change Adaptation for Land and Water Management." Draft Concept Paper. Copenhagen: DANIDA.

De la Torre, Augusto, Pablo Fajnzylber, and John Nash. 2009. *Low Carbon, High Growth. Latin American Responses to Climate Change. An Overview*. World Bank Latin American and Caribbean Studies. Washington, DC.: World Bank

Engle, N. L. 2007. "Adaptive Capacity of Water Management to Climate Change in Brazil: A Case Study Analysis of the Baixo Jaguaribe and Pirapama River Basins." Master's thesis, School of Natural Resources and Environment, University of Michigan. http://hdl.handle.net/2027.42/50490.

Estrada, Daniela. 2006. "Latin America: Family Farms—Durable but Fragile." Inter Press Service (IPS) News Agency. http://ipsnews.net/news.asp?idnews=34996.

Fay, Marianna, ed. 2005. *The Urban Poor in Latin America*. World Bank Report. Washington, DC: World Bank.

Fernandez, B. P., and L. B. Graham. 1999. "Sustainable Economic Development through Integrated Water Resources Management in the Caribbean." Paper Presented at the II Water Meeting, Montevideo, Uruguay, June 15–18.

Formiga-Johnsson, R. M., and K. E. Kemper. 2005. *Institutional and Policy Analysis of River Basin Management—The Jaguaribe River Basin, Ceará, Brazil*. Policy Research Working Paper 3649, World Bank, Washington, DC.

Gleick, P., with W. C. G. Burns, E. L. Chalecki, M. Cohen, K. K. Cushing, A. S. Mann, R. Reyes, G. H. Wolff, and A. K. Wong. 2002. *The World's Water—The Biannual Report on Freshwater Resources.* Washington, DC: Island Press.

Guerrero, V., and F. Garcia-Leon. 2003. "Proposal for the Decentralization of Water Management in Mexico by Means of Basin Councils." In *Water Policies and Institutions in Latin America,* ed. C. Tortajada, B. P. F. Braga, A. K. Biswas, and L. Garcia. 244–59. New Delhi: Oxford University Press.

IADB (Inter-American Development Bank). 1999. *Strategy for Agricultural Development in the Latin American and Caribbean Region.* Washington, DC: Inter-American Development Bank.

IPCC (Intergovernmental Panel on Climate Change). 2001. *Special Report on Emission Scenarios.* Geneva: IPCC.

———. 2007a. *Climate Change 2007: The Physical Science Basis.* Contribution of Working Group I to the Fourth Assessment Report of the IPCC. Geneva: IPCC.

———. 2007b. *Climate Change 2007: Impacts, Adaptation, and Vulnerability.* Contribution of Working Group II to the Fourth Assessment Report of the IPCC. Geneva: IPCC.

———. 2007c. *Climate Change 2007: Synthesis Report.* Contribution of Working Groups I, II and III to the Fourth Assessment Report of the Intergovernmental Panel on Climate Change. Geneva: IPCC. http://www.ipcc.ch/publications _and_data/publications_ipcc_fourth_assessment_report_synthesis_report.htm.

———. 2008. "Technical Paper on Climate Change and Water." IPCC-XXVIII/Doc.13 (8.IV.2008). WMO/UNEP, Switzerland.

IRC (International Water and Sanitation Center). 2006. *An Integrated Water Resources Management Primer.* http://www.irc.nl/page/10433.

Jiménez, B. 2003. "Health Risks in Aquifer Recharge with Recycled Water." In *State of the Art Report—Health Risks in Aquifer Recharge Using Reclaimed Water,* ed. R. Aertgeerts and A. Angelakis. 16–122. WHO/SDE/WSH/03.08. Copenhagen, Denmark: WHO Geneva and WHO Regional Office for Europe.

Kaser, G. 1999. "A Review of the Modern Fluctuations of Tropical Glaciers." *Global and Planetary Change* 22: 93–103.

Kronik, Jakob, and Dorte Verner. 2010. *Indigenous Peoples and Climate Change in Latin America and the Caribbean.* Washington, DC: World Bank.

Lemos, M. C. 2007. *Drought, Governance, and Adaptive Capacity in Northeast Brazil: A Case Study of Ceará.* Occasional paper for UNDP 2007/2008. Human Development Report Office. New York: United Nations Development Program.

Lemos, M. C., and J. L. F. Oliveira. 2004. "Can Water Reform Survive Politics? Institutional Change and River Basin Management in Ceará, Northeast Brazil," *World Development* 32 (12): 2121–37.

Maskrey, A., Gabriella Buescher, Pascal Peduzzi, and Carolin Schaerpf. 2007. *Disaster Risk Reduction: 2007 Global Review.* Consultation ed. Prepared for the Global Platform for Disaster Risk Reduction First Session, Geneva, June 5–7.

Mendelsohn, R. 2008. "Impact of Climate Change on the Rio Bravo River." Background paper for De la Torre, Fajnzylber, and Nash, 2009. World Bank Latin American and Caribbean Studies, World Bank, Washington, DC.

Miller, K. 2006. *Land under Siege. Recent Variations in Sea Level through the Americas.* Faculty of Engineering, University of the West Indies, St. Augustine, Trinidad and Tobago, West Indies.

Milly, P. C. D., K. A. Dunne, and A. V. Vecchia. 2005. "Global Pattern of Trends in Streamflow and Water Availability in a Changing Climate." *Nature* 438: 347–50.

National Environmental Advisory Board/Ministry of Health and the Environment. 2000. "Initial National Communication on Climate Change. St. Vincent and the Grenadines." General Secretariat, Organization of American States, Washington. DC.

National Research Council. 2003. *Cities Transformed: Demographic Change and Its Implications in the Developing World.* Panel on Urban Population Dynamics, Committee on Population, Division of Behavioral and Social Sciences and Education. Washington, DC: National Academies Press.

Nohara, D., A. Kitoh, M. Hosaka, and T. Oki. 2006. "Impact of Climate Change on River Runoff." *Journal of Hydrometeorology* 7: 1076–89.

OECD (Organization for Economic Cooperation and Development). 2003. *OECD Territorial Reviews: Mexico.* Paris: OECD.

———. 2007. *Ranking Port Cities with High Exposure and Vulnerability to Climate Extremes: Exposure Estimates.* December. OECD Environment Directorate. Paris: OECD.

Ohno, A, A. Marui, E. S. Castro, A. A. Reyes, D. Elio-Calvo, H. Kasitani, Y. Ishii, and K. Yamaguchi. 1997. "Enteropathogenic Bacteria in the La Paz River of Bolivia." *American Journal of Tropical Medical Hygiene* 57(4): 438–44.

Oki, T., and S. Kanae. 2006. "Global Hydrological Cycles and World Water Resources." *Science* 313: 1068–72.

Porto, M. 1998. "The Brazilian Water Law: A New Level of Participation and Decision Making." *International Journal of Water Resources Development* 14 (2): 175–82.

Quijandria, B., A. Monares, and R. U. de Peña Montenegro. 2001. *Assessment of Rural Poverty, Latin America and the Caribbean.* Santiago, Chile: International Fund for Agricultural Development.

Ravallion, Martin, Shaohua Chen, and Prem Sangraula. *New Evidence on the Urbanization of Global Poverty.* Background paper for *World Development*

*Report: Agriculture for Development* (2008). Development Research Group, World Bank, Washington, DC.

Ringler, Claudia, Mark W. Rosegrant, and Michael S. Paisner. 2000. *Irrigation and Water Resources in Latin America and the Caribbean: Challenges and Strategies*. EPTD Discussion Paper No. 64. Environment and Production Technology Division. Washington, DC: International Food Policy Research Institute.

Smit, B., I. Burton, R. J. T. Klein, and J. Wandel. 2000. "An Anatomy of Adaptation to Climate Change and Variability." *Climatic Change* 45 (1): 223–51.

Taddei, R. 2005. "Of Clouds and Streams, Prophets and Profits: The Political Semiotics of Climate and Water in the Brazilian Northeast." PhD diss. Graduate School of Arts and Sciences, Columbia University. http://citeseerx.ist.psu.edu/viewdoc/summary?doi+10.1.1.85.4893.

Tendler, Judith. 1997. *Good Governance in the Tropics*. Baltimore, MD: Johns Hopkins University Press.

Tortajada, C. 2000a. *Sustainability of Water Resources Management in Mexico*. Mexico City: Third World Center for Water Management.

———. 2000b. *River Basins: Institutional Framework and Management Options for Latin America*. WCD Thematic Reviews, Institutional Processes V5. Mexico City: Third World Center for Water Management.

———. 2006. *Water Governance with Equity: Is Decentralization the Answer? Decentralization of the Water Sector in Mexico and Intercomparison with Practices in Turkey and Brazil*. Occasional paper for UNDP 2006. Human Development Report Office. New York: United Nations Development Program.

UNDP (United Nations Development Program). 2006. *Beyond Scarcity: Power, Poverty, and the Global Water Crisis, Human Development Report 2006*. New York: Oxford University Press.

UNEP FI (United Nations Environment Program, Finance Initiative). 2004. *Challenges of Water Scarcity—A Business Case for Financial Institutions*. UNEP Finance Initiative. Stockholm: International Water Institute.

UN Population Division. 2006. *World Urbanization Prospects: The 2005 Revision*. Department of Economic and Social Affairs. Urban and Rural Areas Dataset (POP/DB/WUP/Rev.2005/1/Table A.6). http://esa.un.org/unup/. New York: United Nations.

UN Water. 2006. *Water— A Shared Responsibility*. United Nations World Water Development Report 2, World Water Assessment Program. Paris: UNESCO/ New York: Berghahn Books.

———. 2007. "Coping with Water Scarcity. Challenge of the 21st Century." Campaign material for World Water Day, March 22. UN WATER/FAO. http://www.worldwaterday07.org.

U.S. Army Corps of Engineers. 2004. *Water Resources Assessment of the Bahamas.* Mobile District and Topographic Engineering Center, United States Southern Command. http://www.sam.usace.army.mil/en/wra/Bahamas/ BAHAMASWRA.pdf.

Vergara, W. 2008. "Retracting Glacier Impacts: Economic Outlook in the Tropical Andes." Latin America and Caribbean Department, April 23, World Bank, Washington, DC.

Vergara, W., A. M. Deeb, A. M. Valencia, R. S. Bradley, B. Francou, A. Zarzar, A. Grünwaldt, and S. M. Haeussling. 2007. "Economic Impacts of Rapid Glacier Retreat in the Andes." *Eos* 88 (25): 261–64.

WBGU (German Advisory Council on Global Change). 2008. "Climate Change as a Security Risk." London: Earthscan. http://www.wbgu.de/wbgu_jg2007 _engl.html.

Wester, P., D. J. Merrey, and M. de Lange. 2003. "Boundaries of Consent: Stakeholder Representation in River Basin Management in Mexico and South Africa." *World Development* 31 (5): 797–812.

WMO/IADB (World Meteorological Organization and Inter-American Development Bank). 1996. *Water Resources Assessment and Management Strategies in Latin America and the Caribbean.* Proceedings of the WMO/IADB Conference, San José, Costa Rica, May 6–11.

World Bank. 2000. "Argentina: Poor People in a Rich Country." Report No. 19992-AR, World Bank, Washington, DC.

———. 2001. "Attacking Brazil's Poverty: A Poverty Report with a Focus on Urban Poverty Reduction Policies." Report No. 20475-BR, World Bank, Washington, DC.

———. 2002. "Mexico Urban Development: A Contribution to a National Urban Strategy." Report No. 22525, World Bank, Washington, DC.

———. 2005. *World Development Indicators 2005.* Washington, DC: World Bank.

———. 2006. "Project Document for Regional Implementation of Adapation Measures in Coastal Zones (SPACC) Project." World Bank Latin America and Caribbean (LAC)/Global Environment Facility, World Bank, Washington, DC.

———. 2008a. *World Development Report: Agriculture for Development.* Washington, DC: World Bank.

———. 2008b. "Poverty Data: A Supplement to *World Development Indicators 2008.*" Washington, DC: World Bank.

———. 2008c. *Equity and Efficiency in the Greenhouse: Responding to Climate Change in Latin America and the Caribbean Volume I: Overview.* Sustainable Development Department (LCSSD) and Chief Economist Office (LCRCE), Latin American and the Caribbean Region. Washington, DC: World Bank.

————. 2008d. "Poverty and Social Impact Analysis of Groundwater Overexploitation in Mexico." September. Latin America and Caribbean Region, World Bank, Washington, DC.

————. 2008e. "Project Appraisal Document on a Proposed Grant from the Global Environmental Facility Special Climate Change Fund for an Adaptation to the Impact of Rapid Glacier Retreat in the Tropical Andes Project." World Bank, Washington, DC.

World Resources Institute (WRI). 1998. *World Resources Report, 1998–99: Environmental Change and Human Health.* http://www.wri.org/publication/world-resources-1998-99-environmental-change-and-human-health.

# CHAPTER 3

# Climate Change, Disaster Hot Spots, and Asset Erosion

## Tine Rossing and Olivier Rubin

Latin America and the Caribbean are among the regions of the world most prone to climate-related hazards. Flooding and landslides are common in the region, with its complex river basin systems and mountainous terrain. Tropical storms and hurricanes, formed in both the Pacific and Atlantic Oceans, are frequent throughout, especially in the Caribbean hurricane belt, which includes Central America. Climatic variability, in the form of severe droughts, floods, and high winds throughout the hemisphere, has been exacerbated by the recurrent El Niño phenomenon (IADB 2000).

It is therefore alarming that growing evidence suggests that climate change and variability will increase the region's exposure to natural hazards and climate-related disasters (IPCC 2007b; Mendelsohn and Williams 2004; Nagy and others 2006; Vergara and others 2007). The long-term effects of global warming include changes in average temperatures and precipitation, deglaciation, and sea level rise. Climatic variability is also on the rise; its effects are already noticeable. Many natural hazards are projected to become more intense and more common, and increased variability will also create more surprises, such as natural disasters occurring in succession or in places where they have never been experienced before.

Though most of the projections of the Intergovernmental Panel on Climate Change (IPCC) are on a global or regional scale, the impact of climate change, particularly through compounded weather-related disasters, is local. To provide information about the areas potentially the most vulnerable to a changing climate, this chapter first briefly reviews the past patterns of natural hazards and weather-related disasters in Latin America and the Caribbean (LAC), given that the geographical areas that have been affected in the past are highly likely to be hit again and perhaps more severely. It explains that hazard hot spots are not necessarily disaster zones because disasters are as much a function of vulnerability—the susceptibility to harm from events—as they are of the events themselves. The chapter then analyzes how climate-related disasters might affect the livelihoods of vulnerable people, using the sustainable livelihoods framework (SLF) introduced in chapter 1. Because not all groups are equally vulnerable, even within a given hot spot country or area, the chapter ends by stressing the need to apply a context-specific framework to ensure a strong match between the actual needs of particular vulnerable groups and interventions proposed to help them recover from, and adapt to, climate-related disasters.

## Natural Hazards and Disasters in Latin America and the Caribbean

Not only do the frequency and types of natural hazards vary across the LAC region, but hazards also have highly diverse socioeconomic impacts. Thus, the trends and patterns of natural hazards only overlap partially with the occurrence of natural disasters.

### Patterns of Natural Hazards

Where and when natural hazards will occur remains unpredictable. Yet based on past events, it is possible to outline some general trends for the LAC region, where different climate zones experience different rainfall patterns and propensities for hurricanes, floods, and droughts.

To date, the incidence of destructive storms has been highest in Mexico and the Central America and Caribbean subregions—areas that are in the pathways of both western Atlantic and eastern Pacific hurricanes and tropical storms. Since 1970, according to the Emergency Events Database (EM-DAT),[1] more than 40,000 Atlantic and Pacific hurricanes and tropical storms have been recorded, an annual average of 1,053 storms.

Floods are the most common natural hazard in the region. They are a function of the climate patterns of rainfall and storms, as well as of hydrology (the shape of riverbeds, intensity of drainage, and debit flow of rivers) and soil characteristics (absorption capacity). In Central

America and the Caribbean, storms that develop along the intertropical convergence zone and the subtropical high-pressure zone dominate the weather. As a result of these patterns, flooding tends to be associated with hurricanes and other tropical storms, which generate very heavy rainfall continuing over several days (Charveriat 2000). In South America, a major factor in the occurrence of floods is the El Niño–Southern Oscillation (ENSO; see Box 3.1).[2]

---

**Box 3.1**

### Flooding in Bolivia Caused by El Niño Phenomenon

In January 2008, serious flooding across Bolivia brought tragedy for hundreds of people. Due to the severity of the situation, the Bolivian government declared a state of national emergency. The flooding affected all of the country's departments to some degree. It is estimated that 80 percent of the people who had been similarly affected in 2007 also suffered from the torrential rain of 2008. The global climate is changing and Bolivia is feeling it.

Natural hazards are recurrent in Bolivia, and their effects often turn into devastating disasters, particularly in the rainy season, which extends from November to March. These disasters are becoming more frequent and severe. From the end of 2007 to the beginning of 2008, heavy rains caused considerable damage in the departments of Cochabamba, Santa Cruz, Chuquisaca, La Paz, and Potosí. According to civil protection figures, by late January 2008, the floods had affected 25,981 families, claimed 30 lives, and left five missing. True to forecasts, the rains continued through to March. The people affected tried to salvage what little they had left, but also feared for their lives because poor road conditions meant that food supplies took a long time to get through.

The civil protection services coordinate emergency operations in the country and use sectoral commissions to deal with issues such as food, health, water and sanitation, and shelters at both the national and department levels. The National Emergency Operations Center distributed around 200 tons of food, including corn flour, rice, sugar, corn, vegetable oil, and salt, between January 24 and the end of March 2008. Blankets, mattresses, spades, picks, hoes, and wheelbarrows were also distributed. In addition to the Bolivian Red Cross, the center cooperates with international organizations such as the Pan American Health Organization, World Vision, Oxfam, and the UN System in Bolivia to carry out needs assessments and evaluations. Further action will be undertaken to promote economic recovery and protect the livelihoods of the most vulnerable families.

*(continued)*

**Box 3.1** *(continued)*

Workers lay sandbags along the Rio Grande, some 50 kilometers (31 miles) east of Santa Cruz, January 25, 2008. The Red Cross reported that heavy rains and flooding in much of Bolivia killed 30 people and caused hardship for nearly 25,000 families.

According to the Bolivian Red Cross, shortly after the disaster struck, Bolivia had a preliminary response plan in place to provide humanitarian assistance to a total of 1,200 families—600 in Santa Cruz, 300 in Potosí, and another 300 in La Paz—in the form of food parcels and hygiene kits. "One of the most serious problems that we encounter is that the secondary roads often are impassable as a result of the floods, which makes it difficult to get supplies and personnel specialized in assessment and response through to where they are needed," said Rubén Romero, disaster management delegate.

*Source:* Red Cross (2008), http://www.ifrc.org/docs/news/08/08013103/; and authors' elaboration based on fieldwork in Bolivia, in March 2008.

A warming climate may contribute to an increase in the frequency and intensity of El Niño phenomena. By itself, El Niño does not create a hazard, but it is estimated to be the direct cause of changes in precipitation patterns that in turn cause floods and droughts. In Latin America and the Caribbean, El Niño causes higher rainfall in Argentina, Ecuador, Paraguay, Peru, and south Brazil (figure 3.1) (Charveriat 2000). In Central America, El Niño leads to excessive rainfall along the Caribbean coasts, while the Pacific coast remains dry. In South America, it causes precipitation to increase on the coasts of Ecuador, the northern part of Peru, and the southern zones of Chile. In Bolivia, Ecuador, and Peru, projections of increasing drought in the mountainous Andean region imply a continuous retreat of glaciers, with subsequent changes in the availability of water and in local biodiversity. In Colombia, Guyana, and Venezuela, projections suggest

**Figure 3.1    Climate Impacts of El Niño Phenomenon in Latin America and the Caribbean, 1998**

IBRD 37782
MAY 2010

*Sources:* Adapted from IPCC 2001. FAO 2002. UNEP 2003. Climate impacts of El Niño Phenomenon in Latin America and the Caribbean (2005). In *UNEP/GRID-Arendal Maps and Graphics Library.*

that precipitation will tend to decline, leading to drought in northeast Brazil. Finally, for Argentina, Paraguay, and Uruguay, projections indicate increasing rainfall, and this shift in the hydrological balance will cause temperatures in southern Brazil to rise.

Hydrological systems also contribute to flood risks. As Charveriat (2000) points out, the major drainage divide is far to the west of the South American continent, along the crest of the Andes. In the mountainous regions west of the divide, the slopes of riverbeds are very steep. During storms, these slopes increase the risk of the most dangerous type of floods—flash floods. The major basins lie east of the Andes, and the main rivers flow to the Atlantic Ocean. The four largest drainage systems—the Amazon, Río de la Plata (Paraguay, Paraná, and Uruguay Rivers), Orinoco, and São Francisco—cover nearly three-fourths of the continent. As flows from tributaries meet, the risk of flooding in the lower parts of these drainage systems is very high, especially when there is sedimentation or when river channels are not well defined (Charveriat 2000).

Though the LAC region has been less affected by slow-onset weather-related disasters—especially droughts—than by the rapid-onset variety—storms and floods—seasonal drought takes place in climates that have well-defined annual rainy and dry seasons. The region's arid (northeast Brazil, Mexico) and cold (south Chile) climate zones have a greater propensity for seasonal drought (figure 3.1). Unpredictable drought is associated with abnormal rainfall patterns and is most characteristic of humid and subhumid climates. The most important cause of this type of drought in Latin America and the Caribbean is El Niño, which results in dryer conditions in northeast Brazil during the Southern Hemisphere summer. During the Southern Hemisphere winter, the climatic effect of El Niño is drier conditions in Central America, Colombia, and Venezuela (Swiss Reinsurance Company 1998; WMO 1999).

Looking ahead at the prospects for natural hazards, the IPCC's global projections indicate that the number of extremely hot days will increase, while the number of very cold days will decrease. Whereas the global trends generally hold up for the LAC region, they conceal great variation across subregions. Average precipitation is projected to rise, but the intensity and frequency of extreme precipitation are also likely to increase in coming decades. The probability that these trends will lead to more flooding and landslides is very high. The midcontinental areas, such as the inner Amazonian and northern Mexico, are projected to become dryer during the summer months, with a greater risk of drought and forest fires. The

IPCC notes that changes in these types of extreme events due to climate change and variability can already be observed.

As regards future small-scale atmospheric phenomena, such as hurricanes, cyclones, and tidal waves, the IPCC projections are much more uncertain. The traditional IPCC models project a consistent increase in precipitation intensity in future storms. That projection is supported by more recent, higher-resolution models that project increases in peak wind intensities with some consistency and a rise in mean and peak precipitation intensities in future tropical cyclones with even greater consistency (IPCC 2007a). Some of these disaggregated models have even projected fewer but more intense storms, due to climate change and variability, with potential to cause much more damage than a greater number of less-intense storms.

Patterns of actual and predicted natural hazards show that the areas most prone to feel the effects of climate change are also the poorest. The areas at highest risk from storms and flooding include most of the Caribbean (especially Cuba, Dominica, the Dominican Republic, Haiti, Puerto Rico, and St. Lucia); Guatemala, Honduras, and Nicaragua in Central America; and the Bolivian, Brazilian, and Peruvian Amazon regions in South America. Areas highly vulnerable to drought include northeast Brazil, the arid Andean region (particularly Bolivia and Peru), and the north of Mexico.

## Patterns and Impacts of Natural Disasters

A natural hazard is a necessary, but not a sufficient, condition for a natural disaster. Though the LAC region is likely to face more intense and more frequent weather-based natural hazards, such as floods, droughts, storms, and extreme temperatures, the extent to which these will pose serious problems depends crucially on local vulnerabilities. A natural disaster can be defined as a temporary event triggered by a natural hazard that overwhelms local response capacity and seriously affects social and economic development (Hodell, Curtis, and Brenner 1995). The sources of risk for such disasters in Latin America and the Caribbean are both natural and man-made. The high economic and social costs associated with climate-related calamities are mainly a result of vulnerability.

Although livelihoods have constantly adapted to change throughout history, global warming will compound existing vulnerabilities. As pointed out by Sokona and Denton (2001), it is very likely that the impacts of climate change and variability may push people beyond their capacity to cope and adapt, particularly because of greater magnitude or frequency of weather-related disasters such as droughts, storms, and floods.

A consensus exists that natural disasters have increased worldwide during recent decades, whether one considers frequency, numbers of people affected, or amounts of economic damage (Red Cross/Red Crescent 2007; UNDP 2004, 2007; UNEP 2007). The rise in natural disasters has been caused almost wholly by a doubling of weather-related hazards.[3] Globally, such disasters (mainly floods, droughts, and storms) increased from about 200 in 1990 to almost 400 in 2006.[4] The number of people they affected worldwide increased from fewer than 50 million annually at the end of the 1970s, to 262 million annually during 2000–04. The brunt of the increase in extreme weather is being borne by the developing world, and within the developing world, by the poorest of the poor.

An analysis of weather shocks in Latin America and the Caribbean in the period 1970–2008 is consistent with this general picture. Table 3.1 illustrates the incidence of the three major types of weather-related disasters during 1970–2008. It shows that floods and storms account for the clear majority of disasters and that the trend is rising substantially, from a little more than 100 in 1970–79 to around 400 in 2000–08.

The type and frequency of natural disasters and their impact vary widely across Latin America and the Caribbean (table 3.2). Most of the weather-related hazards occur in South America, which is also the subregion where they cause the greatest number of fatalities (46 percent of the regional total) and which has the greatest share of affected population (70 percent). Central America follows closely behind with 41 percent of the region's fatalities due to weather disasters, while the Caribbean is a distant third at 13 percent.

When one takes account of total land area and population size, however, the picture changes (table 3.3). Judged by fatalities per capita,

**Table 3.1    Frequency of Weather-Related Disasters (Floods, Droughts, and Storms) in the LAC Region, 1970–2008**

| Years | Droughts | Floods | Storms | Total weather-related disasters |
|---|---|---|---|---|
| 2000–08 | 28 | 226 | 148 | 402 |
| 1990–99 | 29 | 149 | 128 | 306 |
| 1980–89 | 19 | 136 | 69 | 224 |
| 1970–79 | 11 | 69 | 34 | 114 |
| Total | 87 | 580 | 379 | 1,046 |

*Source:* Authors' elaboration based on EM-DAT data. http://www.emdat.be.

Table 3.2    Impact of Weather-Related Disasters (Floods, Droughts, Storms, and Extreme Temperatures) on People in the LAC Region, by Subregion, 1970–2008
*Percentages*

|  | South America | Caribbean | Central America |
|---|---|---|---|
| Frequency | 42.1 | 28.5 | 29.4 |
| Deaths | 46.2 | 13.1 | 40.7 |
| Affected | 70.1 | 17.0 | 12.9 |
| Homeless | 50.6 | 32.4 | 17.0 |
| Injured | 97.1 | 0.6 | 2.3 |
| Damage (US$) | 31.4 | 33.0 | 35.6 |

*Source:* Authors' elaboration based on EM-DAT data; http://www.emdat.be.
*Note:* According to EM-DAT, affected populations are those requiring immediate assistance during a period of emergency for basic survival needs such as food, water, shelter, sanitation, and medical assistance. Individuals are also considered affected when a significant number of cases of infectious diseases appear in a region or a population that is usually free from the disease. http://www.emdat.be.

Table 3.3    Weather-Related Disaster Exposure Indicators in Latin America and the Caribbean, by Subregion, 1970–2008

|  | South America | Caribbean | Central America |
|---|---|---|---|
| Frequency per square kilometer (%) | 0.3 | 13.4 | 6.2 |
| Fatalities per million inhabitants | 122.0 | 323.1 | 1,043.6 |
| Population affected per thousand inhabitants | 249.2 | 665.3 | 524.3 |

*Source:* Authors' elaboration based on *Wikipedia.org* and United Nations Population Division 2006.
*Notes:* Frequency per square kilometer is calculated based on cumulative events between 1970 and 2008, divided by total square kilometers. Deaths per million inhabitants are calculated by dividing cumulative deaths 1970–2008 by 2005 population estimates. Affected population per thousand inhabitants is calculated by dividing the cumulative affected population 1970–2008 by 2005 population estimates. From a social perspective, calculating damages as a percentage of gross national product (GDP) is a highly imperfect measure for the severity of the disaster. For most vulnerable people, the impact on GDP from a disaster will in most cases be negligible. That is why much higher damage occurs in developed nations (even relative to GDP) in comparison to developing nations.

Central America appears the most vulnerable region with 1,043 fatalities per million population, followed by the Caribbean (323 fatalities per million) and South America (122 fatalities per million). The Caribbean is by far the hardest-hit region when land size is taken into account. Thus, although weather-related disasters are more frequent in South America, they cause more fatalities per capita in Central America. One reason may be that relatively more infrastructure is in place in South America than in Central America.

Individual countries' vulnerability relates closely to their size, governance, and economic wealth. Brazil—the largest country in South America—had by far the largest number of disasters between 1970 and

2008, totaling 2.7 a year, but it experienced among the lowest rates of fatalities per capita. Peru and Colombia experienced moderately fewer disasters but suffered significantly greater damage and fatalities. In the Caribbean, Cuba and Haiti ranked the highest in disaster frequency, but Cuba had only 3 percent of the fatalities that Haiti experienced. Dominica, St. Lucia, the Dominican Republic, and Puerto Rico had the next-highest loss of lives. In Central America, Mexico had many more disasters than other countries but relatively few fatalities. Guatemala, Nicaragua, and Honduras, with moderate numbers of disasters, ranked highest in fatalities per capita. Belize and Panama have the best record in Central America, with low numbers of disasters and fatalities in both absolute and relative terms (Charveriat 2000) (for an example, see the account of flooding in Belize in chapter 10).

Mortality rates are an important indicator of the erosion of human assets from climate-related disasters, but the number of survivors affected is at least as significant. EM-DAT defines "affected populations" as those requiring immediate assistance during a period of emergency to obtain basic survival needs such as food, water, shelter, sanitation, and immediate medical assistance. As table 3.4 shows, for each person who died in a climate-related calamity in the LAC region during 1970–2008, another 1,600 were adversely affected to the extent that they required assistance to survive. The table also shows that whereas droughts led to a relatively small loss of life, they affected more than 60 million people between 1970 and 2008, while floods affected 51 million, and storms 31 million.[5]

Table 3.4    People Killed or Affected by Weather-Related Disasters in the LAC Region, 1970–2008 (thousands)

| | Fatalities | | | | People affected | | | |
|---|---|---|---|---|---|---|---|---|
| | Droughts | Floods | Storms | All weather-related disasters | Drought | Floods | Storms | All weather-related disasters |
| 2000–08 | 0.05 | 6.6 | 5.6 | 12.3 | 3,406.7 | 10,721.2 | 15,409.6 | 29,537.5 |
| 1990–99 | 0.01 | 34.3 | 21.8 | 56.1 | 16,664.4 | 6,199.0 | 9,103.1 | 31,966.4 |
| 1980–89 | 0.02 | 7.0 | 2.1 | 0.1 | 24,165.0 | 24,842.8 | 4,121.7 | 53,129.5 |
| 1970–79 | – | 2.7 | 10.6 | 13.3 | 16,377.0 | 9,752.7 | 2,747.0 | 28,876.7 |
| Total | 0.08 | 50.6 | 40.1 | 81.8 | 60,613.1 | 51,515.7 | 31,381.4 | 143,510.1 |

*Source:* Authors' elaboration based on EM-DAT data.
*Note:* According to EM-DAT, affected populations are those requiring immediate assistance during a period of emergency for basic survival needs such as food, water, shelter, sanitation, and medical assistance. Individuals are also considered affected when a significant number of cases of infectious diseases appear in a region or a population that is usually free from the disease. http://www.emdat.be.

Morbidity data measure the numbers of people injured in disasters. Slow-onset droughts caused no reported injuries during the period 1970–2008 in the LAC region, but during the same period, floods and storms each reportedly injured about 35,000 people (table 3.5). Notably, the number of people reported injured is smaller than the number of people dying from weather-related disasters in the period. The reason may be either that injuries are underreported or that injuries are mainly caused by first-round effects (the impact itself), whereas mortality can be caused by second-round effects such as epidemics and famine.

### Links between Natural Hazards and Disaster Impacts

As stated earlier, the extent to which a natural hazard will pose a serious challenge and translate into an actual disaster depends crucially on local vulnerabilities. The above discussion illustrates that the impact of natural hazards varies widely across subregions. The rate at which natural hazards translate into actual disasters also varies widely across subregions, as illustrated by the figures for disaster occurrence, fatalities, affected populations, and damages. A comparison of weather-related hazards and actual disasters across Latin America and the Caribbean shows only a very limited relationship between number of hazards and numbers of people killed, affected, and injured—statistics that are good indicators of whether a hazard has turned into a disaster. For example, in the cases of both flood and storm hazards, only a weak association appears between hazard frequency and disaster-related deaths.

As a case in point, consider that a flooding hazard of more than 0.4 floods a year closely follows the Amazon River through Bolivia, Peru, and Amazonas in Brazil, but those floods affect and kill relatively few people.

**Table 3.5    People Injured by Weather-Related Disasters in the LAC Region, 1970–2008 (1,000s)**

|  | Droughts | Floods | Storms | All weather-related disasters |
|---|---|---|---|---|
| 2000–08 | 0 | 2.92 | 4.63 | 7.54 |
| 1990–99 | 0 | 5.59 | 14.89 | 20.48 |
| 1980–89 | 0 | 23.14 | 9.46 | 32.59 |
| 1970–79 | 0 | 2.87 | 8.13 | 11.00 |
| Total | 0 | 34.52 | 37.09 | 71.61 |

*Source:* Authors' elaboration based on EM-DAT data. http://www.emdat.be
*Note:* "People injured" are defined by EM-DAT, as those suffering from physical injuries, trauma, or an illness requiring medical treatment as a direct result of a disaster.

In other words, in those locations, the hazards only rarely become disasters. Paradoxically, flooding is less common in the parts of the region that have the most deaths from flooding: the northern parts of South America and Mexico and extensive parts of Central America and the Caribbean islands. The explanation lies in predictability. Flooding occurs regularly along the Amazon, and the indigenous peoples living nearby have adjusted their way of living to its seasonal occurrence. Only when flooding is higher than usual—or fails to materialize—does it cause problems. People living in areas that do not regularly experience floods have no social memory of them, and what to do in response, and are therefore more likely to be caught by surprise. Then such hazards can turn into disasters in which people are killed or suffer injury or material losses.

Similarly, while the most frequent storm hazards arise from the Pacific Ocean along the coast of Mexico, the highest mortality rates from storms that became disasters are caused by the storms in the Atlantic Ocean, which strike the Caribbean islands as well as Honduras and Nicaragua. Although weather-related hazards are more frequent in South America, they cause proportionately more fatalities in Central America, and particularly in the Caribbean, simply because a larger proportion of the population there is exposed to them. The data comparisons also emphasize that disaster occurrence and impact are related to the level of socioeconomic development: poorer countries and areas tend to suffer proportionately more fatalities and injuries.

## Impact of Natural Disasters on Household Assets of the Poor

Throughout this book the findings described repeatedly emphasize the need to allocate more resources to improving the resilience of communities to the impact of climate change, rather than merely to address post-disaster adaptation. Indeed the cost of disaster reduction is much less than the cost of recovery after a disaster. It is estimated that for every United Kingdom pound (£) spent on adaptation (disaster mitigation), £4–10 can be saved on recovery (WWF 2006). In St. Lucia, recovery following the Black Mallet landslide (1999–2000) cost the government US$18 million; the implementation of drainage channels that prevent landslides costs less than 2 percent of that (Anderson and Holcombe 2008). Clearly, good public policies to help vulnerable groups cope, based on anticipated occurrences, are needed not only ex-post (after an event), but also ex-ante (before an expected but sudden disaster strikes, or in preparation for the effects of gradual climate change). Humanitarian efforts to protect people from adverse effects of climate change should try to support existing coping strategies so as to strengthen a community's resilience rather than focus solely on post-disaster recovery measures.

In addition to the need to focus efforts on risk reduction and the need to be proactive rather than just reactive, the question must be asked: Are there socioeconomic determinants of natural disasters? This section discusses how climate-related disasters can affect the asset base of the poor in the LAC region, to compound their existing vulnerability by seriously constricting and often destroying their livelihoods.

### Poverty, Vulnerability, and Resilience of the Poor

Living in poverty makes people more vulnerable to natural hazards. The human and economic impact of natural hazards and shocks depends on the resilience of a system to absorb sudden disturbances. As explained by Adger (2006), vulnerability to environmental change does not exist in isolation from the wider political economy of resource use. This book shares Adger's understanding of vulnerability as the susceptibility to harm from exposure, sensitivity, and lack of capacity to adapt to environmental change. As illustrated in this section, the effects of climate shocks and variability are felt most profoundly among the poor.

In Latin America and the Caribbean, 17 percent of the population—about 100 million people—live in poverty, and 8 percent, or 45 million, are extremely poor. Poverty is not evenly distributed. In Bolivia, Honduras, and Nicaragua more than 30 percent of the population live on less than US$2 a day, but in Chile and Costa Rica less than 10 percent of the population is poor. And while the number of people living on less than US$2 a day has decreased or stabilized for the region as whole, poverty has increased in El Salvador, Guyana, and Honduras, which are among the poorest countries in the region (World Bank 2008c).

The concentration of poverty is a central determinant of vulnerability. Because people cluster densely in urban settings, where 65 percent of the region's poor live (Ravallion, Chen, and Sangraula 2007), residents are at increased risk of being trapped in a large-scale disaster zone when a natural disaster strikes. Poor housing conditions and lack of infrastructure pose great challenges to limiting the spread of disease and delivering emergency assistance. Rural disaster zones develop because of widespread poverty (45 percent of the rural population in Latin America and the Caribbean is poor) and high dependence on natural resources. Residents are particularly vulnerable to the erosion of the natural resource base on which their livelihoods are built. Their dependence on land as a source of food and income, coupled with lack of physical and financial adaptive capacity, means that poor farmers are also at increased risk of harm from slow-onset disasters.

The majority of low-income settlements in both urban and rural areas in the LAC region tend to be in inhospitable areas prone to flooding, landslides, or drought. Such low-income settlements, at constant risk from hurricanes and flooding, can be found in many of the Caribbean and Central American countries. For example, in St. Vincent many households in Georgetown and on the outskirts of Kingston worry that their homes will crumble during the next storm in consequence of their perilous location on steep slopes. And in Belize City, people live in constant fear that the next storm that passes through will bring severe flooding, as occurred during Hurricane Hatti, when water levels rose to 10–12 feet in some of the city's poorest districts.[6] The poor in urban settlements in the region's cities and megacities have their livelihoods destroyed year after year because they reside in areas prone to mudslides and flooding during the rainy season.

Poor people are constrained in their choice of settlement by property rules and landowners. They may choose hazardous areas voluntarily, if those areas seem to improve their access to resources or increase their income-generating possibilities. Hence, environmental hazard might be outweighed by the perceived benefits of living in such hazard-prone areas as fertile volcanic slopes, plains nourished by flood alluvia, riverbanks, or other areas shunned by formal settlements because of their high exposure to risk but which nonetheless offer access to cities and hence income-earning opportunities (Rapp 1991, as cited by Main and Williams 1994). In rural areas, poverty is also a direct cause of environmental degradation (chapter 4), as the extremely poor—who constitute a high percentage of the rural poor—are often forced to deplete their surrounding natural resources for survival (Echeverria 1998). A strong connection also exists between poor-quality housing and low income, especially where low-income people cannot get mortgage financing. Lack of land ownership and lack of rental possibilities exacerbate these links.

It is typically the poor who bear the brunt of weather shocks and who recover most slowly. For example, when Hurricanes Wilma and Stan affected the Yucatan peninsula and Chiapas, in Mexico, in 2005, they caused relatively greater damage to the assets of the poor than to those of the nonpoor. More than 50 percent of the losses from Hurricane Wilma in Yucatan (mostly luxury properties in Cancun) were insured (Carpenter 2006). By contrast, almost all the losses from Hurricane Stan were uninsured, primarily affecting assets of the poor in marginal urban slums (such as Las Americas in Tapachula) and subsistence farming regions (such as Escuintla, Mapastepec, and Cacahoatan). Assets may be recovered quickly or not, depending on the severity and frequency of the disasters and the strategies used to cope. But the effects on the poor in those cases were significant, and they were much longer lasting in the poor areas of Chiapas than they were in the Yucatan peninsula (Zapata-Marti 2006).

In the longer term, the poor are perhaps even more vulnerable if weather shocks recur frequently and their asset stocks are depleted. In that case, households and individuals must forgo opportunities to conserve or accumulate assets for the future, with consequences for the human capital and wealth of future generations.

Recovery of productive assets from natural disasters is possible, and happens regularly, but it is much less easy for the poor. To cope with shocks, poor households that lack credit and insurance may be forced to dispose of their productive assets or adopt erosive coping mechanisms such as withdrawing children from school—which can lead to lifelong losses in learning. If the subsistence base of a household is eroded, that compromises the future recovery or prospects of family members. In Honduras after Hurricane Mitch in 1998, the poorest households appeared to be forced into long-term poverty traps, as they had to dispose of valuable assets needed for recovery. As a result, those households struggled for two to three years or longer, after the shocks, to recover their previous asset levels (De la Fuente 2007).

Governance structures are crucial in determining the livelihood impacts of natural hazards. The most vulnerable groups, who are often the least vocal, mobilized, or empowered entities in decision-making processes, often suffer disproportionately from weak, unresponsive local institutions. Often they also face both poorer infrastructure services and less access to public services—leaving them more exposed to hazards and with more limited access to government-provided safety nets. Community participation, voice coalition, and local governance relating to natural resources are effective ways to mobilize social capital to preserve communities' physical and cultural assets.[7]

Because of their low income level, the poor tend to be more risk averse than richer income groups who have more savings or other assets. However, they are usually less likely to be informed about the risks they incur because they are typically not included in emergency preparedness measures. Nor can the poor usually count on protective infrastructure against weather shocks. Early warning systems are often absent or malfunctioning in rural poor areas that are at risk (Charveriat 2000). When Hurricane Felix struck Nicaragua and Honduras in 2007 it affected roughly 198,000 people, most of them from indigenous Miskito communities. According to Miskito testimonials, authorities sailed by only a few hours before the storm hit, yelling out warnings. At that time, several lobster fishers were already out at sea with no communication devices.[8] In general, countries in the region have less access to information and communications than developed countries, ranging from Barbados (24th rank) to Haiti (159th) (World Bank 2000/2001). Because of the great inequality and the nature of the goods involved, a very unequal distribution of access to information and communications can be expected among income groups, with the poor having little access.

Even if low-income groups had access to information, they might not engage in risk reduction. Resettlement, retrofitting of housing, or insurance coverage might appear too costly to them. Because most of their income is allocated to immediate survival, the low frequency risk of a catastrophic natural disaster might not warrant a change in behavior. Furthermore, as noted above, low-income households are less likely to be able to cope with the effects of a disaster, causing their further impoverishment and feeding future vulnerability. Vicious circles might therefore develop in particularly hazard-prone areas affected by recurrent disasters, such as the coastal areas of northeast Brazil, Ecuador, Nicaragua, or Peru.

In summary, the poor are usually the most immediately and greatly affected by climate shocks and variability because they lack the assets needed to cope with them. Communication and organization networks are often last to be activated for the poor, leaving them without timely information about how to reduce their risk and protect their assets.

## Effects of Natural Disasters on Assets
This section addresses the potential impact of weather-related disasters on natural, human, physical, financial, and social assets.

**Natural assets.** Natural assets comprise the natural resource stock from which households derive resources useful to their livelihood. It is beyond the scope of this chapter to analyze the intricate causal relationship between weather-related natural disasters and natural resources per se.[9] But box 3.2 illustrates how climate change and variability will compound existing stresses on vital natural assets.

Box 3.2

## Natural Asset Erosion in Northeast Brazil

Northeast Brazil is considered a desertification hot spot, with up to 75 percent of the area at risk. This large "drought polygon" has Brazil's highest rural poverty rate (76 percent) and the largest concentration of rural poor in Latin America. More than half the households in the region are considered food insecure.

The region has suffered major droughts primarily driven by ENSO in the years 1911–12, 1925–26, 1982–83, 1997–98, and 2005–06. Zhao and colleagues (2005) reported that the largest area with clear vulnerability to climate variability in the LAC region is northeast Brazil and further, that the potential crop yield impacts of climate change and variability for Brazil are among the most severe for all regions.

Contrary to common perception, the occurrence of droughts cannot simply be attributed to lack of rain. Many other factors give rise to the droughts threatening people's livelihoods. The rain often does not fall when crops need water. Rainfall can be highly irregular in both time and space—there can even be substantial precipitation differences within the same town. In some cases most of the rain evapotranspires. Of the region's 900-millimeter average annual rainfall, only 200 millimeters remains after the normal 3,000 hours of annual sunshine has caused the evapotranspiration of 700 millimeters. The remaining water is absorbed in the topsoil, which, despite being fertile, is shallow and impermeable, inhibiting efficient water use.

The risk of drought-related, long-term disasters is likely to increase, as climate projections for northeast Brazil indicate a strong likelihood of higher temperatures and decreased precipitation. Moreover, climate change will induce greater variation in the amount of rainfall and the locality where it occurs. That variation is believed to be a greater threat to vulnerable farmers than a reduction in average rainfall. Such changes will also augment the need for effective irrigation.

Much therefore depends on whether farmers possess the resources necessary to extract the topsoil water and distribute it effectively.

*Sources:* Authors, based on Lemos and others 2004; World Bank 2008b; and IBGE 2004.

*Human assets.* "Human capital" describes people's education, skills, knowledge, ability to work, and health status. The direct impacts of weather-related calamities on an individual's integrity include fatality, injuries, sickness, and violence. As Charveriat (2000) points out, "Fatalities are a direct and immediate effect of the disaster, as buildings collapse and flows of water, mud, and debris carry people away." She also highlights how fatalities constitute a lasting shock to a household's welfare: as a result, other household members "are left widowed or orphaned at a time when community and extended family solidarity might not be available."

Health is fundamental to human labor, the primary asset of the poor, and can be damaged in many ways by climate-related disasters. Both flooding and drought augment the risk of weather-related and vector-borne diseases. Heavy rains can flood latrines, contaminating water and exposing populations to cholera bacteria. Poor households are disproportionately at risk from this disease, which is strongly related to overcrowding and lack of safe water and sanitation. Vector-borne diseases are likewise prevalent with weather-related climate shocks, particularly with warmer temperatures and flooding. The two most devastating such diseases in the region, malaria and dengue fever, are extremely debilitating and lead to high rates of mortality. With more pronounced El Niño cycles, warming and floods are likely to further encourage the spread of malaria (McMichael, Woodruff, and Hales 2003; Vos, Velasco, and De Labastida 1999). As noted in chapter 6, studies in Venezuela and Colombia have reported the association of malaria outbreaks with El Niño events (Bouma and others 1997). Warmer temperatures speed the maturation of the malaria parasites carried by mosquitoes; for instance, at 20°C they take 26 days to incubate, but at 25°C they take only half that time (Carvajal 2007).

Weather-related disasters affect human health not only by furthering the transmission of diseases but also through greater food insecurity, caused by the erosion of crucial environmental and physical assets (Sachs and Malaney 2002). In adults, temporary hunger reduces body mass, immunity, and productivity, but the results are rarely permanent. In children hunger can stunt growth, impede brain development, or cause death, especially in infants between 12 and 24 months (De la Fuente 2007).

Food insecurity can also affect the education of children and youth. Short-term undernutrition or malnutrition can impede learning and school attendance, so that children may fail to complete a grade. With an increase in food insecurity the incidence of child labor often rises, as families try to bolster their income to ensure survival (box 3.3). Among poor households in Mexico observed over the period 1998–2000, child labor increased as a response to droughts and other climatic hazards, a common pattern for the age 15–18 group. De la Fuente (2007) points out that natural disasters often reduce school enrollment and grade completion. He also notes that floods can damage or destroy transport and schooling

Box 3.3

## Social Capital Is Important in Natural Disaster Management

The importance of social capital in natural disaster management can be high-lighted by a comparative study from a field visit for this book to the subsistence communities of Mecapaca and Palca, Bolivia. During December 2007–January 2008, Bolivia was hit by torrential rains caused by ENSO, which resulted in severe flooding in large parts of the country. Although the subsistence communities the researchers visited all suffered great losses, they dealt with them in different ways.

The Huayhuasi community of subsistence farmers is located in Mecapaca. As a result of the rain, severe flooding from the nearby river devastated the commu-nity, washing away the majority of its crops, damaging some houses beyond repair, and sweeping away others altogether. Once the floods receded, a moon-scape of debris, mud, rocks, and stones remained, and it was estimated that it would take the community two years to clear the land before it could be cultivat-ed again, impairing food security and income in the intervening period. Howev-er, group interviews revealed very strong bonding social capital in the community. Those hardest hit by the floods—who had lost everything—were being helped by others who were less hard hit. During group interviews the participants were

*Photo:* Tine Rossing

*(continued)*

**Box 3.3** (continued)

asked the hypothetical question: "If you could ask for three things, what would they be?" Invariably a request for helping those hardest hit led the list, even when those people were not represented in the group being interviewed. Hence, even though some people essentially lost their livelihood, the strong bonding social capital in the community enabled them to remain in the village, rather than be forced to migrate to La Paz to find other income opportunities.

Another subsistence farmer community, Palca, was also severely affected by the floods but appeared to react differently. Floods destroyed a century-old system of irrigation canals dug into the mountainside. As a result, the community was cut off from its crucial water supply, which is fed by glacial melt above them. To restore the water supply and return the farms to their former productive capacity, the infrastructure would need repair, involving installation of new plastic pipes worth US$10,000, a project that could have been implemented quite quickly. However, the farmers appeared very despondent and did not think it would be possible to mobilize the assistance and funds for the repair. Although they were collectively dependent on the irrigation system, they were all acting as individuals. They also lacked the level of organization to jointly seek assistance or loans to finance the replacement of the pipes. The community therefore appeared to lack sufficient bonding social capital, as well as linking social capital to its municipality government. Instead the coping mechanism that many households adopted was to send the young to La Paz in search of work, so that they could send money home to the family members left behind. Also, small children were taken out of school to work in the fields alongside the elderly in the community.

It appeared that because Palca was so destitute, with so few assets and options available, their supply of social capital was quickly exhausted. In contrast, the community in Mecapaca, situated at a lower altitude and in flatter terrain, could grow a variety of crops and therefore had more diversified income opportunities. Nevertheless, that community was also very conscious of the threat to its sustainability if its young people left for La Paz, and it put much effort into providing opportunities for young people locally as incentives for them to remain in the community.

*Source:* Authors.

infrastructure and thereby delay access to schooling opportunities for children and youths. He concludes that stunted growth or brain development and lower education levels among the population tend to lead to fewer socioeconomic opportunities and fewer financial assets.

***Physical assets.*** Climate-related disasters also lead to partial or total loss of physical assets, particularly assets needed to generate income. "Physical assets" denote productive assets such as land, tools, equipment, and work animals, and household assets such as housing and household services or stocks, such as livestock, food, and jewelry. The term also denotes the basic infrastructure for transport, buildings, communications, and so on.

The loss of agricultural income-generating assets, which are of particular importance to the poor, might be temporary or permanent. Floods tend to make land unsuitable for agricultural production until waters recede, and hurricanes might wash out arable land or permanently increase its salinity through storms and flash floods. Loss of perennial crops, such as forests or banana trees, also has long-term consequences for the ability to generate income. For example, St. Vincent lost two years of banana production as the result of two hurricanes.[10] In Nicaragua in 1998, Hurricane Mitch affected around 80,000 hectares of agricultural land. Of that land, 63,000 hectares were cultivated by small-scale farmers, of whom at least 56,000 suffered losses, which affected about 300,000 poor people who depended on them (UN-ECLAC 1999). Relatively wealthy farmers were also affected, but disastrous consequences were felt mainly by the poorer sections of the population.

Both pastoralists and subsistence households often suffer a blow from the loss of livestock during weather-related disasters. Live animals are an important physical asset, but also a valuable financial asset because they are often traded instead of cash and serve as collateral for credit. Loss of these animals—because they perish or have to be sold—increases the vulnerability of their owners. That is particularly the case with poor households who own few animals and may take years to recover their loss. For instance, to recover from Hurricane Mitch in 1998, poor people had to use a much greater share of their financial assets than wealthier households. For many, that included a large portion of their livestock. By depleting the productive assets of the poor, the weather shock created conditions for an increase in inequality in the future (De la Fuente 2007).

Land and housing are also important to the poor, not just for shelter but also as productive assets. They can generate income through rental opportunities and provide space for home-based production activities. They also provide collateral for credit and other risk management instruments in case of emergencies, for consumption purposes, or to protect productive assets (De la Fuente 2007). People who are deprived of housing and land ownership are usually forced to settle illegally in marginal areas and are often at risk from flooding or landslides. Thus limited access to quality housing and land usually translates into less capacity to withstand damage. It also means that authorities often do not provide the services and infrastructure needed to cope with climate hazards.

Where housing and land are permanently affected by climate hazards, it becomes very difficult for poor households to improve their lives and reduce their long-term vulnerability. In Honduras, when Hurricane Mitch destroyed people's homes, they lost their shelter and their tools and materials for making a living. Many farmers in one ladino village in the Noroiente Region of the country also lost fertile land, as flooding washed away the productive topsoil and deposited debris and rocks all over the fields (De la Fuente 2007). The same happened to the Huayhuasi community in Mecapaca, Bolivia (box 3.3).

*Financial assets.* The poor by definition have meager financial assets, with most of their wealth tied up in their housing and means of production such as land, livestock, and labor. They typically lack access to affordable credit, let alone farm, health, disability, employment, or house insurance. They therefore have very few purely financial options when a shock affects their household assets. In the case of a large covariant shock such as a natural disaster, if many affected households sell their belongings in a shallow market, prices fall significantly. Financial savings thus represent a better asset than physical assets, provided that macroeconomic stability and a reliable banking system exist (World Bank 2005).

It is difficult to assess the impact that weather-related shocks have on employment. The impact depends primarily on the degree of destruction of income-generating assets and how long flows of goods and services are disrupted. If alternative sources of employment are available neither within nor outside a disaster area, the frictional unemployment resulting from a climate shock could reduce income over the long term. The effects would be compounded if reconstruction efforts are limited in scale and slow in execution. Vos, Velasco, and De Labastida (1999) refer to a situation in Ecuador during El Niño in 1997–98, when about 12,000 laborers on banana and sugarcane plantations in the lowlands temporarily lost their jobs. Similarly, in post–Hurricane Mitch Honduras in 1998, according to one estimate, the unemployment rate reached 32 percent (Charveriat 2000).[11]

For farmers who cannot raise their productive yield or enlarge their cropping area, climate shocks can lead to a decline in income, as discussed in chapter 4. That decline then reduces the ability to create a financial pool with which to withstand hazards, creating a vicious circle of low coping capacity, low savings, lack of climate-adaptive instruments (such as crop insurance or a reserve fund), and vulnerability to disasters.[12]

*Social assets.* Research shows that communities endowed with social capital and social networks are more likely to cope better with adverse situations. The findings also highlight the important role such assets play in securing immediate consumption needs and protecting the other assets of poor communities facing adverse climate events (Moser 1996; Narayan 1997). Social capital is based on relations of trust, reciprocity, and exchange; the evolution of common rules; and the functioning of networks (Adger 2003). Interactions between genders, intrahousehold relations, participation in associations and organizations, and interhousehold relations, including politics and markets, are vital to the formation of social assets.

The way in which individuals and groups within a society interact with one another to share risk, provide mutual assistance, and take collective action will influence their vulnerability to climate shocks and variability (Adger 2003). As discussed further in chapter 10, communities use different kinds of social assets to improve livelihood adaptation and resilience.

Social assets are often under stress during natural disasters. For example, if an entire community is severely affected, each community member might be more focused on his or her own survival than on the well-being of others. Generally, social capital tends to be depleted by long-term or continual shocks, if people lack the needed time and effort to restore it. When poverty is also a factor, the stresses on social capital tend to be even greater, leaving little time for association and social action. Relationships with extended families are likely weakened, leading to ineffective reciprocal exchanges with neighbors, less cooperation, and even resentment.

The depletion of social assets is particularly troublesome for the poorest. Because they tend to be marginalized from society they are unlikely to get help from outside, including government agencies, during a crisis, and so they need to be able to rely on help from one another. For low-income groups that are socially excluded, relations with kin, close friends, and neighbors ("bonding social capital," explained in chapter 10) are crucial for adaptive capacity, but with covariate climate-related risks such as floods or storms, the whole community tends to suffer (Adger 2003). Moreover, if one disaster follows another, the capital embodied in these relationships may become exhausted, and such socially excluded communities may not be able to cope and recover.

Higher-level, formal institutions such as local and state governments also participate in promoting and facilitating social capital. But if the state does not step in and help its weakest citizens in the face of growing climate risks, relationships with individuals and informal institutions outside the immediate community ("bridging social capital," as explained in chapter 10) may be crucial in recovering from a disastrous event; they allow the pooling of risk with people who are not being devastated by the same calamity at the same time. Remittances, for example, have been crucially important during natural disasters. Without such bridging social capital, people may be forced to migrate. Conflict and instability then often follow, making the most marginal sections of society more vulnerable.

## Policy Perspectives

Climate-induced changes in resource flows—especially through erosion of key assets—will affect the viability of some livelihoods unless measures are taken to protect and diversify them through adaptation. Not all groups are affected equally, even within a given natural disaster hot spot, however. Hence there is a rationale for applying a context-specific vulnerability analysis using an asset-based framework to ensure a strong match between the needs of particular vulnerable groups and whatever interventions are planned to assist them.

As part of such asset-based vulnerability analyses, it may be helpful to distinguish between the natural hazards caused by increased climate variability of the type we can already observe and the projected long-term trends (30 years and up) for climate change, which might cause more severe natural hazards. Such a distinction has implications for the analysis of asset erosion and vulnerability, particularly with respect to notions of resilience, adaptation, and development. The distinction, in turn, will affect how hazard-risk management is considered for and by the poorest groups. In other words, is it more important to manage the climate variability that is already being felt, or to concentrate on the long term, to increase resilience and prepare people to adapt to the natural hazards to come 30 years from now?

Governance structures are pivotal in determining the impact of natural hazards on livelihoods. The most vulnerable groups often suffer disproportionately from weak, unresponsive local institutions. Community participation, voice coalition, and local governance relating to natural resources should thus be promoted; they are effective ways to mobilize social capital to preserve livelihood practices.

Another key conclusion is that a strategy for assisting the poor in conserving or rebuilding their assets after a disaster must consider the importance of social capital. Although often overlooked or underestimated, social capital is the foundation that allows all other assets to be generated and allocated appropriately. Social capital and its part in disaster resiliency demand study in greater depth. Chapter 10 addresses these issues in more detail.

## Notes

1. Emergency Events Database (EM-DAT), http://www.emdat.be.

2. The periodic El Niño events are a natural phenomenon that has occurred for centuries. Ocean and atmospheric conditions in the Pacific tend to fluctuate between El Niño (warming) and La Niña (cooling). The fluctuations are semiregular and tend to appear every three to six years. A more intense phase of either event may last for about a year.

3. Some of this increase is likely to be due to improved reporting, and some can be attributed to population movements and development—more people live in exposed locations than before, and more material damage occurs because of development where previously there was none or very little.

4. This finding is based either directly on data from the Center for Research on the Epidemiology of Disasters (CRED) or on simple tabulations thereof.

5. These outcomes derive from the nature of the various phenomena in the region (for instance, droughts affecting parts of Brazil) and their respective rates of occurrence. The fact that windstorms affected fewer people is linked to the fact that these phenomena generally affect smaller geographical areas than do floods or droughts.

6. Information obtained during fieldwork in Belize and St. Vincent in 2008.

7. Thanks to Nicolas Perrin of the World Bank for drawing our attention to the need for including the aspect of governance at this stage. These aspects are also addressed in great detail by the World Bank project, led by Nicolas Perrin and Eija Pehu, supporting area-based development initiatives to enhance adaptive capacity and maintain the resilience of local actors and institutions to climate change, and by a research initiative put forward by Arun Agrawal and Nicolas Perrin to shed light on the role of social capital in addressing climate hazards.

8. An early-warning system could have prevented the deaths of several Miskito fishermen, as well as brought women and children to safety before the hurricane hit (Vianica 2007; MADRE 2007; Rodriguez 2007).

9. Thorough analyses have already been undertaken by others (such as the earlier IPCC reports and the Millennium Ecosystem Assessment).

10. Observation made on field visit to St. Vincent, June 2008.

11. A significant discrepancy appears between these figures and the official estimates of unemployment from November 1999, which registered 89,000 unemployed, or only 3.7 percent, compared with 3 percent before Mitch (Boletin del Comisionado 1999). One explanation for this discrepancy could be that much of the frictional unemployment was absorbed between March and November 1999.

12. A survey of subsistence farmers in Chiapas, Mexico, indicated that the respondents tended to depend more on relatives living in the same community and aid from the government than on financial instruments such as insurance and credit. Hence the challenge of integrating financial prevention measures into the region's adaptive strategy.

## References

Adger, W. Neil. 1999. "Social Vulnerability to Climate Change and Extremes in Coastal Vietnam." *World Development* 27 (2): 249–69.

———. 2003. "Social Capital, Collective Action, and Adaptation to Climate Change." *Economic Geography* 79 (4): 387–404.

———. 2006. "Vulnerability," *Global Environmental Change* 16: 268–81.

Anderson, M. G., and L. Holcombe. 2008. "Community-Based Landslide Risk Reduction: Proof of Concept." Presentation at the World Bank, Washington DC, June 16.

Boletin del Comisionado, *La Prensa*, November 16, 1999.

Bouma, M. J., S. R. Kovats, S. A. Goubet, H. St. J. Cox, and A. Haines. 1997. "Global Assessment of El Nino's Disaster Burden." *Lancet* 1435 (8): 350.

Carpenter, G. 2006. "Tropical Cyclone Review 2005." Instrat Briefing, January. http ://gcportal.guycarp.com/portal/extranet/popup/pdf/GCBriefings/Tropical_ Cyclone_Review_2005.pdf.

Carvajal, Liliana. 2007. "Impacts of Climate Change on Human Development." Background Paper to UNDP 2007, *Fighting Climate Change: Human Solidarity in a Divided World, Human Development Report 2007/08*. New York: Oxford University Press.

Charveriat, C. 2000. *Natural Disasters in Latin America and the Caribbean: An Overview of Risk*. Working Paper 434, Research Department, Inter-American Development Bank, Washington, DC.

De la Fuente, A. 2007. *Climate Shocks and their Impacts on Assets*. Occasional paper for UNDP 2007, *Fighting Climate Change: Human Solidarity in a Divided World, Human Development Report 2007/08*. New York: Oxford University Press.

Echeverría, R.G. 1998. "Rural Poverty Reduction." Bank Strategy Paper, Inter-American Development Bank, Washington, DC.

Hodell, D. A., J. H. Curtis, and M. Brenner. 1995. "Possible Role of Climate in the Collapse of Classic Maya Civilization." http://www.ncdc.noaa.gov/paleo/drought/drght_mayan.html.

IADB (Inter-American Development Bank). 2000. *Development beyond Economics: Economic and Social Progress in Latin America, 2000 Report.* Washington, DC: Inter-American Development Bank.

IBGE (Instituto Brasileiro de Geografia e Estatistica). 2004. "IBGE Releases Previously Unseen Profile of Food Security in Brazil." http://www.ibge.gov.br/english/presidencia/noticias/noticia_visualiza.php?id_noticia=600&id_pagina=1.

IPCC (Intergovernmental Panel on Climate Change). 2007a. *Climate Change 2007: The Physical Science Basis.* Contribution of Working Group I to the Fourth Assessment Report of the IPCC. Geneva: IPCC.

———. 2007b. *Climate Change 2007: Impacts, Adaptation, and Vulnerability.* Contribution of Working Group II to the Fourth Assessment Report of the IPCC. Geneva: IPCC.

Lemos, M. C., D. Nelson, T. Finan, and R. Fox. 2004. "The Social and Policy Implications of Seasonal Forecasting: A Case Study of Ceara, Northeast Brazil." NOAA Report, University of Arizona.

MADRE News. 2007. "Hurricane Felix Tears through Nicaragua; Local Communities Destroyed." September 5. http://www.madre.org/index/press-room-4/news/hurricane-felix-tears-through-nicaragua-local-communities-destroyed-16.html.

Main, H., and S. W. Williams, eds. 1994. *Environment and Housing in Third World Cities.* New York: John Wiley and Sons.

McMichael, A.J., R. E. Woodruff, and S. Hales. 2003. "Climate Change and Human Health: Present and Future Risks." *Lancet* 859 (69): 367.

Mendelsohn, Robert, and Larry Williams. 2004. "Comparing Forecasts of the Global Impacts of Climate Change," *Mitigation and Adaptation Strategies for Global Change* 9: 315–33.

Moser, C.O. 1996. *Confronting Crisis—A Comparative Study of Households.* Environmentally Sustainable Development Studies and Monograph Series No. 8. Washington, DC: World Bank.

Nagy, Caffera, Aparicio, et al. 2006. "Understanding the Potential Impact of Climate Change and Variability in Latin America and the Caribbean." Report prepared for the *Stern Review on the Economics of Climate Change.* http://www.hm-treasury.gov.uk/media/6/7/Nagy.pdf.

Narayan, D. 1997. *Voices of the Poor: Poverty and Social Capital in Tanzania.* Washington, DC: World Bank.

Ravallion, M., S. Chen, and P. Sangraula. 2007. "New Evidence on the Urbanization of Global Poverty." *Population and Development Review* 33 (4): 667–701.

Red Cross/Red Crescent. 2007. "Climate Guide." Red Cross/Red Crescent Climate Centre. http://www.climatecentre.org/downloads/File/reports/RCRC_climate guide.pdf.

Rocha, J., and I. Christoplos. 2001. "Disaster Mitigation and Preparedness in Nicaragua after Hurricane Mitch." Report for research project. NGO Natural Disaster Mitigation and Preparedness Projects: An Assessment and Way Forward. ESCOR Award No. R7231.

Rodriguez, O. 2007. "Felix Survivors Describe Being Caught at Sea." Associated Press. September 8. http://www.msnbc.msn.com/id/20661143/.

Sachs, Jeffrey, and Pia Malaney. 2002. "The Economic and Social Burden of Malaria." *Nature* 415: 113–25.

Sokona, Y., and F. Denton. 2001. "Climate Change Impacts: Can Africa Cope with the Challenges? *Climate Policy* 1:117–23.

Swiss Reinsurance Company. 1998. *Floods: An Insurable Risk? A Market Survey.* Zürich: Swiss Reinsurance Company.

UNDP (United Nations Development Program). 2004. *Reducing Disaster Risk. A Challenge for Development. A Global Report.* Bureau for Crisis Prevention and Recovery. New York: United Nations Development Program.

———. 2007. *Fighting Climate Change: Human Solidarity in a Divided World, Human Development Report 2007/08.* United Nations Development Program. New York: Oxford University Press.

UN-ECLAC (Economic Commission for Latin America and the Caribbean). 1999. "Nicaragua: Assessment of the Damage Caused by Hurricane Mitch, 1998." LC/MEX/L.372, April 19.

UNEP (United Nations Environment Program). 2007. *Global Environment Outlook.* Geneva: United Nations Environment Program.

UNEP/GRID. 2010. Climate Impacts of El Niño Phenomenon in Latin America and the Caribbean, *Arendal Maps and Graphics Library*, http://maps.grida.no/go/graphic/climate_impacts_of_el_ni_o_phenomenon_in_latin_america_and_the_caribbean (accessed April 4, 2010).

United Nations Population Division. 2006. *World Population Prospects: The 2006 Revision.* Department of Economic and Social Affairs. New York: United Nations. Dataset in digital form. http://www.un.org/esa/population/publications/wpp2006/wpp2006.htm.

Vergara, W., H. Kond, E. Perez, J. M. Mendez Perez, Victor Magana Rueda, M. C. Arango Martinez, J. F. Ruiz Murcia, G. J. Avalos Roldan, and E. Palacios. 2007. "Visualizing Future Climate in Latin America: Results from the Application of the Earth Simulator" Latin America and the Caribbean Region, Sustainable Development Department, World Bank, Washington, DC.

Vianica. Hurricane Felix Information Center. 2007. "Hurricane Felix and the Miskito Cays." http://www.vianica.com/projects/1/

Vos, R., M. Velasco, and E. De Labastida. 1999. *Economic and Social Effects of El Niño in Ecuador.* Washington, DC: Inter-American Development Bank.

WMO (World Meteorological Organization). 1999. *The 1997–1998 El Niño Event: A Scientific and Technical Retrospective.* No. 905. Geneva: World Meteorological Organization.

World Bank. 2000–01. *World Development Report 2000/2001: Attacking Poverty.* Washington, DC: World Bank. http://go.worldbank.org/L8RGH3WLI0.

———. 2005. *The Urban Poor in Latin America.* Ed. Marianne Fay. Directions in Development. Washington, DC: World Bank.

World Bank. 2008a. "At a Glance: Poverty Numbers in LAC." http://www.worldbank .org/lacpoverty

World Bank. 2008b. *Equity and Efficiency in the Greenhouse: Responding to Climate Change in Latin America and the Caribbean Volume I: Overview.* Sustainable Development Department (LCSSD) and Chief Economist Office (LCRCE), Latin American and the Caribbean Region. Washington, DC: World Bank.

World Bank. 2008c. "Poverty Data: A Supplement to *World Development Indicators 2008.*" Washington, DC: World Bank.

World Bank. 2008d. *World Development Report: Agriculture for Development.* Washington, DC: World Bank.

WWF (World Wildlife Fund). 2006. *Up in Smoke? Latin America and the Caribbean—The Threat from Climate Change to the Environment and Human Development.* Third Report from the Working Group on Climate Change and Development. London: New Economics Foundation.

Zapata-Martí, Ricardo. 2006. *Los efectos de los desastres en 2004 y 2005: La necesidad de adaptación de largo plazo.* Punto Focal de Evaluación de Desastres, ECLAC, Mexico City.

Zhao, Y., C. Wang, S. Wang, and L. V. Tibig. 2005. "Impacts of Present and Future Climate Variability on Agriculture and Forestry in the Humid and Sub-humid Tropics," *Climatic Change* 70: 73–116.

# CHAPTER 4

# Agrarian Livelihoods and Climate Change

## Jørgen E. Olesen

This chapter analyzes social vulnerability to climate change in agrarian communities in Latin America and the Caribbean (LAC), with a particular focus on the dryland communities in which more than half of the region's rural poor live. Climate change implies higher temperatures and an intensification of the hydrological cycle, which will lead to both more intense rainfall and more frequent droughts (IPCC 2007a). The drier conditions are projected to occur mainly in the currently dry tropics and subtropics and may lead to salinization and desertification of agricultural lands. These conditions are likely to have particularly negative effects on the rural poor, affecting both food security and sources of income.

Because climate change is only one of the factors affecting the livelihoods of poor rural people, it is important to look at how the economic, social, and environmental context of production influences farmers' strategic choices. The first section of this chapter briefly describes the structure of rural poverty in the LAC region and the agrarian production systems on which poor rural communities directly and indirectly depend. Using the sustainable livelihoods framework (SLF; explained in chapter 1) helps to clarify the vulnerability context within which these communities live. Poor farmers' ability to use good husbandry practices for the land on which they depend is severely constrained by their lack of other types of assets. In many parts of the region, processes of land degradation already under way severely undermine the resilience and adaptability of local communities in the face of climate change and variability. The second part of the chapter reviews how climate change and variability are likely to affect the productive assets, livelihood, and food security of the rural poor. The third section addresses options for improving resilience and adaptability to climate change at the community level and at the individual farm level and draws implications for the role of institutions, including local government, and the for orientation of government policy. The final section highlights areas where further research is needed.

## Rural Poverty and Production Systems

The largest rural populations in the LAC region are in Brazil, Mexico, Colombia, Peru, and Guatemala. In several countries of the region, more than 30 percent of the population lives in rural areas (World Bank 2007a). The highest rural poverty rates are in Bolivia (where 82 percent of rural people live below the national poverty line), Colombia (79 percent), Guatemala (75 percent), Honduras (70 percent), and Peru (72 percent).

The rural poor in Latin America and the Caribbean tend to live in arid and semiarid regions, on mountain slopes or plateaus, and in tropical rainforests. These are fragile environments that in many cases suffer from environmental degradation, partly due to overpopulation (box 4.1).

The areas most densely populated by the rural poor are the arid and semiarid subtropical parts of the region, which cover more than 9 million square kilometers in northeastern Brazil, northern Mexico, northeastern Venezuela, the Pacific coastal and central areas of Honduras and Nicaragua, and northern Peru and Brazil. Substantially more than half of the region's rural population lives in dryland conditions where the mean annual ratio of precipitation to potential evapotranspiration is significantly less than one (World Bank 2002). These areas are subject to environmental degradation and have generally low productivity. Livelihood systems are often complex, diverse, and risk prone (Morton 2007). The limited availability of resources and the general projection that rainfall will become more erratic for drylands as climate changes suggest that communities in these areas are particularly vulnerable to climate change and variability and will increasingly feel the effects.

Box 4.1

## Climate Change Effects and Adaptation in the Bolivian Highlands

Some 200 indigenous territories are located in the highlands of Bolivia. Around 92 percent of the land within the titled territories is experiencing a high level of degradation and is at risk from desertification, partly caused by unsustainable land use due to overpopulation (Bosque 2008). Each year some 16,000 new small-scale farmers, many of them indigenous people, struggle to set up a living with too little land available. Less than 10 percent of the cultivated land in the highlands and valleys is irrigated, which means a high reliance on rainfall. Unsustainable land management practices leading to soil degradation include (a) expansion of the agricultural frontier of quinoa, (b) reduced area for the breeding of llamas, (c) loss of communal norms to maintain a balance in agricultural activities, (d) disappearance of traditional vegetation cover, and (e) contamination from the mining industry.

Climate change speeds up the degradation of natural resources that are already stressed by overpopulation and unsustainable management. That degradation can have undesired impacts on the livelihood of the indigenous population and undermine development efforts in the region. Among the principal climate changes that will affect the indigenous communities negatively are (a) higher temperatures and the disappearance of the Andean glaciers, (b) less and more variable rainfall, and (c) more frequent droughts and floods. More frequent droughts are one of the main dangers in the highlands, since they will accelerate desertification.

Less-predictable climate and weather conditions also make it more difficult for farmers to know when to begin sowing. Early frost and heavy rainfall may ruin part of the harvest, and stronger winds out of season further degrade the soil in the dry season.

*Source:* Bosque 2008.

## Agrarian Production Systems among the Rural Poor

Several types of agrarian production systems, identified by the primary activity of the family, its location, and the ethnic origin of its members, occur among the poor in Latin America and the Caribbean (table 4.1).

The largest group in the population of rural poor (33 percent) is the indigenous communities of the Andean region in Bolivia, northern Chile, Colombia, Peru, and Venezuela and the numerous indigenous communities and ejidos of Mexico.[1] Some 10 percent, or around 60 million of the

Table 4.1    Principal Production Systems among Agrarian Rural Poor

| Rural group | Main countries affected | Primary activity | Secondary activity |
|---|---|---|---|
| Small livestock farmers | All LAC | Cattle, sheep, and goat raising and wool | Milk and cheese production |
| Small crop farmers | All LAC | Cash and staple crops | Seasonal wage labor |
| Small mixed livestock and crop farmers | Brazil, Mexico, Peru, Bolivia | Staple and cash crops, livestock raising | Seasonal wage labor |
| Subsistence farmers | Brazil, Mexico, Peru, Bolivia | Seasonal paid employment | Staple crops |
| Landless farmers | Brazil, Mexico, Peru, Ecuador, Venezuela | Seasonal or permanent paid employment | Crops |
| Rural day laborers | Brazil, Mexico, Chile | Seasonal or permanent paid employment | |
| Andean herders | Peru, Bolivia | Alpaca and llama raising | Seasonal wage labor |
| Indigenous communities | Mexico, Peru, Bolivia | Crop farming | Livestock raising |

Source: Quijandria, Monares, and de Peña Montenegro 2001.

region's population, are indigenous, and most of them live in rural areas. Most rural indigenous families live in extreme poverty, generally having little or no schooling, few or no productive resources, limited knowledge of production, few work skills, little or no access to basic and rural production resources, and little or no political voice.

Small farmers make up 27 percent of the rural poor. They possess (with or without title) small plots of land in arid or semiarid regions, on hillsides, or on the fringes of irrigated valleys. They farm on ecologically fragile lands subject to climatic variation and uncertainty. In many cases, they raise crops and livestock on hillsides, where their activities lead to loss of natural vegetation and soil erosion. These small farmers combine agricultural production with seasonal off-farm work. Men take responsibility for preparing the land and for harvesting, while women and children tend the livestock. When men migrate in search of seasonal employment, the women take over all farm activities.

Subsistence or landless farmers account for 20 percent of the rural poor and are among the poorest of rural people. This group only has access to leased land, whether as tenants or sharecroppers. Families depend on seasonal or permanent off-farm employment for most of their income and consume most of their farm production themselves. Any surplus—depending on weather conditions—is sold on local markets or used in backyard production of poultry and pigs.

The number of rural day laborers has increased with the development of intensive export-oriented agriculture in the LAC region, which has created a considerable demand for skilled and unskilled labor. A large number of poor workers from both rural and urban areas earn income from harvesting, sorting, processing, and packaging fruits and vegetables.

Herders of the Andean highlands raise alpacas and llamas on natural grasslands at altitudes of more than 3,500 meters. Herders living at lower altitudes also raise sheep and small cattle. Most of the herder population is located in Peru and Bolivia, but there are also small groups of herders in northern Argentina and Chile. All are members of indigenous groups, and draw their principal income from the sale of alpaca and llama wool (Kronik and Verner 2010). They have some additional income from production of jerky (dried, salt-cured meat) and handicrafts.

### Sources of Vulnerability

Climate change and variability are but two of many sources of vulnerability for the rural poor. Like others worldwide, the LAC region's poor rural households suffer social and economic exclusion, often linked to ethnicity and gender; lack of access to basic services (health, education, housing); and very low levels of income and other assets. Poor rural communities are often geographically isolated, with unreliable systems of communication and services. Many are untouched by conditional cash transfer programs. To safeguard themselves, they create deeply rooted forms of reciprocity to maintain social and family relationships, and they depend on extended family networks, or social bonding capital (defined in chapter 1), for social and economic survival. Their livelihoods depend on fragile natural resources that in many cases are already being rapidly degraded. Particularly in drylands and other marginal lands, climate change is likely to speed up the degradation.

Particularly important processes that affect the natural resources used by poor farmers and herders are the conversion of rangelands or silvopastoral systems to arable lands and the shortening of fallow periods, leading to land degradation and overall lower productivity. Eventually the degraded lands may become desert, incapable of supporting life. These processes are often driven by increasing population pressure; the expansion of commercial, export-oriented agriculture; or the introduction of biofuel crops, pushing farmers or pastoralists who lack property rights onto more marginal land. Continuing rapid deforestation imperils the livelihoods of many indigenous people and threatens biodiversity resources. Government policies and structural change in the agriculture sector also appear to be making the rural poor more vulnerable in many cases. These factors are briefly reviewed below, followed by an assessment of how climate change and variability are likely to hasten and compound their effects.

### Soil Degradation and Desertification

Increasingly, land degradation has reduced the productivity of ecosystems in LAC (Bai and others 2008). Erosion, the main cause of land degradation in the region, affects more than 14 percent of the territory in South America and 26 percent in Central America (Oldeman 1994). Water and wind are both major causes of erosion (figure 4.1), and with the expectations for more intensive rainfall and longer droughts, erosion from both causes is projected to increase considerably (Meadows and Hoffman 2003).

Nutrient depletion is another serious issue. Largely driven by intensification of land use (UNEP 2002), nutrient depletion has exacerbated poverty—which, in turn, has contributed to greater environmental degradation and land deterioration.

Desertification is land degradation in arid, semiarid, and dry subhumid areas. Its several causes include climatic variation and human activities (Sivakumar 2007). It is common in drylands, and the areas of high vulnerability to desertification in Latin America and the Caribbean coincide to some extent with the areas where water and wind erosion are also causing land degradation.

Salinization is a particularly significant form of soil degradation, because it is difficult to treat and can lead to desertification. Salinization from irrigation occurs where the irrigation water carries nutrients and solutes that accumulate in the soil when the irrigation is insufficient to leach excess nutrients. Salinization is estimated to affect 18.4 million hectares in the region, particularly in Argentina, Brazil, Chile, Mexico, and Peru (UNEP 2002).

Desertification processes are likely to intensify considerably in areas where climate change reduces total rainfall or makes rainfall patterns more erratic. Since a large proportion of the rural poor in the LAC region live under dryland conditions, combating desertification will become a major issue for improving rural livelihoods in conditions of climate change.

**Figure 4.1** Water and Wind Erosion and Desertification as Sources of Land Degradation in Latin America and the Caribbean

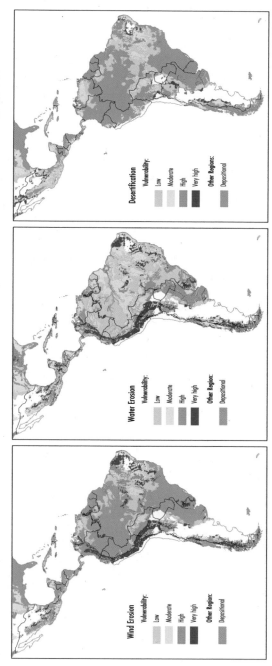

**Wind Erosion**

Vulnerability:
- Low
- Moderate
- High
- Very high

Other Region:
- Depositional

**Water Erosion**

Vulnerability:
- Low
- Moderate
- High
- Very high

Other Region:
- Depositional

**Desertification**

Vulnerability:
- Low
- Moderate
- High
- Very high

Other Region:
- Depositional

IBRD 37783
MAY 2010

*Source:* Adapted from UNEP 2002.

*Note:* Areas outside vegetation zones (dry, cold, glacier) are not classified as erosion zones, and deposition areas are places with accumulation of eroded soil.

Land users allow their land to become degraded for many reasons. Some of them reflect societal perceptions that land has a low value (Eswaran, Lal, and Reich 2001). Another is that soil degradation is a slow, barely perceptible process that leaves many people unaware that their land is degrading. Creating awareness and building a sense of stewardship are key elements in any effort to reduce degradation. Appropriate technology is only a partial answer. The main solution lies in the behavior of farmers, who need to be aware of the different sources of risk to the quality of the land they farm.

## Deforestation

Excessive deforestation has been taking place in tropical areas of Latin America and the Caribbean for the past several decades. Among its causes are population growth, extensive logging, an expanding agricultural area, and lagging agricultural yields. The forested area is declining in all LAC countries except Uruguay, mostly at rates of 0.5 percent to 1.5 percent annually (Velarde 2004). At the current rate, LAC forests are projected to shrink by 9 percent between 2000 and 2020, and the areas under permanent pasture, permanent crops, and arable crops are projected to increase by 2 percent, 25 percent, and 13 percent, respectively.

In many tropical countries, most of the deforestation results from the actions of poor subsistence cultivators. In the Amazon, however, those farmers contribute only about 30 percent of deforestation. The other 70 percent is caused by the encroachment of cattle ranches (Nepstad and others 2008), largely as an indirect result of an expansion of soybean farming. Most soybean cultivation takes place outside the rainforest in the neighboring cerrado grassland ecosystem and in areas that have already been cleared. But as cattle farms and the lands of some subsistence farmers are converted to soybean cultivation, cattle and subsistence farmers turn to forest clearing to obtain new land. Studies have also shown a strong association between logging and future clearing for settlement and farming.

## Impacts of Policy and Structural Change in Agriculture

Changes in agricultural policies have worked to the detriment of small farmers in much of the region. Structural adjustment policies introduced in the 1990s led to the elimination or severe reduction of state services to support rural production. For commercial agriculture, a private market grew up for technical services and proved very effective in increasing yields and acreage of major export crops. But small and medium-sized farming operations and small rural enterprises were left with no technical assistance and services.

Another trend that has been causing difficulties for small producers is the vertical integration of food supplies. Over recent decades, a handful of vertically integrated, transnational corporations have gained increasing control over the global trade in food, its processing, and sale. The biggest chains, most of them owned by multinational giants, now control 65 percent to 95 percent of supermarket sales in Latin America and the Caribbean (FAO 2007). While the expansion of supermarkets could present an opportunity for small farmers to reach new markets, it also presents a substantial risk of greater marginalization and even deeper poverty (Berdegue and Ravnborg 2007). For example, between 1997 and 2001, more than 75,000 Brazilian dairy farmers were delisted by the 12 largest milk processors during consolidation of the market. Most of them presumably went out of business. Similar structural change in agriculture is happening elsewhere in the region. It need not lead to increased poverty, if society can provide other and possibly better job opportunities and sources of livelihood to replace income from agriculture.

## Likely Impacts of Climate Change on Agrarian Societies

The vulnerability of agrarian societies to climate change and variability depends closely on the societies' access to assets of different types. Climate change primarily affects the natural and physical asset bases, making farming more difficult and unpredictable. Some of the major effects will come from increased frequency and severity of droughts and floods that degrade farmland and damage farm property as well as public infrastructure such as roads and irrigation channels. Other effects will be seen in changes in crop seasonality and yield caused by higher mean temperatures. The rural poor are particularly vulnerable to these physical losses, since they lack other assets, in particular human and financial capital, with which to offset them. They may have some bonding social capital, as noted above, but as livelihood sources are eroded, so too will be their social capital—for example when family members migrate away, as has happened during droughts in Mexico (box 4.3). Alternatively, social capital may simply become exhausted after a succession of severe weather events. In this section, the probable impacts of climate change on agriculture are reviewed, followed by an examination of some of the broader social impacts and the prospects for food security.

## Impacts on Agriculture

Climate change and variability affect cropping systems directly and indirectly through a range of pathways (Tubiello, Soussana, and Howden 2007). The effects are mediated through the farmer's management of the interactions between crops and their growth environment, which depend crucially on available resources, including climate, soil, water, nutrients, genetic diversity, and machinery or labor.

The projected increase in greenhouse gases will affect agro-ecosystems directly, mainly by increasing photosynthesis at higher atmospheric concentrations of carbon dioxide ($CO_2$), and indirectly by altering climate, with consequent effects on the functioning of ecosystems (Olesen and Bindi 2002). Increasing temperature affects plant development. With warming, the start of active growth is advanced and plants develop faster, which reduces crop duration. An analysis of global production shows that cereal crop yields are reduced with increasing temperatures (Lobell and Field 2007). The effects of higher temperatures on yields will be greatest in tropical crops such as coffee that in the present warm tropics perform close to the optimum, so that warming will often lead to relatively large yield reductions (box 4.2). However, extreme events such as cyclones, droughts, and floods cause greater damage to commercial crops than changes in mean climate variables alone (Porter and Semenov 2005).

Simulations of crop yield changes under climate change show large differences across the region (De la Torre, Fajnzylber, and Nash 2009). Considering the direct effect of higher atmospheric concentrations of $CO_2$, models using the Intergovernmental Panel on Climate Change's (IPCC) A2 emission scenario suggest that by 2080, crop yield changes will range from reductions of 20 percent to 30 percent in such countries as Bolivia, Mexico, Peru, and Venezuela, to increases of up to 5 percent in Argentina (Cline 2007; Parry and others 2004).[2] For smallholders, a mean reduction in maize yields can be expected by 2050 (Jones and Thornton 2003). With a reduction in land suitable for growing coffee in Brazil, and the stresses caused by higher temperatures and lower rainfall for coffee producers in Mexico (Gay and others 2006; IPCC 2007b), cultivation of coffee will shift to other parts of the region, such as Argentina (box 4.2).

Climate variability and climate change affect livestock production in two ways: indirectly, by influencing the productivity of grassland and hence the quantity and quality of fodder, and directly by causing more heat- and disease-related stress and death (Zhao and others 2005). In

Box 4.2

## Two Aspects of Coffee Production in Mexico

Gay and others (2006) used data from coffee production in the mountainous Mexican state of Veracruz to predict how coffee production might respond to changes in economic and climatic variables. The model includes economic variables, along with mean seasonal temperature and precipitation and their variability. The results show that coffee production responds significantly to changes in seasonal temperature patterns and also to changes in the minimum wage, as coffee production is labor intensive.

Temperatures in Veracruz are already above the optimum for coffee production. Higher mean temperatures and lower precipitation could cause a reduction of up to 34 percent in coffee production there by 2020. That would have important economic and social implications, since many of the rural poor obtain their income as laborers on the coffee plantations, and some small-scale farmers also grow coffee for sale in the marketplace. Importantly, in the past coffee producers in Veracruz have shown a limited capacity to adapt to climatic and economic stress. They are constrained by lack of money and poor access to credit; by the fact that coffee plantations are a long-term investment; by limited land availability; by little government support; and by tradition.

Coffee in Mexico has traditionally been grown under a diverse tree canopy in what is known as an "agroforestry" system, in which the canopy trees also provide fruit, nuts, and lumber that are additional sources of income. Studies in Chiapas have shown that shade trees increase the resilience of coffee growing to climate change (Lin, Perfecto, and Vandermeer 2008), primarily by reducing maximum temperatures and increasing the soil's moisture retention. The effects of Hurricane Stan in Chiapas in 2005 also showed that shade cultivation of coffee provided significant protection against landslides (Philpott and others 2008). However, in an effort to boost coffee production, supported by government incentives, many farmers have in recent years been abandoning traditional shade-growing techniques, leaving themselves more vulnerable to the effects of climate change.

*Source:* Gay and others 2006; Philpott and others 2008.

regions that are currently warm and dry, climate change will mostly affect both those aspects negatively.

Reduced water availability will particularly affect grain crops and livestock production in Central America (Mexico, Costa Rica, and Panama), the Andes, and parts of Argentina, Brazil, and Chile (box 4.3). In drier

**Box 4.3**

## Droughts in Mexico

Mexico has a long and varied experience with drought, from descriptions in early historical chronicles to contemporary climatic data and disaster declarations (Liverman 1990). More than 85 percent of the country is arid or semiarid, and inter-annual rainfall varies widely. Biophysical factors are the main reasons for vulnerability in the northern and north-central regions, where rainfall is most variable, and in the highlands, where the timing of rainfall is critical. In these regions, vulnerability is likely to increase as a result of deforestation, overgrazing, and the projected climate changes. Social reasons for vulnerability vary greatly among regions and population groups. More than half of Mexico's cropland is operated by ejidos, a form of cooperative land tenure, and drought losses are generally considerably higher on ejidos than on private farms, for several reasons. After the Mexican Revolution, the ejidos mostly received the less-productive and drier land. In many regions they do not have much irrigated land and have problems obtaining credit. Two policy changes effectively diminished the supply of water for small and traditional communal farmers, who also have less access to credit: the 1992 reform of the water law led to higher water prices (Wilder and Lankao 2006), and the North American Free Trade Agreement (NAFTA) encouraged export-oriented farmers to switch into more water-demanding crops (Liverman 2000).

Mexican communities have developed many traditional technologies for coping with drought, such as sophisticated irrigation systems, growing adapted maize varieties, and relying on traditional famine foods (such as cactus, agave, and mesquite) in drought years (Eakin 2000). More recent adaptation schemes that rely on irrigation have shown themselves vulnerable to multiyear droughts, though attempts to make irrigation more efficient have been beneficial. Experience from the 1996 drought in northern Mexico has shown wide variation in the responses of individual farmers to drought, with farmers planting and then praying for rain or government relief, and others deciding to abandon their land and migrating to find work in other regions of Mexico or the United States. Small landholders were disproportionally affected, especially those who were in debt or who farmed or ranched more fragile land (Liverman and Rosenberg 1996).

*Source:* Liverman 1990; 2000; Liverman and Rosenberg 1996.

areas, which largely coincide with the areas showing risk of desertification, climate change and variability will likely speed up the salinization and desertification of agricultural lands, compounding the effects of reduced rainfall.

The demand for water for irrigation is projected to increase in a warmer climate, heightening the competition for both surface water and groundwater between residential and agricultural uses and among different farmer groups. About 14 percent of the people of the LAC region lack access to a safe water supply, and 63 percent of those live in rural areas (IADB 2004). In many parts of Latin America and the Caribbean, further growth in demand from an increasing population, along with the expected drier conditions in many basins where water is already scarce, will produce a net increase in the number of people experiencing water stress. The vast majority of them will be among the poor, including those living in rural areas, where the impact will be particularly severe in already arid and semiarid areas.

More frequent intense rainstorms and floods will affect infrastructure such as roads and irrigation systems, which influence market access and production capacity. They may also threaten existing food and feed storage systems, with immediate effects on food security and safety.

### Social Impacts

Climate change will affect the livelihood sources of farmers, agricultural laborers, and other rural workers whose jobs depend on the health of agriculture (IAASTD 2009). Climate change also interacts with other pressures on livelihoods, such as the continued deforestation that reduces the livelihoods of many indigenous people and threatens biodiversity resources.

As noted above, climate change will affect the suitability of locally adapted crops, giving rise to cultivation of new crop species and cultivars. Because many rural people obtain most of their food from local agriculture and forestry, local diets will need to take stock of the new crops being grown. In the Andes, for example, other crops may replace potatoes, and in some of the drier and warmer regions maize may need to be replaced with more drought-tolerant crops such as sorghum or millet. Such new crops may be less favored for human consumption, and that may lead to more imports of food from outside. The local agricultural products may then be used instead for animal feed—which may support the livelihoods of farmers but increase the poverty of other rural people.

In many parts of the region, the effects of climate change on ecosystems have begun to compromise the traditional livelihoods and lifestyles of indigenous peoples who depend on them (box 4.4).

Changes in biodiversity, resulting from climate change and exacerbated by human pressures on natural ecosystems, will likely reduce the availability of many plant species used in modern as well as traditional medicine.

**Box 4.4**

## Indigenous Peoples across Latin America and the Caribbean Experience the Negative Social Impacts of Climate Change

The greater people's dependence on natural resources, the greater are the negative effects they experience from climate change, and the more likely they are to be increasingly affected in the future. Like other people who depend closely on natural resources, indigenous peoples are threatened by unforeseen seasonal variations, rising temperatures, droughts, more and more intense precipitation (rain and hailstorms with potential to result in floods), frosts, storms, and hurricanes.

But what sets indigenous peoples apart is the intimate ways in which they use and live by natural resources. These livelihood strategies are developed and maintained through social and cultural institutions, practices, and knowledge. Indigenous peoples' knowledge systems are based on experimenting with nature and are juxtaposed with a stock of knowledge developed over time and passed on through generations. Their ability to predict and interpret natural phenomena, climatic conditions, and the like, has not only been vital for survival and well-being, but has also been instrumental in the development of cultural practices, social structures, trust, and authority. Because their traditional ways of life are already under pressure from other societal trends and actors—such as expansion of the agricultural frontier, advancing colonization, political unrest, and pressure from the military and police, as well as illegal coca cultivation— their capacity to adapt to climate change events has in many cases been stretched to the limit.

In the Andean region, the climatic changes cause glacier retreat; variations in seasonality and in the number and intensity of rain- and hailstorms; frosts during unusual periods of the year; and droughts. All of these affect the Aymara and Quechua peoples by putting food security at risk, to the detriment of social stability, health, and psychological well-being. High-elevation Andean herders' livelihoods are threatened by the rapid melting of the glaciers; once the glaciers are gone they will face water scarcity in dry periods. Also particular to the high mountains are the effects of unexpected hailstorms and frosts during the crop-growing season. In the lower Andes, the melting of glaciers has different effects, including floods and soil erosion.

In the Caribbean and Mesoamerica, the increased intensity of extreme events endangers entire ecosystems. The gradual warming and acidification of the oceans are threatening the viability of, for example, sea turtles and coral reefs,

*(continued)*

**Box 4.4** *(continued)*

which some communities rely on for their livelihoods. Expectations are for more powerful hurricanes and a continuing upward trend for intense precipitation followed by extended periods of drought resulting from higher temperatures. Indigenous people interviewed in Nicaragua said that even now 20 years after Hurricane Joan, they have yet to fully recover the abundance of forest resources that existed before the hurricane (especially lumber and wildlife). The failure of forest resources to recover was closely linked to the expansion of the agricultural frontier that was facilitated by large tracts of forest being felled by the hurricane, allowing easy access to indigenous territories. In contrast, after 5 years yields were close to levels prior to the event for fishing and perennial crops such as fruit trees; yields recovered similarly after just one year for rice, beans, banana, plantain, cassava, and maize. Indigenous peoples affected by climate change and variability include the Miskitu, the Sumu-Mayangna complex (Ulwas, Twaskas, and Panamaskas), and the Rama.

In the lowland forests of the Amazon region, drought is exacerbating the effects on indigenous peoples of the ongoing deforestation. The threats from climate change compound the effects of other dangers such as expansion of the agricultural frontier, advancing colonization, political unrest, and illegal coca cultivation. The very great diversity of fish fauna in the Amazon provides an important food resource for indigenous communities. The successful reproduction of fish is dependent on the great rainy season that causes the river to rise. Normally, it rains through April and May, when the river floods the lower and higher alluvial plateaus and inundates extensive areas of the forest. During this time, most wild fruits, fed by the water, ripen and fall into the rising waters. They attract fish, which come to the surface and disperse all along the flooded forest floor to feed on the plentiful food and to lay their eggs. However, the river no longer rises and falls as before. In 2005 in the northern region, the river flooded and fish hatched normally, but then suddenly, the river receded before the fish were sufficiently mature, killing them. On the Amazon River (southern regime), flooding has not been sufficient since 1999, and that has directly affected fish reproduction. Abnormal river levels also directly affect the reproduction of turtles, which require a drop in the river level at precisely the time they are ready to lay their eggs on the beaches emerging out of the retreating river.

Indigenous peoples possess a strong awareness of complex seasonal rhythms. The annual succession of seasons is of utmost importance for indigenous peoples. It orders the timing of the horticultural cycle and the ritual practices that

*(continued)*

**Box 4.4** *(continued)*

help prevent illnesses and promote human well-being, and it is crucial for the reproduction of wildlife. Its disruption causes food insecurity and undermines the array of
solutions provided by cultural institutions and authorities.

According to key people interviewed in different areas, the natural signs that
they now perceive are alarming. Seasons have become irregular, the once regular
flood and ebb of rivers are now out of synchrony with the fall of wild fruits, and
heat is increasing. A recurrent lament voiced during all research field visits was
that seasonal variation has become so unpredictable that the cultural adaptation
strategies developed to tackle the normal span of variation no longer provide the
necessary security.

Without secure livelihood strategies, communities tend to break down. Some
indigenous peoples are compelled to change their livelihood so dramatically that
they lose conditions vital for the development and reproduction of their culture.
When climatic conditions become impossible to predict, the elders and traditional
leaders—who are the experts within traditional knowledge systems—lose credibility. When traditional authorities cannot guarantee abundance and prosperity,
their status falls, and people look elsewhere for solutions to their problems, both
by turning to other bodies of knowledge and by migrating. This has been spoken
of among several peoples as leading to the end of life as indigenous peoples.
Not only do individual family members leave, but whole families are now
uprooting. Some communities are left ghostlike; in others, only children and the
elderly remain. This phenomenon is seen in communities across the region from
Argentina to Mexico. An example is the response to flood damage to irrigation
channels in Palca, Bolivia, which is described in box 3.3 in chapter 3.

*Source:* Kronik and Verner 2010.

Together the increased climate-related risks to agriculture, property,
infrastructure, and ecosystems are likely to have negative effects on health
by impeding access to safe water sources and sufficient food (FAO 2007),
as discussed further in chapter 6.

## Impacts on Food Security

Essentially all quantitative assessments show that climate change and
variability will adversely affect food security in Latin America and the
Caribbean (Schmidhuber and Tubiello 2007). "Food security" is defined

as a "situation that exists when all people, at all times, have physical, social, and economic access to sufficient, safe, and nutritious food that meets their dietary needs and food preferences for an active and healthy life" (FAO 2002). Though the proportion of undernourished people in the region has generally been decreasing over the past two decades, hunger and malnutrition remain a major problem among the rural poor, particularly among those with little or no access to land.[3]

Projections of the biophysical effects of climate change and variability on hunger in the LAC region suggest that an additional 1 million people could be short of food by 2020 (Parry and others 2004). And in Brazil, simulations indicate that an average reduction of 18 percent in agricultural output could lead to an increase in rural poverty of 2 to 3.2 percentage points (De la Torre, Fajnzylber, and Nash 2009).

Climate change may also affect food security indirectly, notably by increasing the demand for bioenergy in an effort to reduce reliance on fossil fuels, affecting world food prices. Increasing the use of energy from biological sources presents both opportunities and risks for food security. It could create new market opportunities for agriculture, foster rural development, and alleviate poverty, not least by improving rural access to sustainable energy. But if not managed sustainably, it could seriously threaten food security, especially for some of the most vulnerable people. Rising demand may raise food prices and increase food price volatility, compounding the food-price effects of more frequent floods and droughts across the world. Higher food prices will have both positive and negative effects on rural poverty, affecting food producers and food consumers in opposite directions. The net effect in most low-income countries, however, including those in Latin America and the Caribbean, is to increase the poverty rate (Ivanic and Martin 2008).

It must be said that the socioeconomic environment is a more important determinant of food security than are the biophysical impacts of climate change (Tubiello and Fischer 2007). Much depends on the socioeconomic status of a country or a population group prior to climate change (FAO 2006). In a poor agrarian society, a failure in agricultural production exacerbates poverty, malnutrition, and hunger for all, not just those whose primary activity is agriculture. It also increases dependency on other sources of income and livelihood. Recurrent production failures erode livelihood assets, exacerbating poverty and eventually leading to migration. Much will therefore depend on implementation of pro-poor policies to combat these effects.

## Adaptation to Climate Change: Institutional and Policy Perspectives

In principle, a wide range of options exists for adapting livelihood strategies to climate change and variability. Strategies for use by households and communities are described first, followed by a look at changes that are needed in agricultural practices and how these may be supported by institutions, policy, and research.

### Community Strategies for Coping with Environmental Risk

Agrarian communities adopt various strategies to deal with climatic variability and climate risk, such as growing maize for subsistence in addition to cash crops, keeping livestock to provide income in case of crop failures, and sending members to work in less-vulnerable areas and send remittances home (Eakin 2005). These strategies are of five broad types (Agrawal 2008).

Mobility pools risk across space. Temporary migration to seek temporary or seasonal employment in rural areas and jobs in urban areas or across borders, coupled with remittances, is a common way of coping with adverse climatic conditions. With recurring droughts and severe degradation of the available land, entire families may need to resettle elsewhere.

Crop storage spreads risk across time. Lack of safe storage facilities is an important and typical characteristic of poor agrarian households (TWN 2006). When well-built infrastructure (for example, safe storage for food and feed grain) is combined with good coordination across households and social groups, crop storage facilities can protect effectively against climatic variability and even total crop failure. However, they are no cure for a permanent erosion of production resources. Many of the food and feed storage systems of smallholder farmers are not robust and will probably be damaged if severe storms become more frequent. Support for climate-proof storage facilities, to be operated by individual farmers or cooperatives, is therefore needed (SDC 2008).

Diversification pools risks across the assets owned by the household or collective. At the farm level, diversification often involves integration of crop and livestock production so as to enhance overall farm income, and it often includes improved use of crop residues and other wastes. Diversification can vary widely, depending on what livelihood strategy is in use, and may involve both farm and nonfarm production assets, consumption strategies, and job opportunities. This is a beneficial and reliable strategy to the extent that climate change and variability affect different assets differently, but when several assets are simultaneously damaged, for instance, by a hurricane, it may be insufficient. It also may not be an option for people who lack sufficient land or a relatively stable social structure or who have insufficient human capital.

Communal pooling distributes risks across households, perhaps through joint ownership of assets or through sharing labor or incomes across households. It may also involve mobilization of resources that are held collectively during times of scarcity. A kind of communal pooling that draws heavily on social capital is the formation of cooperatives, to improve farmers' access to markets and improve the provision of inputs to offset a less-favorable climate, for example, by organizing water harvesting and irrigation facilities.

Exchange can substitute for the first four of these strategies if the rural poor have access to markets. The sale of farm products for cash allows people to purchase other assets they need or to save for a time of greater need. For people not fortunate enough to hold such savings, a crisis such as a drought may force them to resort to exchange of assets such as livestock or household items for basic necessities to survive. To be successful, however, exchange often requires considerable specialization and institutionalization of exchange relations. Buying insurance to cover crop or property losses from droughts or floods would fall into this category of specialized and institutionalized exchange. The use of insurance against crop failure varies considerably among LAC countries. On average across countries, 6 percent of cropland is covered by insurance. The figure in Uruguay is 20 percent (Edmeades and others 2008). In Mexico, the use of insurance declined considerably when the agricultural insurance market was privatized, and smallholder farmers have had difficulty paying insurance premiums. A more extreme climate with heavy storms, in combination with mismanagement of slopes and rivers, will threaten infrastructure that is critical for ensuring market access. Rural development therefore needs to give priority to infrastructure that can withstand more intense rainstorms.

These strategies typically draw on several of a household's available assets simultaneously. Agrawal and Perrin (2008) note that the most common adaptive strategies are diversification and communal pooling, and a combination of diversification plus exchange.

### Changes in Agricultural Practices

As climate change increases the incidence and severity of natural disasters, resilience-building activities, including agro-ecological management, will become more important in helping vulnerable communities adapt to a changing environment. Communities with a more secure, sustainable natural asset base are better equipped to deal with sudden shocks and disruptive trends. Increasing the resilience of agro-ecosystems generally entails improving, or at least maintaining, the natural resource base.

Resource-poor land users are often forced to overexploit their natural capital to make a living, however, and they frequently lack the resources to replace what is used up (Thomas and others 2007). Studies among different agrarian groups in Mexico have shown that managing climatic risk—both present and future—is not always a priority of farm households, despite frequent and even increasing losses to climatic hazards (Eakin 2005). And case studies of agricultural adaptation have shown that nonclimatic factors are often bigger determinants of individual farmers' strategies than climatic factors, emphasizing that poor farmers will likely need assistance to adapt to climate change in an optimal way (Ziervogel and others 2006).

In principle, farmers have a range of options to improve agricultural management to cope better with climate change and variability. Most of them are extensions of current practices (Howden and others 2007):

- Altering the timing and location of cropping activities
- Altering inputs such as crop varieties, fertilizer rates, or irrigation
- Using technologies that "harvest" and conserve water and soil moisture
- Managing water to prevent flooding, waterlogging, erosion, and nutrient leaching under increased rainfall
- Improving crop protection practices, including changing crop rotations
- Diversifying farm activities, for example, by integrating crop and livestock production
- Using weather forecasting to reduce production risk (box 4.5)

Findings from Central American countries emphasize the importance of conserving entire hillside and watershed ecosystems rather than individual plots of land (box 4.6). Farming should use longer-term agro-ecological strategies to maintain the natural resource base and thus the economic viability of people's livelihoods.

Changes in climate variability may be particularly difficult for many farmers to adapt to, and adaptation strategies to cope with variability may differ from those dealing with changes in mean climate. Adapting to increased variability may require measures to avoid periods of high stress or steps to make the system more resilient, for example,

Box 4.5

## Forecasting El Niño–Southern Oscillation Events

El Niño-Southern Oscillation (ENSO) is the major cause of interannual variation in temperature and rainfall in large parts of Latin America and the Caribbean. Particularly strong El Niño years (such as 1982–83 and 1997–98) are associated with increased risks of floods and droughts, and with climate change these events - could become more intense and frequent.

There is therefore great interest in increasing the ability of climate models to forecast ENSO events (Eakin 2005). Agriculture is a key sector for the potential use of climate forecasts to plan production strategies. Such climate forecasts have been used in northeast Brazil since the early 1990s and have been successful in reducing production losses, but the broad adoption of this technique has been held back because failure to forecast some weather events eroded its credibility.

New studies have focused on factors that influence the perception, communication, interpretation, and use of forecasts (Orlove, Broad, and Petty 2004). Their results show that poor and less-educated people may be just as able as wealthier and more educated people to use forecasts. In fact, in the Peruvian and Bolivian Andes, indigenous populations have used the visibility of stars at a certain time of year to predict rainfall, and the ability to see high clouds seems to be a simple indicator of a La Niña year. Recently in Tlaxcala, Mexico, based on strong stakeholder involvement, ENSO forecasting was successfully used to switch crops (from maize to oats) during an El Niño event.

With ENSO events projected to cause more droughts and floods in Latin America and the Caribbean, the use of climate forecasts will be of increasing importance for avoiding their negative consequences. This will require increased institutional support for delivering and communicating information suitably targeted to different farmer groups.

*Source:* Eakin 2005; Orlove, Broad, and Pretty 2004.

adding diversity to the crop rotation and improving soil and water resources.

Changes in farming practices, like the broader changes in family and community livelihood strategies that were outlined in the preceding section, will need to draw on and strengthen the assets available. Changing the timing and location of agricultural activities draws on social capital, using available labor to make better use of rainfall variations. That often means intensive periods of soil cultivation, planting, and harvesting,

**Box 4.6**

## Increasing Resilience through Agro-ecological Practices

In October 1998 Hurricane Mitch clearly showed the human and ecological vulnerability of Central America to climate-related disasters. The storm caused massive floods and landslides, killing more than 18,000 people and destroying thousands of homes, along with infrastructure, crops, and animals. The people most vulnerable were those living and farming on hillsides and near riverbanks (IISD 2003). With little access to credit, land titles, or technical assistance to help diversify and enhance their livelihoods, these farmers had had little incentive to invest in sustainable farming practices. Clear-cutting of forests for timber, ranching, and farming, along with widespread burning, had led to massive losses of protective vegetative cover, leaving hillsides barren and unable to absorb or retain water. During Hurricane Mitch, heavy rainfall led to massive runoffs on these degraded hillsides, carrying away enormous amounts of topsoil, along with rocks and vegetation.

Based on data collected from Honduras, Nicaragua, and Guatemala, it has been found that farms using agro-ecological practices such as soil and water conservation, cover cropping, organic fertilizer use, integrated pest management, and reduced or zero grazing were more resilient to erosion and runoff and retained more topsoil and moisture.

*Source:* Holt-Gimenez 2001.

engaging the members of the household and possibly other members of the community. Altering farm inputs draws on human and financial capital, since improved knowledge is often needed concerning how best to combine inputs to meet the demands of a changed climate, and some of the needed changes in inputs require investments or bring higher costs. Likewise, introducing water-harvesting methods and investing in irrigation schemes involve mobilizing a range of assets, including human, social, and financial ones.

Since the most vulnerable farming systems are those located in drylands, it is especially important to focus on measures that improve the efficiency of water use and reduce soil degradation, in particular desertification. Notably, changing from subsistence farming to high-value irrigated crops does not necessarily reduce the uncertainty of household livelihoods, since such systems are often very labor intensive and product prices often extremely volatile (Eakin 2005).

Irrigation is vital to agriculture in several regions of Latin America and the Caribbean, but it often makes very inefficient use of water. For example in Peru, where 85 percent of the irrigated area is watered by surface irrigation, nearly two-thirds of the water provided in this way is lost to plants through a combination of surface runoff, deep percolation, and evaporation. Even worse wastage rates occur in Mexico, where 94 percent of irrigation is done with surface irrigation (Edmeades and others 2008). Climate change will reduce the water available for irrigation in many parts of the LAC region. But there is considerable scope for improving water use, both through adopting more efficient irrigation systems and through switching to less-water-intensive cropping systems.

Technologies such as multistory agroforestry systems (for example, growing coffee under shade trees) and some organic farming practices can effectively satisfy both short-term demands and long-term requirements for sustainability because they help to protect the soil and allow it to store more carbon. Because such systems can also help to reduce emissions of greenhouse gases, they may be attractive for additional funding through initiatives for reducing carbon emissions. To achieve this would require policy and economic interventions, supported by local institutions, to close the gap between the short-term goals of resource-poor farmers and the longer-term demands of the broader community—and society in general—for resilient and sustainable farming systems.

### Institutional and Policy Dimensions

Enhancing the ability of households and communities to deal with climatic variability and risks brings many benefits. Enhanced coping capacities lead to longer-term improvements in livelihoods (Moser and Satterthwaite 2008). It is of paramount importance to incorporate climate change considerations in future development initiatives and to consider how institutional processes could support doing so.

The structures and processes that can help or hinder the rural poor in adapting to the effects of climate change and variability include the institutions, organizations, policies, and legislation that shape livelihoods. It is the task of these structures and processes to facilitate adaptation, to build capacity, and to remove obstacles to ensuring long-term sustainable livelihoods in conditions of climate change. This task requires partnerships, capacity building, involvement of a wide range of stakeholders, and above all political awareness and political will (IISD 2003).

All adaptation practices occur in institutionally rich contexts, and the success of adaptation depends on specific institutional arrangements

(Agrawal and Perrin 2008).We define "institutions" in the broad sense, as formal or informal systems of social interactions that (a) regulate access to production factors (land, labor, capital, and information), (b) facilitate transactions, and (c) permit economic cooperation and organization (Wiggins and Davis 2006).

In drylands, because of the importance of variability and risk, the dominant institutions tend to be flexible enough to allow for variable and complex use of resources (Morton 2007). Markets for produce and consumer goods that operate in dryland societies may function flexibly through institutions such as shopkeeper credit and long-distance trade networks linked by specific ethnic groups, within which trust and information can flow. Markets function alongside informal institutions for allocating resources, which also tend to operate flexibly. Examples include traditional patterns of loaning livestock and sharing their produce, rights to use wells of different kinds, seasonal use of fallows and crop residues, collective labor parties, families as institutions for regulating migration, and networks for sharing information.

Smallholder farmers in marginal areas experiencing climatic variability and climate change often distrust government and other external institutions (Young and Lipton 2006). In contrast, they tend to view community institutions as providing reliable support, information, and expertise in times of need and risk. Local governments should thus seek to support household and community initiatives. The ability of local governments in Latin America and the Caribbean to perform in this manner has in general been improved by changes that have led to greater democracy and more decentralized decision making.

It is part of the role of central government to help local governments develop asset-based adaptation strategies. For the agricultural sector, these strategies need to focus on helping smallholder farmers to improve water use and maintain soil fertility. It is also very important for central government to support transport infrastructure and agricultural research and extension, since such facilities are critical for effective adaptation to climate change. In many of the region's poorest countries, such support and development have been neglected over the past 20 years (Binswanger 2007).

In particular, greater political focus is needed on supporting local governance in vulnerable dryland societies—whether governance is provided by formal or informal institutions (Eakin and Lemos 2006). Policies affecting land, soil, water, nutrients, agrochemicals, energy, genetic diversity, research capacity, information systems, culture, infrastructure, and market access should seek to ensure that sustainable, long-term solutions, rather than short-term economic gain, are pursued. Since the impacts of climate change and variability on poor farmers and agrarian communities are highly locally specific and difficult to predict, flexible policies are needed in which the local government and organizations are given the resources and the decision power to act in the interest of the local population.

Land and water management will be increasingly critical to livelihoods in agrarian societies. Policies to promote and support management practices that combat land degradation and desertification should receive much more attention. In some cases that may require reduction of population pressure by resettling the inhabitants of affected areas. Where water can be made available, new investments in canals, dams, and water-saving technologies, as well as new institutional arrangements (water rights, water user associations, water pricing) need to be promoted. These arrangements must ensure greater efficiency in water use, with highest priority given to the highest-value uses, and must include consideration of the social consequences.

International donors should support central governments in developing communities' adaptive capacity in three ways: (a) examine whether current priorities are appropriate or should be modified to enhance resilience to climate change and variability, (b) identify how donors could support national structures (including financial, advisory, and regulatory capacity) that assist in developing adaptive capacity, and (c) directly support innovation at local and community levels to reduce short- and long-term risks associated with climate change and climatic variability.

## Research Perspectives

The current research focus on studying the effects of extreme events on poor rural people in developing countries needs to move toward studying how climate change and variability affect the long-term sustainability of agricultural systems in marginal environments. Increased attention should also be given to how complex agrarian farming systems can adapt to changes in climate and climatic variability. Research should also address the coping capacity of different regions and different social groups, particularly with respect to how migration and livelihood diversification (and support for diversification) may assist adaptation to climate change. Such research will not only improve knowledge of impacts, but most importantly, it will help in building adaptive capacity at all levels within the farming community.

A particular need exists to strengthen the interactions between research and local knowledge, to ensure that local environmental knowledge is up to date for conditions of climate change and variability (World Bank 2009). Many of the farming communities in the LAC countries, including indigenous groups, rely on such local environmental knowledge, and their capacity to cope with, and adapt to, climate change and variability depends on whether that knowledge can be quickly updated to reflect the changing climate. This emphasizes the need for better education in indigenous communities and the need to direct research and knowledge development and dissemination more toward them. It may also be advantageous to involve indigenous groups in developing new ways of learning that better incorporate both traditional knowledge and science-based results, and that can compensate for poor conventional learning skills stemming from high rates of illiteracy.

## Notes

1. The ejido is a traditional system of communal landholding that combines collectively used areas with individually assigned plots of land.
2. The A2 emission scenario envisages a mean temperature change of 3.4°C by 2090–99 relative to 1980–99; see IPCC 2001.
3. For many of the extremely poor, improving nutrition is the first priority if more resources become available in the household (World Bank 2007b).

## References

Agrawal, A. 2008. "The Role of Local Institutions in Adaptation to Climate Change." Paper presented at Workshop on Social Dimensions of Climate Change, Social Development Department, World Bank, Washington, DC.

Agrawal, A., and N. Perrin. 2008. *Climate Adaptation, Local Institutions, and Rural Livelihoods.* IFPRI Working Paper W08I-6. Washington, DC: International Food Policy Research Institute.

Bai, Z. G., D. L. Dent, L. Olsson, and M. E. Schaepman. 2008. "Proxy Land Assessment of Land Degradation." *Soil Use and Management* 24: 223–24.

Berdegue, J. A., and H. M. Ravnborg. 2007. *Agricultural Development for Poverty Reduction: Some Options in Support of Public Policy Interventions.* Copenhagen: Danish Institute for International Studies.

Binswanger, H. P. 2007. "Empowering Rural People for Their Own Development." *Agricultural Economics* 37: 13–27.

Bosque, H. 2008. *Diagnotico en Tierras Altas de Bolivia, el marco de la Temática de la "recuperación de suelos y sistemas tradicionales de producción" en tierras comunitarieas de origen*. La Paz, Bolivia: Viceministerio de Tierras.

Cline, W. R. 2007. *Global Warming and Agriculture*. Washington, DC: Center for Global Development.

De la Torre, A., P. Fajnzylber, and J. Nash. 2009. *Low Carbon High Growth: Latin American Responses to Climate Change. An Overview*. World Bank Latin American and Caribbean Studies. Washington, DC: World Bank.

Eakin, H. 2000. "Smallholder Maize Production and Climate Risk: A Case Study from Mexico." *Climate Change* 45: 19–36.

———. 2005. "Institutional Change, Climate Risk, and Rural Vulnerability: Cases from Central Mexico." *World Development* 33 (11): 1923–38.

Eakin, H., and M. C. Lemos. 2006. "Adaptation and the State: Latin America and the Challenge of Capacity-Building under Globalization." *Global Environmental Change* 16: 7–18.

Edmeades, S., W. Janssen, C. Dengel, A. Horst, A. Bucher, D. Behr, S. Georgieva, and D. Arias. 2008. "Climate Change Aspects in Agriculture." Latin America and Caribbean Region, World Bank, Washington, DC. http://go.worldbank.org/Q1YTC1WII0.

Eswaran, H., R. Lal, and P. F. Reich. 2001. "Land Degradation: An Overview." In *Responses to Land Degradation*, ed. E. M. Bridges, I. D. Hannam, L. R. Oldeman, F. W. T. Pening de Vries, S. J. Scherr, and S. Sompatpani. Proceedings of the 2nd International Conference on Land Degradation and Desertification, Khon Kaen, Thailand. New Delhi, India: Oxford University Press.

FAO (Food and Agricultural Organization). 2002. *The State of Food Insecurity in the World, 2001*. Rome: FAO.

———. 2006. *The State of Food Insecurity in the World, 2006*. Rome: FAO.

———. 2007. *The State of Food and Agriculture. Paying Farmers for Environmental Sciences*. Rome: FAO.

Gay, C., F. Estrada, C. Conde, H. Eakin, and L. Villers. 2006. "Potential Impacts of Climate Change on Agriculture: A Case Study of Coffee Production in Veracruz, Mexico." *Climatic Change* 79: 259–88.

Holt-Gimenez, E. 2001. "Measuring Farmers' Agroecological Resistance to Hurricane Mitch." *LEISA Magazine*, April.

Howden, S. M., J. F. Soussana, F. N. Tubiello, N. Chhetri, M. Dunlop, and H. Meinke. 2007. "Adapting Agriculture to Climate Change," *Proceedings of the National Academy of Sciences* 104: 19691–96.

IAASTD (International Assessment of Agricultural Knowledge, Science, and Technology for Development). 2009. *Agriculture at a Crossroads: Global Report*. Washington, DC: Island Press.

IADB (Inter-American Development Bank). 2004. "Financing Water and Sanitation Services: Options and Constraints." Seminar of the Inter-American Development Bank. Salvador, Bahía, Brazil.

IISD (International Institute for Sustainable Development). 2003. *Livelihoods and Climate Change*. Winnipeg: IISD.

IPCC (Intergovernmental Panel on Climate Change). 2001. *Special Report on Emission Scenarios*. Geneva: IPCC.

————. 2007a. *Climate Change 2007: The Physical Science Basis*. Contribution of Working Group I to the Fourth Assessment Report of the IPCC. Geneva: IPCC.

————. 2007b. *Climate Change 2007: Impacts, Adaptation, and Vulnerability*. Contribution of Working Group II to the Fourth Assessment Report of the IPCC. Geneva: IPCC.

Ivanic, M., and W. Martin. 2008. *Implications of Higher Global Food Prices for Poverty in Low-Income Countries*. Policy Research Working Paper 4594, World Bank, Washington, DC.

Jones, P. G., and P. K. Thornton. 2003. "The Potential Impacts of Climate Change on Maize Production in Africa and Latin America in 2055." *Global Environmental Change* 13: 51–59.

Kronik, Jakob, and Dorte Verner. 2010. *Indigenous Peoples and Climate Change in Latin America and the Caribbean*. Washington, DC: World Bank.

Lin, B. B., I. Perfecto, and J. Vandermeer. 2008. "Synergies between Agricultural Intensification and Climate Change Could Create Surprising Vulnerabilities for Crops." *BioScience* 58: 847–54.

Liverman, D. M. 1990. "Vulnerability to Drought in Mexico: The Cases of Sonora and Puebla in 1970." *Annals of the Association of American Geographers* 80: 49–72.

————. 2000. "Adaptation to Drought in Mexico." In *Drought: A Global Assessment*, ed. D. Wilhite, 35–45. New York: Routledge.

Liverman, D. M., and J. Rosenberg. 1996. *Preliminary Assessment of the 1996 Drought in Sonora*. Tucson, Arizona: Latin America Area Center.

Lobell, D. B., and C. B. Field. 2007. "Global Scale Climate-Crop Relationships and the Impacts of Recent Warming." *Environmental Research Letters* 2: 014002.

Meadows, M. E., and T. M. Hoffman. 2003. "Land Degradation and Climate Change in South Africa." *Geographical Journal* 169: 168–77.

Morton, J. F. 2007. "The Impact of Climate Change on Smallholder and Subsistence Agriculture." *Proceedings of the National Academy of Sciences* 104: 19680–85.

Moser, C., and D. Satterthwaite. 2008. "Pro-poor Climate Adaptation in the Urban Centers of Low- and Middle-Income Countries." Paper presented at

Workshop on Social Dimensions of Climate Change. Social Development Department, World Bank, Washington, DC.

Nepstad, D. C., C. M. Stickler, B. Soares-Filho, and F. Merry. 2008. "Interactions among Amazon Land Use, Forests and Climate: Prospects for Near-Term Forest Tipping Point." *Philosphical Transactions of the Royal Society* B363: 1737–46.

Oldeman, L. R. 1994. "The Global Extent of Soil Degradation." In *Soil Resilience and Sustainable Land Use*, ed. D. J. Greenland and T. Szaboles. Wallingford, U.K.: Commonwealth Agricultural Bureau International.

Olesen, J. E., and M. Bindi. 2002. "Consequences of Climate Change for European Agricultural Productivity, Land Use, and Policy." *European Journal of Agronomy* 16: 239–62.

Orlove, B. S., K. Broad, and A. M. Petty. 2004. Factors that Influence the Use of Climate Forecasts. *Bulletin of the American Meteorological Society*, doi: 10.1175/BAMS-85-11-1735.

Parry, M. L., C. Rosenzweig, A. Iglesias, M. Livermore, and G. Fischer. 2004. "Effects of Climate Change on Global Food Production under the SRES Emissions and Socioeconomic Scenarios." *Global Environmental Change* 14: 53–67.

Philpott, S. M., B. B. Lin., S. Jha, and S. J. Brines. 2008. "A Multi-scale Assessment of Hurricane Impacts on Agricultural Landscapes Based on Land Use and Topographic Features." *Agriculture, Ecosystems, and Environment* 128: 12–20.

Porter J. R., and M. A. Semenov. 2005. "Crop Responses to Climatic Variation." *Philosophical Transactions of the Royal Society, London* B360: 2021–35.

Quijandria, B., A. Monares, and R. U. de Peña Montenegro. 2001. *Assessment of Rural Poverty, Latin America and the Caribbean*. Santiago, Chile: International Fund for Agricultural Development.

Schmidhuber, J., and F. N. Tubiello. 2007. "Global Food Security under Climate Change." *Proceedings of the National Academy of Sciences* 104: 19703–08.

SDC. 2008. Central America: Fighting Poverty with Silos and Job Creation. Berne, Switzerland: Swiss Agency for Development and Cooperation.

Sivakumar, M. V. K. 2007. "Interactions between Climate and Desertification." *Agriculture and Forest Meteorology* 142: 143–55.

Thomas, R. J., E. de Pauw, M. Qadir, A. Amri, and M. Pala. 2007. "Increasing the Resilience of Dryland Agro-ecosystems to Climate Change. *SAT eJournal* 4 (1): 1–37.

Tubiello, F. N., and G. Fischer. 2007. "Reducing Climate Change Impacts on Agriculture: Global and Regional Effects of Mitigation, 2000–2080." *Technological Forecasting and Social Change* 74: 1030–56.

Tubiello, F. N., J. F. Soussana, and S. M. Howden. 2007. "Crop and Pasture Response to Climate Change." *Proceedings of the National Academy of Sciences* 104: 19686–90.

TWN (Third World Network). 2006. Globalization, Liberalization, Protectionism: Impacts on Poor Rural Producers in Developing Countries. Rome: IFAD.

USDA (U.S. Department of Agriculture). 2003. Global Desertification Vulnerability Map. http://soils.usda.gov/use/worldsoils/mapindex/desert .html.

UNEP (United Nations Environment Program). 2002. *Global Environment Outlook – 3*. Nairobi, Kenya: United Nations Environment Program.

Velarde, S. J. 2004. *Socio-economic Trends and Outlook in Latin America: Implications for the Forestry Sector to 2020*. Rome: FAO.

Wiggins, S., and J. Davis. 2006. *Economic Institutions*. IPPG Briefing Paper 3. Improving Institutions for Pro-Poor Growth Consortium. Manchester, U.K.: University of Manchester.

Wilder, M., and P. R. Lankao. 2006. "Paradoxes of Decentralization: Water Reform and Social Implications in Mexico." *World Development* 34: 1977–95.

World Bank. 2002. *The Dryland Predicament: Natural Capital, Global Forces*. Washington, DC: World Bank.

———. 2007a. "The Invisible Poor. A Portrait of Rural Poverty in Argentina." Washington, DC: World Bank.

———. 2007b. "Chapter 6: Sustainable Fisheries through Improved Management and Policies." In *Republic of Peru Environmental Sustainability: A Key to Poverty Reduction in Peru*. Environmentally and Socially Sustainable Development in Latin America and the Caribbean. Washington, DC: World Bank.

———. 2009. "Building Response Strategies to Climate Change in Agricultural Systems in Latin America." Washington, DC: World Bank.

Young, K. R., and J. K. Lipton. 2006. "Adaptive Governance and Climate Change in the Tropical Highlands of Western South America." *Climatic Change* 78: 63.

Zhao, Y., C. Wang, S. Wang, and L. V. Tibig. 2005. "Impacts of Present and Future Climate Variability on Agriculture and Forestry in the Humid and Sub-humid Tropics." *Climatic Change* 70: 73–116.

Ziervogel, G., A. Nyong, B. Osman, C. Conde, S. Cortés, and T. Downing. 2006. *Climate Variability and Change Implications for Household Food Security*. Working Paper No. 20. Assessments of Impacts and Adaptation to Climate Change. Washington, DC: START Secretariat.

# CHAPTER 5

# Coastal Livelihoods and Climate Change

## Sara Trab Nielsen

This chapter analyzes social vulnerability to climate change and variability within coastal communities in Latin America and the Caribbean (LAC). The region's 64,000-kilometer coastline is one of the most densely populated in the world (Cavallos 2005; Sale and others 2008). Its coastal states have more than 521 million residents, of whom two-thirds (348 million) live within 200 kilometers of the coastline.[1] More than 8.4 million people in Latin America and the Caribbean live in the path of hurricanes, and roughly 29 million live in low-elevation coastal zones[2] where they are highly vulnerable to sea level rise, storm surges, and coastal flooding (McGranahan, Balk, and Anderson 2007; UNEP 2007). Human activities such as overfishing, marine pollution, and development of the coast have already eroded the natural resources in many coastal areas to a level where they can hardly withstand the impacts of climate change. Climate change and variability are likely to compound the damage, however, putting severe pressure on coastal communities, both directly as the result of storms and rising sea levels, and indirectly through further degrading natural resources.

Coastal communities at greatest risk from climate change and variability are generally those that rely on natural resources for a living, occupy marginal lands, and have limited access to the livelihood assets that are necessary for building resilience to the impact of climate change.[3] They include nature-dependent communities, especially those that rely on coastal tourism and on fisheries. They also include much of the region's large population of urban slum dwellers, who live in places such as low-elevation areas or steep slopes threatened by coastal flooding and erosion.

The first section of this chapter establishes the vulnerability context within which coastal communities live.[4] Next, three examples are presented to suggest how these communities are likely to be affected by climate change and variability, taking the view that resilience and adaptability depend on access to assets—which in turn depends greatly on local as well as national governance. The third section of the chapter reviews options and makes recommendations for building resilience and adapting livelihoods where needed. Much will depend on societal choices about the protection of the environment, the protection of the poor, land rights for marginal communities, and the degree of overall regard for incorporating climate change considerations into policies. Overall, it is argued that climate change considerations need to be incorporated in future development initiatives. Implementing policies and developing ideas to safeguard natural resources—improving natural capital—will benefit coastal communities, fostering the health of the natural environments they rely on for their livelihoods. Healthy natural environments are much better able to withstand climate change impacts. The chapter concludes by highlighting some needs for further research.

## Vulnerability Context of Coastal Communities

The coastal population of Latin America and the Caribbean is already highly vulnerable as a result of environmental deterioration. The region's wide variety of coastal natural resources has long supported subsistence livelihoods, fishing communities, and tourism, but in the past 100 years, those resources have deteriorated immensely in many areas as a result of population pressure, poverty, and destructive human activities such as overfishing, marine pollution, and unrestricted, unsustainable development of the coast (McField and Kramer 2007). Environmental problems such as declining fish populations; changes in water clarity, quality, and availability; ecosystem destruction in coral reefs and mangroves; erosion; and a host of other environmental problems inevitably affect people's livelihoods.

Population growth has put severe pressure on the coastline. As population pressure increases in coastal areas, building and development activities to support livelihoods lead to widespread conversion of coastal lands (Cavallos 2005; UNEP 2007). The clearing of vegetation causes severe erosion, which will worsen with intensifying hurricanes and more frequent torrential rainstorms.

Overfishing poses another—and possibly one of the biggest—non-climatic threats to fish stocks (McField and Kramer 2007). Though most fish populations are overharvested, catches in both South America and the Caribbean have been steadily increasing since the 1950s (FAO 2007). Overfishing has become an issue particularly as new fishing techniques and equipment have been introduced, including gill nets, lobster traps, and trawlers.[5] In Belize, for example, field interviews conducted for this book revealed that several artisanal fishers noted a drastic drop in near-shore fish stocks after the introduction of gill nets to the region by foreigners in 1981. As fish stocks decline, fishermen have to spend more days at sea, traveling farther to obtain their needed catch and finding it more difficult to make ends meet.[6] In addition, a consistent open-access regime, without clearly delineated fishing rights for communities or individual catch limits based on the natural productivity of the fish, has worsened the situation significantly.

Marine pollution remains one of the biggest environmental threats to the health of ecosystems and water supplies. In the Gulf of Mexico, industrial and agricultural runoff has resulted in the world's second-largest anoxic "dead zone"—covering 70,000 square kilometers. In Chile, the bays of Valparaiso and Concepcion receive a total of 244 million metric tons annually of untreated effluent, mainly from copper mines, pulp and paper mills, fish processing, and oil refineries. The most widespread problem in the LAC region, however, is discharge of wastewater, 80 percent to 90 percent of which is untreated before reaching the sea (Cavallos 2005; Hinrichson 2008; Nelleman and Corcoran 2006; Sale and others 2008). Other coastal pollution comes from waste discharged by ships and runoff from industrial development.

Coral reefs are especially vulnerable to the effects of climate change and variability. The causes of coral reef decline are a combination of local anthropogenic factors, such as overfishing, dredging, sedimentation, and marine pollution, and climate-related ones, such as bleaching events, that leave the corals more vulnerable to disease, acidification, hurricanes, and El Niño–Southern Oscillation (ENSO) events. Reliable data for coral reef loss are hard to come by, in part because coordinated monitoring of these ecosystems across their range is a relatively recent initiative, having only taken place over the last few decades, making long-term comparisons impossible. Also, in some cases corals that were presumed lost, following bleaching events, have revitalized after some years, making it difficult to pronounce with certainty the extent of permanent loss. One meta-analysis of 65 studies covering 263 sites in the Caribbean found that between 1977 and 2001 the average hard coral cover on reefs was reduced by 80 percent, from about 50 percent to 10 percent. Table 5.1 shows the status of coral reefs in Latin America and the Caribbean as reported by the Global Coral Reef Monitoring Network in 2008. According to that report, around 36 percent of coral reefs are effectively lost or are at a critical stage, and 24 percent are threatened. Only 39 percent appear not to be under significant threat. These findings are consistent with those of Burke and Maidens (2004), who identified more than 80 percent of the reefs in Jamaica, Haiti, and the Dominican Republic as threatened by human activities, with one-third under very high threat. However, these data are associated with great uncertainty and specifically do not consider the future threat posed by climate change. Recent World Bank studies using the Intergovernmental Panel on Climate Change (IPCC) business-as-usual scenario (and a low-temperature sensitivity scenario) predict that coral reefs in the Caribbean will disappear completely between 2060 and 2070 under these external impacts, assuming no adaptation measures are implemented.

More than half of the region's inventory of mangroves is classified as endangered. Mangrove loss is primarily due to clearing and changes in hydrology associated with reclamation for aquaculture and coastal development. Mangroves are threatened by sea level rise in areas where the coastline is built up, making their landward retreat impossible, and by

**Table 5.1    Status of Coral Reefs in Latin America and the Caribbean**

| Region | Coral reef area (km²) | Effectively lost reefs (%)[a] | Reefs at critical stage (%)[b] | Reefs at threatened stage[c] | Reefs at low threat level (%)[d] |
|---|---|---|---|---|---|
| North Caribbean | 9,800 | 12 | 13 | 30 | 45 |
| Central America | 4,630 | 14 | 24 | 22 | 40 |
| Lesser Antilles | 1,920 | 13 | 31 | 22 | 34 |
| S Tropical America | 5,120 | 13 | 40 | 17 | 30 |
| Total | 21,470 | 13 | 23 | 24 | 39 |

*Source:* Wilkinson 2008.
*Note:* Coral reef area from the *World Atlas of Coral Reefs* (2001). Estimates determined using considerable coral reef monitoring data, anecdotal reports, and the expert opinion of hundreds of people associated with the Global Coral Reef Monitoring Network. Assessments should be regarded as indicative because of insufficient data for many regions.
a. Reefs "effectively lost" are 90 percent lost and unlikely to recover soon.
b. Reefs at "critical stage" have 50 percent to 90 percent loss of corals.
c. "Threatened" reefs, are reefs showing moderate signs of damage: 20 percent to 50 percent loss of corals.
d. "Low threat"–level reefs are under no immediate threat of significant losses (except from global climate change).

changes in salinity gradients due to changes in freshwater flows to estuaries. However they are far more threatened by local human pressure than by climate change. In the face of expected increases in the severity of tropical storms, the loss of mangroves makes coastal zones more vulnerable to the force of those storms. In the two decades 1978–98, mangrove coverage declined by 65 percent around Mexico and by 20 percent around Ecuador, Guatemala, and El Salvador. According to the United Nations Environmental Program, mangroves, particularly in Ecuador and Honduras, are being destroyed to make way for aquaculture, most of it using environmentally unsustainable approaches They are being lost in Mexico through coastal development, particularly along the big tourist corridors of the Riviera Maya and the Costa Maya along the Yucatan (see box 5.1).

**Box 5.1**

## Impacts of Climate Change on Livelihoods Dependent on Mangroves and Coral Reefs

Climate change and variability will significantly decrease the ability of coral reefs and mangroves to absorb shocks from waves and wind and to provide essential ecosystem services. Coral bleaching events caused by rising sea surface temperature will become more frequent, while acidification—caused by heightened carbon dioxide ($CO_2$) levels—will lower calcification rates, ending coral productivity.

Mangroves in Southern Belize
*Photo:* Sara Trab Nielsen.

*(continued)*

**Box 5.1** *(continued)*

Acidification occurs once availability of carbonate ions in the ocean starts to decrease because more $CO_2$ is dissolved in the ocean. Carbonate ions are vital skeletal building blocks for reef-building corals. Once carbonate levels fall below a certain threshold, corals lose their ability to calcify, eroding coral reef structures and the multitude of services they provide (Hoeg-Guldberg and others 2007; McField and Kramer 2007; Wilkinson and Souter 2008).

In 1998 and again in 2005, increases in sea surface temperature in the Caribbean caused severe bleaching of corals, weakening and killing many. For mangroves, the possible change in the saltwater/freshwater balance will alter productivity, and temperature changes are expected to decrease photosynthesis and impair root structures and seedlings. Already, large storms have eliminated 10 mangrove forests in the past 50 years in the Caribbean (McLeod and Salm 2006), and in Antigua and Barbuda a 3–4-millimeter rise in sea level has caused a 2 percent annual rate of mangrove loss. If sea level rise continues at this rate, other things equal the island would have no mangroves left by 2075 (Murray n.d.)

Millions of people in communities that rely on mangroves and coral reefs for fisheries and tourism will be affected by their deterioration, which will wipe out the amenities that these ecosystems provide, such as absorbing carbon dioxide and acting as filtering systems for nutrients. The deterioration will decrease coastal protection from storms and floods, diminish the outstanding biodiversity that people come to view, and decrease the availability of the fish that tourists—and locals—consume. Tourist communities throughout the Caribbean and in the Pacific count on coral reefs to attract snorkelers, scuba divers, and other tourists.

**Potential Value of Lost Economic Services of Coral Reefs, 2040–60 (2008 US$ millions, Assuming 50 percent Loss of Corals in the Caribbean)**

|                     | *Low estimate* | *High estimate* |
|---------------------|----------------|-----------------|
| Coastal protection  | 438            | 1,376           |
| Tourism             | 541            | 1,313           |
| Fisheries           | 195            | 319             |
| Biodiversity        | 14             | 19              |
| Pharmaceutical uses | 3,651          | 3,651           |
| Total               | 4,838          | 6,678           |

*Source:* Vergara and others 2009.

*(continued)*

**Box 5.1** *(continued)*

In addition, small communities of poor artisanal fishers will feel the impacts because of their reliance on near-shore, demersal fish species, whose numbers may decline as climate change takes its toll on the reefs and mangroves on which these species rely on for spawning and hiding grounds.

According to the World Rainforest Movement (WRM 2002), annual productivity per hectare of mangroves is estimated at 1,100 to 111,800 kilograms of fishery catches, and it is 10 to 370 kilograms per hectare for corals. The annual value of mangroves varies from US$900 to $12,400 per hectare. These numbers only take into account fisheries and none of the other services that mangroves provide, including recreation and tourism. If they are included in calculations, products and services of mangroves are valued from US$ 200,000 to $900,000 per hectare annually (McLeod and Salm 2006). The accompanying table shows the potential loss to coastal industries from a major loss of coral.

A recent World Bank study found that around 65 percent of all species in the Caribbean rely to some extent on coral reefs for survival. Thus if corals collapse, so may the species and overall ecology in the area that depends on them (De la Torre, Fajnzylber, and Nash 2009). Studies have shown that some species of mangroves and coral reefs have reactive adaptation capacities; thus speculations exist whether adaptation over time is possible. However, due to the severe pressure from human activities, most scientists agree that prolonged climate trends will impair mangrove and coral productivity to a point of no recovery unless serious measures are taken to protect them.

*Source:* Author.

Mangroves and coral reefs play vital roles in supporting the livelihoods of millions of people who depend on them for fisheries and tourism, and they help to weaken the force of storm surges. Coral reefs and mangroves are most abundant throughout the Caribbean Sea, the Pacific Central American coast, and the northern coast of South America where they act as nurseries to a large range of marine species. They are vital to the attractiveness of these coasts as tourist destinations and hold great potential for pharmaceutical uses. For centuries, coastal populations in Latin America and the Caribbean have relied on them for survival. Any decline—or further decline—caused by climate change will significantly affect people's livelihoods.

## Likely Impacts of Climate Change and Variability on Coastal Societies: Three Examples

Climate change and variability are likely to exacerbate the damage to natural resources from the trends discussed in the previous section. The incidence of climate-related disasters in the LAC region was about 2.4 times higher in 2000–05 than in 1970–90 (UNEP 2007). As detailed in chapter 1, that trend is likely to continue, with more intense storms, hurricanes, and ENSO events, as well as changes in sea surface temperatures, ocean currents, and ocean pH levels (acidification), and a rise in sea level.

As made clear by the U.K. Department for International Development's (DFID) sustainable livelihoods framework (SLF; see chapter 1), the resilience and adaptability of coastal communities in the face of climate change depend heavily on their access to assets. Poverty remains a major problem in the LAC region despite recent improvements, especially in rural areas and in urban slums. Though poverty rates among coastal populations are lower than national averages (McField and Kramer 2007; World Bank 2001), they vary according to location and livelihood. For example, poverty rates are relatively high on the northeastern coast of Brazil, with 30 percent to 50 percent of the population below the national poverty line in the extensive slum districts of coastal cities and among coastal subsistence communities. On the southeastern coast of Brazil, which is more urbanized and benefits from tourism, poverty levels are much lower, at 9 percent to 14 percent. Similarly, tourist-reliant regions such as Cancun, Mexico, and Belize District, Belize, have lower poverty rates than, say, the Toledo district in Belize, which relies on fisheries (McField and Kramer 2007).

This section analyzes three selected community structures—fisheries, tourism, and urban slums—taking into account poverty and access to assets, to assess their resilience and adaptability to the effects of climate change and variability.[7]

### Artisanal Fishing Communities

Fisheries play an especially vital role in small coastal communities, providing jobs as well as an important source of protein in diets. Regionwide, fisheries employ some 3 million people.[8] Overall, fish contribute 21 percent of people's protein consumption in the LAC region and up to 36 percent in small island states in the Caribbean (FAO 2008). In addition, employment in fisheries has grown in nearly all LAC countries since 1970, largely because of the increasing global demand for fish. It is growing particularly in poorer countries, whose fishing fleets largely comprise smaller rather than larger, industrial boats—notably in Bolivia (7.9 percent a year, on average, from 1970 to 2006) and in Nicaragua and the Dominican Republic (both 6.3 percent a year from 1970 to 2005) (table 5.2).

The commercial fishing industry comprises both large industrial vessels and small-scale artisanal boats. Industrial fleets tend to be responsible for exports and the production of canned and packaged goods, while small-scale fleets supply local markets for fresh fish (FAO 2008). Artisanal fishing predominates in poorer countries such as Belize, Bolivia, and Nicaragua, as well as in small island states, while industrial fishing predominates in larger economies such as Argentina, Brazil, Chile, and Peru.[9] Peru's is the biggest fishing industry in the LAC region and one of the largest in the world. The industry, including harvesting and processing—largely into fish meal—contributes about 25 percent of Peru's gross domestic product (GDP) and accounts for nearly half of the human consumption of protein. Because of the importance of the industry in Peru, and because it employs a vast number of the poor, Peru is expected to be one of the countries worst hit by climate change from a fisheries perspective (box 5.1). Along with Colombia and República Bolivariana de Venezuela, Peru was also among the 33 countries whose national fishery economies are most vulnerable to climate change, according to research by Allison and others (2009).[10]

The focus of this chapter is on small-scale and artisanal fisheries because of their importance in the local markets, the predominance of the small-scale industry in poorer countries, and the vulnerability of small-scale, poor fishermen. Small-scale fishermen are among the groups most vulnerable to climate change (Allison and others 2009). The great majority of small-scale fishermen in developing countries are poor and many live on the margins of society. In Latin America and the Caribbean, the countries with the greatest concentrations of poor people employed in small-scale fisheries include Bolivia, Brazil, Colombia, Haiti, and Nicaragua (table 5.3). Several fish species are highly heat sensitive and therefore change their migration patterns in response to sea surface temperatures (Anderson 2000; Kawasaki 2001; Perry and others 2005). Whereas large-scale fishing fleets with bigger boats can withstand much

Table 5.2　Overview of the Fishing Industry in Select LAC Countries

| Country | Gross value (% of GDP) | Per capita protein consumption (%) | Employment in fisheries (1,000s) | | | Employment in fisheries as % of national employment | Avg. annual growth rate 1970–2006 (%) | Primary fishery practice |
|---|---|---|---|---|---|---|---|---|
| | | | Primary sector | Secondary sector | Artisanal | | | |
| Argentina | 0.1 | 5.3 | 17 | n.a. | n.a. | 0.15 | 5.4 | Industrial |
| Belize | 5 | 14.9 | 2.1 | 0.1 | n.a. | 0.00 | 1.9 | Artisanal (coastal) |
| Bolivia | 0.1 | 3.8 | 26.8 | n.a. | 3.6 | 0.64 | 7.9 | Artisanal (Inland) |
| Brazil | 0.4 | 6.6 | 790 | 250 | 53.9 | 1.16 | 1.8 | Industrial |
| Colombia | 3.9 | 12.5 | 200.4 | 33.2 | 26 | 1.29 | 6.3 | Artisanal (coastal) |
| Jamaica | 0.3 | 25.1 | 16.2 | n.a. | n.a. | 1.46 | 2.5 | Industrial |
| Mexico | 0.8 | 15.1 | 257.7 | n.a. | n.a. | 0.60 | 2.2 | Artisanal |
| Nicaragua | 1.5 | 14.2 | 26.5 | 12.2 | 13.4 | 1.85 | 6.3 | Artisanal (coastal) |
| Peru | 1.1 | 47.9 | 99 | 45 | n.a. | 1.57 | 2 | Industrial |
| St. Lucia | n.a. | 31.6 | 2.4 | 0.1 | 2.5 | 4.18 | -0.2 | Artisanal (coastal) |
| Trinidad and Tobago | 0.1 | 27.4 | 10.7 | 2 | n.a. | 2.21 | 3.2 | Artisanal (coastal) |
| Total LAC | 2.9 | 14 | 2,306.20 | 798.3 | 672,815 | 1.17 | n.a. | Industrial/Artisanal |

*Sources:* Author's calculations based on most recent data from FAO 2007; ILO and the World Bank, and e-mail communication with FAO 2008.
*Note:* n.a. indicates data not available.

**Table 5.3   Sensitivity to Climate Change Based on Poverty Level and Number of Fishermen**

| Sensitivity | Country |
|---|---|
| Highest | Haiti |
| Medium-high | Nicaragua, Bolivia, Colombia, Honduras, Paraguay, Guyana, Ecuador, Brazil, Guatemala, Peru |
| Medium-low | El Salvador, Caribbean Small Island States (SIDS), Cuba, Venezuela, R.B. de, Uruguay, Argentina, Belize, Panama, Chile, Mexico |
| Low | SIDS: St. Kitts and Nevis, Antigua and Barbuda, Barbados, Bahamas, Guadeloupe |

*Source:* Allison and others 2005.

*Note:* Table based on table A1-1 in Allison and others 2005, which rates countries' sensitivity to climate change based on number of fishermen working in the industry and level of poverty. More poverty combined with a higher percentage working in fisheries makes a country more sensitive. The index is one of four developed to describe and understand the degree to which fisheries production systems are sensitive to climate change taking into consideration poverty, number of fishermen, economic dependence, and food security.

tougher conditions and travel farther to obtain their catch, small-scale artisanal boats can travel only about 2 kilometers offshore.[11] Hence artisanal fishers rely highly on fishing grounds such as mangroves, coral reefs, and other vital ecosystems close to the shore, which are the areas that will be most severely affected by climate change and variability.

Slow-onset climate change and climate-induced disasters will significantly affect fishermen's limited access to assets. Destruction of fragile infrastructure and equipment (physical capital) during climate change–related disasters, such as hurricanes and floods, has already proven detrimental to many fishermen's livelihoods (box 5.2). In Belize in 1998, Hurricane Mitch caused more than US$1.2 million in losses when it destroyed fishing grounds such as mangroves and coral reefs in the north and damaged equipment, keeping fishermen on land for months without any alternative livelihood option (Allison and others 2005). In Antigua and Barbuda in 1995, Hurricane Luis destroyed about 16 percent of the fishing fleet and damaged another 18 percent, causing a loss of roughly one-third of the fishing capacity and a 24 percent drop in annual revenues from fishing (Murray n.d.). In Jamaica in 1988, Hurricane Gilbert caused losses of about 90 percent of fish traps in small communities. During such times fishermen are forced to stay on land, or they have to take jobs on other boats. However, rather than receiving a regular pay from their hire, they are often paid in fish, which affects their ability to build credit and savings.[12]

Box 5.2

## ENSO Impact on Peruvian Fisheries

Peru has one of the largest fishing industries in the world, accounting for about 8 percent of the marine fishing industry worldwide. It is the country's second-biggest foreign currency earner after mining, contributing about US$1.1 billion to $1.7 billion a year to the national economy and accounting for roughly 4 percent of rural employment. About a quarter of Peru's protein consumption comes from fish. The poor in particular in Peru are dependent on fish for protein and as a source of income, as many of them not only work as small-scale fishermen throughout the year but also take seasonal jobs within the large-scale industry.

The Peruvian fishing industry will be affected by climate change through several pathways, but the most worrisome is the possibility of intensifying and more frequent El Niño–Southern Oscillation (ENSO) events. ENSO can cause worsening storm conditions making fishing more dangerous and causing more destruction of equipment. It may also change sea surface temperatures in the short term. Most of all, ENSO is known to reverse the upwelling on which the Peruvian industry most relies.

Peru has twice been hit by severe ENSO events affecting the fishing industry. In the early 1970s, ENSO conditions changed the natural balance of upwelling that pelagic fish rely on, causing a severe drop in the available fish stock. Peruvian industrial fishing, which relied solely on anchovy for its catch, collapsed. Artisanal fishers, who depended on a variety of species, handled the crisis better. The industrial fleet has since diversified its catch, though anchovy still accounts for roughly 70 percent. In 1982–83, another ENSO year caused losses of more than US$2 billion for the anchovy fleet, damaging infrastructure and reducing the fish catch volume by 45 percent. This ENSO event came with rougher-than-usual weather conditions, which kept small-scale fishermen on land for weeks.

As ENSO events are expected to become more severe, with rougher weather events, the Peruvian fish industry is likely to continue to be hit, possibly resulting in layoffs and losses of livelihood. Small-scale artisanal fishermen especially will feel the impact. Most of them are among the 54 percent of people in Peru who live below the poverty line; they receive little education and health care and lack access to local institutions and insurance.

*Sources:* World Bank 2007; FAO 2008; DFID 2004; Allison and others 2005.

Furthermore, because fish are so important as a source of protein, losses in fisheries may jeopardize nutrition or even food security in communities where fish is a key part of the diet. Though many fishing villages supplement their diets with farming, their farming practices will not fully offset what may be lost in fisheries. Moreover, with the expected impacts of climate change on farming, many communities may be hit from two sides (see chapter 4).

Fishermen's vulnerability to climate change depends greatly on the quality of national and international governance of available assets such as access to insurance and land and sea rights. Small coastal and artisanal fisheries in Latin America and the Caribbean have long faced challenges related to the globalization of trade, privatization of access rights to fisheries, and poorly functioning natural resources protection systems. They also often lack democratic representation, which prevents small-scale fishery organizations from participating effectively in national decision making (Allison and others 2005).

First, a lack of property rights to fisheries causes particular difficulty. Though some communities have relied on an area for fishing for centuries, the common-property nature of most fisheries leaves the area open to anyone. For example, property rights to protect mangroves from conversion to shrimp farms and other aquaculture are rare, and so small-scale fishermen have often been overrun by large-scale industrial aquaculture. Aquaculture is a fast-growing industry in Latin America and the Caribbean, but poor local fishermen have benefited little from it and have often lost their rights to fish in certain areas because of it. In Placencia, Belize, for example, local communities are fighting to protect their mangroves from the intrusion of large-scale shrimp farmers. The same is occurring in Cispata Bay, Colombia, where pollution from aquaculture has severely affected water quality in nearby agricultural zones that support roughly 2,500 families, none of whom obtains any financial benefits from the farms.

Second, small-scale fishermen often lack access to sufficient insurance. Particularly, they are unable to obtain any insurance that protects against losses following a disaster. Insurance is too expensive because of the high risk associated with their work and the poor safety of their boats (Allison and others 2005; McGoodwin 2001).

Finally, because many small-scale fishing communities are remote from urban areas, early warning systems have been lacking. Fishing communities, particularly those of indigenous origin, receive little early warning before a disaster strikes. In 2007, Hurricane Felix struck Nicaragua and Honduras, wiping away farmlands, destroying vital fishing grounds, and affecting the lives of about 198,000 people. As described in chapter 4, those affected were primarily from Miskito communities that had received little warning before the hurricane hit.

Overall, the fishing communities of Latin America and the Caribbean are seriously threatened by climate change. It is important to note, however, that with better planning, improved management of natural resources, and greater access to assets, resilience to long-term climate change and short-term disasters can be built within poor fishing communities.

## Tourism

Scholars and researchers agree that the tourism industry in Latin America and the Caribbean will be damaged by climate change and variability if no adaptation measures are implemented. Overall, researchers expect a northward shift in the tourist industry away from the region (Ehmer and Heymann 2008; UNWTO 2008; see figure 5.1). Increased threats from natural disasters, torrential rainfall, and elevated temperatures—possibly coupled with higher travel costs reflecting government policies to mitigate global warming—may prompt travelers to vacation closer to home.[13] Unfortunately, little research has been conducted on the causal relationship between climate change and tourism or on how climate change might affect people working in the industry. Even so, it is clear that because of the climate-sensitive nature of tourism and the increasing reliance on it in the LAC region, safeguarding tourism against climate change is strongly called for.

Overall, roughly 40.7 million people in the LAC region are employed in primary tourist-related activities, and 23.5 million in secondary ones.[14] Though people working in tourism usually have higher standards of living and more education than those in fishing and farming communities, the industry employs many illegal immigrants and is so closely connected to other industries that employ the poor that it deserves much attention in the debate on climate change and livelihoods (Aguayo, Exposito, and Nelida 2001; ECLAC 2003b). For example, in the Galapagos thousands of illegal migrant workers from mainland Ecuador have been arriving for decades to work as maids, waiters, cleaners, and shop assistants (Carrol 2008). Poor people who work in any industry related to tourism, or impoverished communities that wish to develop tourism, may be affected by climate change and variability. Tourism is one of the main development tools used to raise the economic status of small communities. Indeed it has been a successful tool, but it has been developed without climate change considerations, making it vulnerable to climate change.

In recent decades, the tourist industry has grown significantly in nearly all LAC countries in Latin America and the Caribbean, in terms of both contribution to GDP and employment.[15] Though the Caribbean and Mexico hold the highest market shares (14.4 percent and 8.2 percent, respectively), poorer countries with small market shares have experienced

**Figure 5.1    Impacts of Climate Change and Variability on the Tourist Sector in Latin America and the Caribbean**

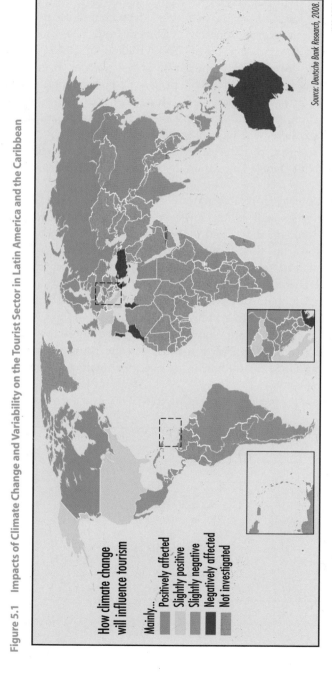

How climate change
will influence tourism

Mainly...

Positively affected
Slightly positive
Slightly negative
Negatively affected
Not investigated

*Source: Deutsche Bank Research, 2008.*

IBRD 37784
MAY 2010

*Source:* Ehmer and Heymann 2008.

the greatest growth in both employment and value (table 5.4). For example, as table 5.4 shows, Nicaragua, with only 0.5 percent of the market, saw an annual growth in tourism of nearly 21 percent from 1990 to 2005, raising the value of the industry from US$12 million to US$207 million. Employment has grown at an annual rate of 8.4 percent in Nicaragua's primary tourist-related activities and 0.9 percent in secondary tourist-related industries (table 5.5). The trend is quite clear; poor countries in particular are benefiting from increased tourism. Unless climate safeguard measures are implemented, these benefits may come under threat.

The tourist industry of Latin America and the Caribbean will feel the impacts of climate change and variability through three pathways.

**Hurricanes.** Along with related natural disasters, such as landslides and flooding, hurricanes will inflict severe damage on infrastructure and buildings and result in short- and long-term closings of stores and businesses, depending on the level of damage and reconstruction costs. Hurricanes will lead to significant erosion of physical capital through damage to hotels and restaurants and to natural capital such as beaches and coral reefs, which, in turn, will lead to erosion of financial capital. In 2005, Hurricane Stan cost the industry in Guatemala millions of dollars in destruction of infrastructure, closings, and the loss of more than 100,000 visitors. In Belize in 2001, Hurricane Iris hit the south as a category 4 storm, causing US$12.5 million in damage to hotels, tour operations,

**Table 5.4    Value of the Tourist Industry in Selected Countries in the Americas, 1990–2005**

| Country | US$millions | | | Average annual growth (%) | Market share (%)[a] |
|---------|------|------|------|-----------|-----------|
| | 1990 | 2000 | 2005 | 1990–2005 | 2004 |
| Belize | 44 | 116 | 133[b] | 8.2 | 0.2 |
| Bolivia | 55 | 68 | 205 | 9.2 | 0.3 |
| Brazil | 1,492 | 1,810 | 3,861 | 6.5 | 3.8 |
| Caribbean | 8,721 | 17,145 | 20,386 | 5.8 | 14.4 |
| Colombia | 406 | 1,030 | 1,218 | 7.6 | 0.6 |
| Mexico | 5,526 | 8,294 | 11,803 | 5.2 | 8.2 |
| Nicaragua | 12 | 129 | 207 | 20.9 | 0.5 |
| Peru | 217 | 837 | 1,308 | 12.7 | 1.0 |

*Source:* UNWTO 2008.
*Note:* Includes values based on directly related tourist activities, such as hotels, restaurants, entertainment, and tours.
a. The United States holds almost 56 percent of the market share.
b. Most recent reported numbers are from 2004.

**Table 5.5    Change in Total Employment in Primary and Secondary Tourism-Related Industries, 1997–2006**

| | Primary tourism-related industries | | | Secondary tourism-related industries | | |
|---|---|---|---|---|---|---|
| | Base year (1,000s) | End year (1,000s) | Average annual growth (%) | Base year (1,000s) | End year (1,000s) | Average annual growth (%) |
| Belize[a] | 15.2 | 25.7 | 6.8 | 7.8 | 13.3 | 7.0 |
| Bolivia[b] | 561.8 | 660.5 | 5.5 | 310.2 | 425.2 | 11.0 |
| Brazil[c] | 9,223.0 | 19,143.0 | N/A | 7,342.0 | 10,119.0 | N/A |
| Colombia[d] | 4,234.5 | 4,420.5 | N/A | 1,715.6 | 2,215.6 | N/A |
| Dominican Republic[a] | 656.0 | 899.1 | 4.0 | 367.5 | 451.9 | 2.6 |
| Jamaica[e] | 206.2 | 272.4 | 3.1 | 135.8 | 194.2 | 4.6 |
| Mexico[e] | 9,442.3 | 12,109.8 | 2.8 | 3,290.6 | 5,763.6 | 6.4 |
| Nicaragua | 233.8 | 481.1 | 8.4 | 175.2 | 189.8 | 0.9 |
| Peru[f] | 1,144.1 | 1,225.1 | N/A | 506.7 | 569.4 | N/A |
| St. Lucia[g] | 14.3 | 16.5 | 2.1 | 9.0 | 14.3 | 6.8 |
| Trinidad and Tobago[a] | 81.0 | 103.5 | 3.1 | 82.9 | 139.1 | 6.7 |
| Total LAC 2000–01[h] | 40.7 million | | | 23.5 million | | |

Source: Author's calculations based on numbers from the International Labor Organization (ILO) Database of Labor Statistics, http://aborsta.ilo.org, 2008.
Note: Figures include persons aged 10 or older unless otherwise indicated. Base year is generally 1997 and end year 2006. However, some countries' base and end years vary as result of data discrepancies and missing data. N/A = calculations not possible because of data discrepancies.
a. End year is 2005.
b. End year is 2000.
c. Prior to 2003, numbers for rural populations in Rondonia, Acre, Amazonas, Roraima, Para, and Amapa were excluded.
d. Persons aged 12 years and older; before 2001 only seven main cities were included; after 2001 the entire country was included.
e. Persons aged 14 and older.
f. After 2001, only the Lima metropolitan area was included; before 2002, all urban areas were included.
g. Base year is 1997, but end year is 2004.
h. Excludes Guyana, Cayman, Guatemala, Paraguay, and Belize due to insufficient data.

restaurants, bars, and gift shops. In the six weeks that followed the event, revenue loss amounted to US$7 million (Jackson n.d.). In the Bahamas— one of the Caribbean's most popular tourist destinations—most shops at the international market in Freeport were still closed in mid-2009, following the landfall of Hurricane Hanna in September 2008.[16] In Grenada, Hurricane Ivan in 2004 caused damage to more than 90 percent of all hotels, totaling US$128 million, or 29 percent of GDP. The hurricane severely hurt ecotourism and cultural heritage sites, accounting for more than 60 percent of job losses in this subsector (Becken and Hay 2007).

*Temperature.* Increases in temperatures may result in long-term visitor losses or shifting patterns of visitors, to which the industry will have to adjust. According to modeled projections by Hamilton, Maddison, and Tol (2005), countries in the LAC region could expect a decrease of at least 10 percent to 25 percent in the number of visitors by 2025 if temperatures rise by 1°C, and as much as 80 percent if temperatures rise by 4°C. Furthermore, increased temperatures and prolonged droughts may worsen water scarcity problems, particularly in the Caribbean, with severe consequences for the tourist industry (box 5.3).

---

**Box 5.3**

## Implications of Water Scarcity for the Tourist Industry: The Case of Barbados

Water is a key environmental resource and asset in tourism. The tourist industry relies heavily on access to an ample supply of clean water for pools, drinking water, showering, washing, and cleaning, not to mention irrigating golf courses and landscapes to keep them green and fresh looking. In fact, the tourist industry is one of the most water-intensive service industries worldwide. Previous research on climate change and tourism shows that the average tourist's water consumption varies between 26 and 550 gallons per day, depending on location. Consumption in city hotels at higher latitudes is much less than consumption in major resorts at lower latitudes, such as those in the Caribbean and Latin America (Hall and Higham 2005). Tourists in Barbados, on average, use 178 gallons of water per day per person (Drosdoff 2009). Such high consumption levels, combined with the climate-induced water scarcity described in chapter 2, in places that lack effective regulation, are likely to affect water availability not only for local residents but also for the tourist industry.

According to Morrison and others (2009), the possible impacts of increased water scarcity on the tourist industry include

- a decrease in the amount of water available,
- an increase in the cost of water,

*(continued)*

**Box 5.3** *(continued)*

• possible operational disruptions, and
• constraints on future growth and license to operate.

Moreover, as long as water regulation remains lacking and infrastructure is inadequate, water use will remain high. For example in Barbados—one of the driest countries in the world—use has increased from an average10 gallons per resident in 1978, to 60 gallons per resident per day today. At the same time, water availability has decreased, posing particular problems for the tourism industry as resorts have difficulty keeping their landscaping and golf courses green. Problems in Barbados are partly due to the fact that 6 of the past 10 years have been abnormally dry. But the greatest contributors are increased water use and poor infrastructure: about half of water production in Barbados is considered unaccounted for due to losses through old pipes and leakage. Though the island country has made efforts to increase the amount of water available by constructing desalination plants, high usage and losses continue. In Barbados, the need to improve water facilities and regulation is clear—to save water not only for the tourist industry but also for the residents (Drosdoff 2009).

*Source:* Authors.

*Disease.* A possible increase in *disease outbreaks* may destroy the reputations of tourist destinations. Combined with possible hikes in ticket prices, that could discourage people from visiting for several years. In 1991, a cholera outbreak in Peru cost the country roughly US$770 million through food embargoes and loss of tourism (WHO 2008). The cholera outbreak was linked to algae blooms fueled by warmer-than-usual coastal waters (Lydersen 2008).

Currently, the tourist destinations most at risk in the LAC region are the Yucatan peninsula and the Caribbean islands, mainly because they are located right in the path of hurricanes. Owing to the relative size of the industry in these countries, increased intensity of hurricanes would have devastating effects on their economies (box 5.4). These areas hold such popular destinations as Cancun, Isla Mujeres, and Cozumel in Mexico, and the Bahamas, Jamaica, and the Dominican Republic, and they are expected to feel increased impact from intensifying hurricanes and eventually rising sea level. Both areas rely heavily on their white sandy beaches and magnificent coral reefs; they have much at stake with intensifying climate change.

**Box 5.4**

## The Impact of Climate Change on Tourism on the Yucatan Peninsula, Mexico

Perched at the outermost part of the Yucatan peninsula and only 7 meters above sea level, Mexico's Quintano Roo district is highly vulnerable to the possible intensification of hurricanes, related natural disasters, and a gradual rise in sea level. With roughly US$20.1 million worth of tourist arrivals, and generating about US$10.8 billion annually from tourist-related activities, the Yucatan is the most visited region in Latin America and the Caribbean. It is estimated that by 2020 Mexico will be the world's second-fastest growing tourist destination. This prediction may not be realized if the effects of climate change and variability worsen the existing impact on the areas' natural resources, which in turn may reduce the number of visitors.

Erosion is already a significant problem in the area as a result of overdevelopment. With the intensification of storms, rising sea level, and loss of the structural integrity of coral reefs, erosion rates will pick up. Since 1980, 11 storms and hurricanes have assailed the peninsula, all causing significant damage to both beaches and hotel infrastructure (Martinez 2008). A 1-meter rise in sea level would result in complete inundation of Cancun, Isla Mujeres, Cozumel, and other nearby tourist destinations.

**Inundation of Yucatan Peninsula with 1-Meter Sea Level Rise**

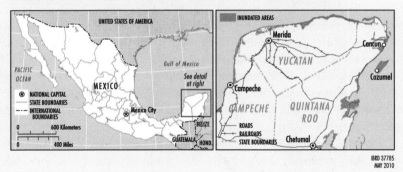

IBRD 37785
MAY 2010

*Source:* World Bank.                    *Source:* Adapted from CReSIS 2008.

Indeed, storms have historically posed the most significant threat to the area. In 1988, Hurricane Gilbert swept away about half of the beaches in Cancun, and in 2005 Hurricane Wilma took roughly 8 miles of sandy beach shore. Wilma also significantly discouraged tourism for the 2006 season and severely damaged

*(continued)*

**Box 5.4** *(continued)*

hotels, which were forced to stay closed for months. The total cost of Wilma was US$17.2 million, and overall more than 16,000 rooms, or 60 percent of capacity, were damaged (Hendershot 2006; Martinez 2008). That year the number of jobs in primary tourist activities fell by 234,200 (ILO 2008). As climate change and variability intensify, the Yucatan peninsula is almost certain to feel similar and more long-term effects.

*Source:* Author compilation.

Because of the integrated nature of the tourist industry, the effects of any economic losses from a natural disaster and a loss of visitors, or from the loss of the pristine environment that tourists hope to experience, are likely to have ripple effects through the economy, causing job losses. Though few data exist for Latin America and the Caribbean on the loss of livelihoods in tourism from natural disasters, studies of similar events elsewhere show significant job losses. For example, the 2004 tsunami in the Indian Ocean destroyed 250,000 jobs. In the three countries where job losses were worst—Thailand, the Maldives, and Sri Lanka—the tourism industries lost roughly US$1.2 billion, US$55 million, and US$201 million, respectively (WTTC 2005).

Figure 5.2 shows how economic losses flow through society to losses in tourism-related livelihoods. It illustrates how erosion of one kind of capital can lead to erosion of other types of capital. In tourism, the impact on people's livelihoods is not as straightforward as in fisheries, being caused more by the overall economic decline that ripples through society once the industry is in decline. The effects of climate change will have indirect impact on livelihoods through erosion of financial capital, leading to a loss of employment and deterioration of social services (human capital).[17]

Because of the climate-sensitive nature of tourism, it is possible that—in a worst-case scenario—a collapse of crucial natural resources such as coral reefs could cause a complete collapse of the tourist industry in the Mesoamerican and Caribbean regions. According to a recent study of the area, the cost of inaction against the effects of climate change in the Caribbean would amount to 22 percent of GDP, and as much as 75 percent or more of GDP in Dominica, Grenada, St. Kitts and

**Figure 5.2    Effects of Climate Change on the Economy, the Tourist Industry, and Livelihoods**

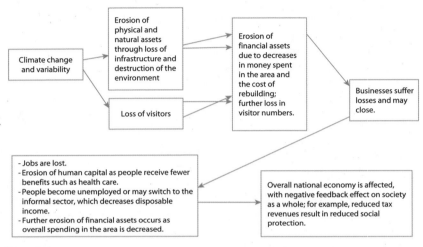

*Source:* Author.

Nevis, and Turks and Caicos, all nations where income from tourism makes up a large share of GDP. Cuba could face an economic impact of 13 percent loss of GDP by midcentury and a 27 percent loss by 2100 (Bueno and others 2008). It is clear that such losses in the tourist industry would significantly affect both the economy and society as a whole in the individual countries.

In many cases, the potential for damage from climate change is greater because of past disregard for environmental sustainability in the development of the tourist industry. The dense concentration of hotels in coastal areas has been pushed by national development policies responding to the preference of developers and visitors for beachside properties (Jackson n.d.). In addition, because beachside tourism has remained a prime choice for visitors, beaches have been poorly protected from development and abuse. Developers have been allowed to create white, sandy beaches even where these are not natural and may cause damage. For example, converting the unique lava stone beaches in St. Vincent and the Grenadines to white, sandy ones may worsen the country's existing erosion problems.[18] Building beachside cabanas in the path of hurricanes, as well as in other low-lying coastal areas, may cause significant safety issues and long-term economic problems in the event of storms, flooding, and rising sea level.

To sum up, the climate-sensitive nature of the tourist industry makes it vulnerable to climate change and variability. It is certain that any losses caused by direct damage or declines in visitors will extend into society at large, resulting in job losses. That said, using tourism as a development tool is not beyond reach, but it will require changes in the way the industry is further developed, so that new ventures fit into the natural environment and embody measures to safeguard resorts from the effects of climate change and variability (box 5.5).

### The Urban Poor

With more than two-thirds of the population of the LAC region living in urban settings, and 60 of the region's 77 biggest cities located on the

---

**Box 5.5**

## The Impact of Climate Change on Ecotourism

Ecotourism is the most vulnerable part of the tourism industry. Not only does it depend strongly on the integrity of natural resources, but people who rely on it as a source of livelihood and income growth are often short of other types of assets. Ecotourism is being pushed in small, poor communities, where it has become one of the preferred alternative livelihoods. It has been the most rapidly expanding part of the tourist industry in Latin America and the Caribbean in recent years, and it is currently growing at roughly 6 percent a year regionwide (Kelly and Sanabria 2008). Ecotourism takes advantage of the abundance of natural resources in the region, such as beautiful coral reefs, unique marine and land species, mountains, and rainforest. If these natural assets erode, so does the base for the ecotourism industry in particular. Because of the threat that climate change poses to ecotourism, further development of the industry calls for significant attention from sector developers, local governments, and politicians, not only from an environmental point of view—because of the impact that climate change is expected to have on natural resources—but also from an equity point of view. Although ecotourism development has been successful in small communities, it has been implemented without any safeguards against climate change and variability. Ecotourism clearly needs specific measures to safeguard it against climate change. Therefore it is strongly recommended that pro-poor, resilience-building mechanisms that incorporate ecosystem-based approaches be implemented.

*Source:* Author compilation.

coast, a discussion of the vulnerability of the urban poor to climate change and variability is imperative.[19] Although the urban poor do not depend directly on natural resources for a living, their sheer numbers in the region, combined with their poor housing conditions and job insecurity, make them another of the groups most vulnerable to climate change and variability, for several reasons. First, a large share of the urban poor use their dwellings for productive activities and hence income generation. Second, about 65 percent of the poor and roughly 60 percent of the extremely poor, some 110 million people altogether, live in urban areas (Ravallion and others 2007). According to the Organization for Economic Cooperation and Development (OECD 2007), Brazil has the greatest number of cities vulnerable to climate change and variability because of their location on the coast and their vast numbers of slum dwellers.[20] Finally, the urban population in the LAC region has been growing steadily and continues to rise. On average, from 2000 to 2005, annual growth rates in the region's urban areas were about 1.9 percent, compared to a 0.2 percent decrease in rural areas. The highest urban growth rates, ranging from 2.1 percent to 2.7 percent, are found in Nicaragua, Colombia, Bolivia, and Brazil (United Nations Population Division 2006).

Climate change is likely to accelerate this already-rapid urbanization. Many of the urban poor have migrated from rural areas in search of a better life after losing their livelihoods, either because of eroding natural and financial capital or because of some directly climate-related factor. It can be expected that greater impacts of climate change in the future will cause further increases in migration by the rural poor (chapter 7).

Owing to their poverty and limited income-earning opportunities, incoming migrants typically settle in shantytowns on marginal lands, such as floodplains, unstable slopes, and low-lying coastal land, which others consider too risky for settlement, and accept jobs in the socially unprotected informal sector. Consequently, the already densely populated urban slums around some of the LAC region's megacities, such as Rio de Janeiro, Bogotá, Quito, Mexico City, Buenos Aires, and São Paulo have experienced some of the highest growth rates in recent decades. This growth of slums is often reflected in the increase in urban poverty rates. In Bogotá, Colombia, for example, the poverty rate grew from 19.4 percent in 1994 to 23 percent in 2003 (UN Habitat 2003).

In the megacities of Latin America and the Caribbean, many of the poor have inadequate access to such basic services as piped water, sanitation, and electricity. Almost half of the urban poor in Argentina have no adequate sanitation, and about half of the urban slum dwellers in Rio de

Janeiro have no sewerage connection. Even when such services are available, many of the urban poor find their cost prohibitive. For example, in Guatemala, where services are in principle available to the poor in most cities, 20 percent to 40 percent of the poor do not connect to them because they find the cost too high.

Few of the urban poor have the opportunity to attend school beyond the primary level, often because the distance they need to travel to reach secondary school is too long. Their lack of sufficient education confines them to work in the informal sector, which offers little or no social insurance or security. Only about 20 percent of the urban poor work in the formal sector regionwide. Only 6 percent of the urban poor in Mexico, and as few as 4 percent in Peru have pension entitlements (Fay 2005).

All in all, the urban poor are very vulnerable both to climate-related disasters and to the effects of ongoing climate change. Intensifying hurricanes, torrential rainfall, landslides, and other climate-related events may cause severe damage to houses or destroy them altogether. Such events also destroy vital roads and utilities, often cutting off poor neighborhoods from help. A lack of water and sanitation infrastructure encourages disease outbreaks, especially outbreaks of diarrheal diseases. Warmer mean temperatures and standing water following hurricanes and floods also raise the prevalence of diseases such as Chagas disease, dengue fever, and malaria in urban areas—which is worrisome, considering the cramped conditions and overall lack of health care in the urban slums. Further, it can be assumed that disease outbreaks and the rebuilding of houses and other infrastructure will keep individuals away from work, perhaps depleting their savings and credit (financial capital). And when adults fall ill, younger family members may enter the workforce, compromising their education and later earning prospects (decreasing their human capital).

Lack of social and cultural capital is likely to worsen the situation for urban slum dwellers. Social networks, familiar ties, and traditions are much more integrated in rural areas, where individuals share the same culture and personal attachment to the community (Verner and Egset 2008). Once migrants arrive in cities, they tend to lose their community and cultural identity and find it difficult to establish a personal network that offers the same level of security. Although some social capital exists in urban areas, Fay (2005) has established that social ties are more individualistic and based less on familial bonding. In addition, the networks that are established tend to be unstable because of the mobility of people in urban areas, high crime rates, and corruption. The lack of social capital may have very negative effects after natural disasters, especially considering the existing high crime rates in megacities.[21]

To sum up, the urban poor are highly vulnerable to climate change and variability because they lack the assets to cope with them, because of the vulnerable lands on which they live, and because of the density of their living quarters. The lack of such assets as clean water, sanitation, electricity, social services, safe land, insurance, and secure employment significantly limits their human, physical, and financial capital, and they lack some of the important social and cultural capital often found in rural areas. During a natural catastrophe, the lack of these assets is likely to worsen their situation, and capital may continue to deteriorate. It is clear that to help the urban poor cope with either short-term disasters or long-term climate change affecting health and work conditions, building their livelihood assets is vital. That implies the governance structures affecting the distribution of assets need to change.

## Adaptation to Climate Change: Institutional and Policy Perspectives

Mitigating the impacts of climate change and variability on coastal communities will depend on societal choices about the protection of the environment, the protection of the poor, land rights for marginal communities, and the degree of overall regard for incorporating climate change considerations into policies.

It is particularly clear that within the three types of communities discussed above, success in building the six types of capital on which people depend for survival depends very much on how assets are governed. A clear need exists to transform governance structures and improve policies to help safeguard the abovementioned livelihoods and build resilience to climate change and variability.

This section provides a few examples of measures that enhance not only the financial prospects of the poor but also the social and political conditions that can improve resilience and adaptability, while protecting the natural environment and hence the potential for long-term, sustainable growth. Empowerment of the poor, promotion of environmentally sustainable development, and improving access to assets are key to successful development and adaptation to climate change.

## Building Environmental Resilience and Alternative Livelihood Opportunities in Coastal Zones

Governments need to pay more attention to the protection and sustainable use of environmental resources in coastal zones. Keeping ecosystems healthy and protecting natural resources by applying solid environmental management principles are imperative in safeguarding and improving resilience to climate change and variability in small communities. Healthy reefs and mangroves provide long-term benefits for society as a whole and can help the coastline withstand rougher climatic conditions.

Often initiatives designed to protect the environment have not sufficiently incorporated management principles that consider protection of livelihoods, and because of that they have aroused opposition from local communities. The potential impacts of climate change make it crucial to strengthen protection of environmental resources, *with the support and assistance of local communities*. Various experiences show that this win-win outcome can be achieved through close cooperation among all stakeholders.

One example is the establishment of marine protected areas (MPAs). MPAs protect coastal marine habitats against local human stressors and enhance the capacity of the surrounding ecosystem to resist or recover from external impacts, including climate change–related ones. They also provide opportunities for alternative livelihoods for people who live in the surrounding area. Using integrated coastal zone management (ICZM), MPAs stress the importance of protecting not only an ocean ecosystem but also the adjacent land. Because of the close interrelation between land and sea in coastal zones, incorporating a concern for both in an MPA is essential (Nelleman and Corcoran 2006). For example, if an MPA is protected through restricted fishing practices and closed seasons, but no similar restrictions are placed on agricultural use or development of the adjacent land, fish species may still be critically harmed by marine pollution and overdevelopment. Currently, MPAs cover only about 0.5 percent of the LAC region's marine area. Most of them are small, allow multiple uses—rather than being "no-take" areas—and do not extend far beyond the coastline (IUCN 2007). Considering the ecological richness and many unique features of the region's coastal habitats, the area protected by MPAs clearly needs to be extended.

When establishing an MPA or related protected zone, it is essential to involve the local population. Generally, MPAs have been established without regard for socioeconomic factors—mainly because they have been looked on as safeguarding ecosystems rather than providing economic benefits for society (Sumaila and others 1999). But recently, as the socio-economic benefits of MPAs and their potential for job creation have been recognized, involving local communities is becoming more common.

Incorporating local communities in the management of the protected area may prove highly beneficial: not only will the managing entity gain increased support from the communities, but many of the communities have an intimate connection to the area and can provide valuable advice and guidance. That is the case in Belize, where eight community-based MPAs have been established thus far. They generally are run by local nonprofit organizations, which hire local people to help protect the area, thereby creating alternative livelihoods in locations where fisheries are failing (box 5.6). This integrated socioeconomic and environmental approach needs to be mainstreamed throughout the LAC region.

Similarly, reforestation of mangroves can be used to generate revenue for small communities while building their resilience to climate change and variability. Mangroves not only function as vital buffers against hurricanes but are also great repositories for $CO_2$. Thus programs to restore mangroves can also help to mitigate climate change itself and potentially provide communities with a vital source of income from the sale of carbon credits through entities such as the World Bank Biocarbon Fund. One such project is being implemented in the Nariva swamp in Trinidad, where local communities are working alongside foresters, ecologists, hydrologists, and engineers to restore native swamp vegetation. At the time of writing, the project planned to apply for credits under the constituents of the clean development mechanism (CDM) (Tropical Re-Leaf Foundation 2008; Vergara 2005). Similar projects might be viable in areas of deteriorated mangroves in Brazil, Mexico, Nicaragua, and parts of the Caribbean— though with the carbon market still in its infancy, the full benefits to local communities are still to be seen, let alone measured and tested.

### Enhancing Land Rights and Using Small-Scale Aquaculture to Protect Fishing Communities

Policies are needed to reduce pressure on current fish stocks and provide improved livelihood options for local fishermen. Developing sustainable aquaculture communities is one way to better manage fisheries in addition to mitigating pressure on artisanal fishermen.

Pro-poor, sustainable, and environmentally friendly aquaculture can potentially provide an important source of livelihood for fishermen losing their jobs. In practice, aquaculture has put significant pressure on coastal

Box 5.6

## Integrated Coastal Management—The Port Honduras Marine Reserve in the Toledo District of Belize

People in the Toledo District in southern Belize have traditionally relied on fisheries for their livelihoods. This district is the poorest in Belize, with a poverty level of 79 percent, and the majority of the region's population is classified as Garifuna. Since the early 1980s, Belizeans have experienced a severe decline in fish stocks, mainly due to overfishing caused by the introduction of gill nets and lobster traps. As a result, in 2000 the Port Honduras Marine Reserve (PHMR), covering roughly 160 square miles, was established, banning all destructive gear and enforcing a strict catch season.

The managing agent for the PHMR is the Toledo Institute for Environment and Development (TIDE), which is headquartered in Punta Gorda. Adjacent to the reserve lies Paynes Creek National Park, which covers about 38,000 acres of mangroves and lush broadleaf forest and is also under the management of TIDE. The marine reserve and the nearby area support the livelihoods of Punta Gorda and Monkey River Village. From the beginning, the local community has supported the reserve, knowing the importance of protecting natural resources. That support has been beneficial for TIDE, which has made every effort to include the community in the management of the two protected areas for three significant reasons: (a) locals have intimate knowledge of the area, (b) they have a personal

*Photo:* Sara Trab Nielsen.

*(continued)*

**Box 5.6** *(continued)*

Rangers in PHMR.
*Photo:* Sara Trab Nielsen.

connection to the area, and (c) as fish resources became scarce, alternative employment for fishermen was needed.

In the past eight years, TIDE has led a successful initiative helping fishermen to diversify their livelihoods. Fishermen are trained as rangers and marine specialists, who help monitor fish stocks and diversity in the reserve. In addition, TIDE Tours was established to promote ecotourism in the region. It offers canoe trips up the Rio Grande, boat trips through the mangroves, fly-fishing, catch and release, and snorkeling tours, all guided by local people who previously worked as fishermen. According to testimonials from several people employed by TIDE, their new jobs as tour guides, taxi drivers, captains, and rangers provide a much more sustainable and reliable livelihood than fishing.

Overall, the reserve is a success. Though TIDE occasionally catches fishermen in the reserve during the off season, the institute generally receives positive support from the community, and species diversity as well as fish stocks have increased in recent years. Even manatees can now be spotted in the reserve.

*Source*: Author.

zones as the result of clearing of vegetation and nutrient enrichment, leading to pollution of surrounding waters, and as noted above, it has tended to provide few benefits for local people. Done sustainably, however, and with local fishermen in mind, aquaculture could be an alternative to declining marine fisheries, helping communities without other livelihood options. Culturally, aquaculture keeps fishermen within their own realm, and it may diminish the problems associated with cultural identity loss when traditional livelihoods die (McGoodwin 2001).[22] Finally, development of small-scale aquaculture helps to provide a supply of high-quality food.

Small-scale aquaculture could use the same principles employed by small-scale fair trade farmers. Such projects would require the establishment of much clearer land rights that support the interests of indigenous and impoverished communities, and they should ensure that the local community keeps the right to manage aquaculture production, without outside influence from big corporations. Guidelines that prevent abuse of natural resources and pollution and protect the surrounding environment need to be incorporated, as do rules that (a) permit only the cultivation of native species, (b) ensure proper nutrient cycling to minimize release, and (c) impose production limits to keep the ponds from overproducing and eventually collapsing.[23] Additionally, the development of aquaculture should not undermine capture fisheries, for example, by relying on juveniles caught in the wild for stocking ponds.

Aquaculture must also be climate-proofed. Protecting aquaculture from the effects of climate change and variability calls for the use of adaptation mechanisms such as building deeper ponds and selecting species for culture that can withstand possible sea level rise, saltwater intrusion, and temperature change. Possibly, climate change and variability will even create more favorable conditions for aquaculture, in particular marine aquaculture: areas where rising sea levels are expected to cause salinization and in some cases inundate freshwater resources may become suitable for development of marine-based aquaculture.[24]

### Preparing for Climate Change and Variability in the Tourist Industry

Because climate change and variability have potential to damage tourism significantly, and because tourism is a main poverty alleviation tool, the industry needs to consider climate implications in project development to safeguard communities that rely on it. As stated by UNWTO Secretary General Francesco Frangialli, "Climate change as well as poverty alleviation will remain central issues for the world community. Tourism is an important element in both. Governments and the private sector must place increased importance on these factors in tourism development strategies and in climate and poverty strategies. They are interdependent and must be dealt with in a holistic fashion" (UNWTO 2008).

That is particularly the case in the development of ecotourism. Several community-based ecotourism initiatives are under way in the LAC region, including in Belize (box 5.6); Paraná, Brazil; Silvestra Agroecotur, Colombia; and the area of Corazón in Nicaragua and Honduras (LCSSD 2008).

Indeed, ecotourism has been a successful form of development and empowerment. In Bolivia, the community of Quechua-Tacana developed the ecotourism resort Chalalan, which takes advantage of the surrounding area's pristine rainforest and the Chalalan lagoon.[25] Today this resort is fully owned by the community and provides jobs for roughly 1,500 local people. The resort has made the community more prosperous, bringing in roughly US$350,000 annually, of which half is reinvested in the community for health care, infrastructure, and education and the remainder divided among families. The initiative in Chalalan is a good example of how building financial assets and investing in physical assets through community activities can enhance human capital. In addition, it shows the benefits from a high level of social capital through trust and cooperation within society.[26]

Further, the Chalalan model is a good example of why tourism should be run by—and benefit—local stakeholders: if tourism is to be a sustainable way to improve lives in Latin America and the Caribbean and build resilience to the effects of climate change, it is highly important that its management and benefits be in the hands of people whose livelihoods depend on the sustainability of the local natural resources. This approach contrasts with the resource-mining mentality that has typified the tourist industry.

Clearly, such success helps raise some capital, but other adaptation strategies must be considered as well to fully climate-proof tourism. New building codes and policies must be developed and implemented that restrict development in near-shore zones or other areas at high risk of damage from climate change. Detailed environmental impact assessments that incorporate climate considerations are necessary to identify risks and assess viability before developing resorts. New hotels and resorts should be built further inland or in non-windward areas away from erosion and landslide-prone slopes. Areas that are not natural to the coastal zone should not be established; for example, beaches should not be developed where they are not natural, and indigenous coastal vegetation surrounding resorts should be maintained to provide protection from storms and hurricanes.

Finally, the tourist industry must consider that beaches and other ecosystems may be altered, seasons may change, and some popular tourist attractions such as sea turtles and coral reefs may disappear. Such changes require not only development of alternative tourist attractions but also a focus on new marketing strategies. Enhancing cultural tourism may be an alternative. All countries and regions within them have their own unique cultural traditions and traits that may be used to attract visitors.

### Adaptation in Coastal Megacities

Thus far it has been rare for urban authorities in Latin America and the Caribbean to adopt measures for climate change adaptation. City authorities need to identify and map where climate hazards are greatest and where the most vulnerable people live and target adaptation measures accordingly. Disaster management plans need to be made or updated, but focus is also needed on providing necessary public infrastructure and upgrading high-risk housing or promoting resettlement away from high-risk neighborhoods. To mitigate the social impacts of climate change and variability, expertise in local governance, in community-driven development, in enhancement of social capital, and in resettlement needs to be applied, together with knowledge about the effects of climate change, so as to give the urban poor a voice in decision making about development.

Sometimes slum dwellers have devised their own safeguard measures, with disregard for their neighbors, often making matters worse for the neighborhood as a whole.[27] To prevent such counterproductive measures and decrease the vulnerability of urban slum dwellers to disasters resulting from climate change events, governments are called on to provide increased protection for the poor. Necessary measures include improving property rights for the poor, improving building codes and their application in practice, and providing funding to strengthen adaptation infrastructure in urban slums. From a structural point of view, building gabion baskets and drainage channels may significantly help to prevent erosion and landslides and manage standing water.[28] As seen in chapter 4, the financial rationale for improving disaster prevention over disaster management is obvious: it is less expensive to dig drainage channels than to provide emergency aid and fund reconstruction following a natural disaster. Improvements in utilities infrastructure, such as that for safe water supply, sanitation, and electricity, may reduce disease as well as reduce mortality during natural disasters. Institutional improvements to protect the poor are also needed, including the extension of eligibility for social security and insurance, which provide monetary relief and access to health care.

Governments implementing relocation programs need to ensure that the programs provide adequate alternative housing at affordable prices in safer areas and should inform residents about the importance of their vacating high-risk areas. Relocation programs have not always succeeded, because they are costly to the people affected, especially if they entail moving away from their means of making a living.[29] In St. Vincent, for example, the government has found that individuals do not accept relocation packages because they are reluctant to leave a house that essentially costs them nothing—even if it lies in a high-risk area. Some families who agree to be relocated leave some of their members behind in the risk-prone house, in fear that someone else may take it.[30] Indeed, to prevent people from remaining in high-risk settlements, or discourage new people from moving in, the areas should be converted to alternative uses precluding habitation, such as urban agriculture, parks, or sports and playing fields. Informing people about the risk associated with staying is crucial to success, as is making sure that the substitute housing is affordable and offers some of the amenities that are lacking in the high-risk location, such as clean water, sanitation, and electricity. Relocating people to neighborhoods with services may indeed cost less in the long term, while protecting people against losing their livelihoods.

Finally, it is vital to improve services that enhance human capital, such as education, health care, and other social services. Providing better education benefits society as a whole. More educated people can help improve the overall standard of living, provided enough jobs are available. Jobs should come with social benefits such as better health services. People in better health are more resilient to disease outbreaks, and public health measures during a disease outbreak help to prevent deaths (chapter 6).

## Research Perspectives

Overall, the impacts of climate change phenomena on coastal livelihoods remain severely understudied. More detailed information is needed on each subregion, country, locality, and sector. In particular, up-to-date time-series data on employment and the value of coastal sectors, as well as an overview of climate-related shocks and trends in the different countries, are vital. International organizations, nongovernmental organizations, and individual governments should continue to fund research and take initiatives to improve the livelihoods of impoverished coastal populations to safeguard them against climate change and variability.

In particular, more detailed research is needed in two sectors:

- *Fisheries.* More detailed information is needed on small-scale aquaculture, to analyze how land should be used, the best locations, the communities in which it would work, and how to safeguard it from climate change and variability. In addition, much more information is needed on the location of vulnerable, small-scale fishing villages, including up-to-date data on the value of the industry and the number of people it employs, including employment in artisanal fisheries. More information is needed on vulnerability within communities and on communities' exact wants and needs. Some fishing communities may be more accepting of outside ideas than others.

- *Tourism.* Given the speed at which the tourist industry is growing in Latin America and the Caribbean, it is crucial that institutions act to research and establish ways to safeguard it. More research is needed on how the industry can cope with climate change and variability and adapt to prevent job losses. Very little research has been done on ecotourism and the impact that climate change and variability may have on it. Because that part of the industry relies so heavily on natural resources, further analysis is needed to understand how those resources will be affected by climate change and variability and how small communities can adapt.

## Notes

1. The complete LAC region has a population of 556 million (Hinrichson 2008; World Bank 2008).
2. Low-elevation coastal zones (LECZ) are the areas along the coast that are less than 10 meters above sea level and will be most hit by climate change events and variability (McGranahan, Balk, and Anderson 2007).
3. The livelihood assets comprise physical, financial, human, social, cultural, and natural capital.
4. Only a little information will be drawn from the agricultural sector since it is covered in chapter 4. The science behind climate change is discussed in appendix A.
5. One of the main issues is that large fish populations and predator species have been replaced in the catch by smaller, less valuable species, or juveniles of the species traditionally caught. Thus it can be assumed that while the volume of

fish caught may increase for a time, fish stocks are being serially depleted as effort transfers from one species to another, down the food chain.

6. Fishers also noted that when gasoline prices began to rise, several of them were forced to stay on land, as they were no longer able to obtain a catch sufficient to cover the cost of their fuel. Information obtained through interviews with Dennis Garbutt from EarthWatch Belize and local fishers in Monkey River, Belize, 2008.

7. Tourism is included in this analysis because of its climate-sensitive nature, the increase in tourism development in smaller communities, and in particular the increase in development of ecotourism.

8. Number based on estimations compiled from FAO databases. The number includes artisanal, inland fisheries, marine fisheries and all people employment in secondary industries such as marketing and processing. However, it excludes the vast number of unregistered fishermen. It is difficult to obtain accurate figures for how many people work in fisheries in the LAC region and worldwide, as millions are unregistered (see Allison and others 2009, 9).

9. Honduras is an exception, as it has a well-developed industrial shrimp fleet and is a major exporter of shrimp.

10. The study conducted by Allison and others (2009) included 132 countries worldwide. Countries were rated based on their possible climate change exposure, their economies' dependence on fisheries, and their ability to adapt to climate change.

11. With the exception of shrimp, the large-scale industrial fleet does not fish shallow-water demersal species, such as those found adjacent to coral reefs. Rather they fish a few pelagic species far off shore. It is important to note that climate change and variability will affect demersal and pelagic species differently.

12. Information obtained from fieldwork in Belize in 2008.

13. Climate change effects such as rising temperatures may simply result in changing tourist seasons overall, for example, with the high season for travel beginning and ending later in the year. Not enough local climate data exist to be able to predict these possible changes and their impact (Ehmer and Heymann 2008).

14. This discussion splits the tourist industry into two categories: (a) primary tourist-related jobs, which are directly related to tourist activities, including jobs in hotels, restaurants, and wholesale and retail trade, and (b) secondary tourist-related jobs, including anything that may be affected by turmoil in the primary industry, such as construction, transport, communications and non-classifiable jobs, usually in the informal sector.

15. The rapidly expanding tourist industry in the LAC region can be separated into three categories: (a) mass tourism such as cruise liners, (b) large scale "sun

and surf" tourism, and (c) sustainable ecotourism, including cultural tourism. "Sun and surf" tourism refers to the classic sort of tourism consisting of long days at the beach, shopping, and nights out, with little regard for the environment; ecotourism consists of visitors who travel to see the local environment and learn about environmental protection. However, the two different industries are starting to overlap, as ecotourism now also includes beach travel in which tourists can stay in sustainably built and kept resorts, and regular hotels are starting to incorporate environmental measures. "Mass tourism" refers mainly to cruise liners that travel from port to port. Because cruise liners provide travelers with all their basic needs, hotels, restaurants, stores, and local guides and communities profit little from the industry, and so mass tourism is not considered in this analysis.

16. Information obtained during fieldtrip to The Bahamas, May 2009.

17. Tourism, more than any other industry, faces indirect impacts from climate change mitigation measures. For example, climate mitigation bid the arrival of carbon taxation systems, which may raise fuel prices and the price of plane tickets. Currently, most airlines give travelers a choice of whether to purchase carbon offsets or not. However, it is to be expected that a global carbon tax will be implemented on transport costs in the near future. That, coupled with rising fuel prices, may drive travel costs to a level where people simply opt out of traveling. That, according to the CHA-CTO (2007) could jeopardize the sustainable livelihoods of about one-third of the Caribbean population.

18. Currently many of these beaches are being destroyed by removal of stones for construction purposes. Most of this industry is informal. Observations from field trip to St. Vincent, 2008.

19. For the purpose of this chapter, "urban poverty" will refer only to megacities and associated slums. However, the general definition of the term "urban" includes everything from small towns to megacities (Fay 2005).

20. In Brazil these cities include Baixada Santista, Belém, Fortaleza, Grande Vitória, Maceió, Natal, Porto Alegre, Recife, and Rio de Janeiro. Other highly vulnerable cities include Barranquilla, Colombia; Lima, Peru; Montevideo, Uruguay; Maracaibo, República Bolivariana de Venezuela; Belize City, Belize; Panama City, Panama; Havana, Cuba; Santo Domingo, Dominican Republic; and Port-au-Prince, Haiti. These cities' vulnerability has been rated based on level of infrastructure, economy, and population size (OECD 2007).

21. That was apparent in the United States when looting and crime soared during the days following Hurricane Katrina.

22. Cultural identity is a vital consideration in initiatives that aim to build new livelihoods. It should be mentioned that some cultures would not accept aquaculture as an alternative. The choice is highly differentiated. For example, in Belize fishing villagers were willing to develop sustainable small-scale aquaculture if land rights were allocated to them, while in other parts of the

Mesoamerican region villages were more open to undertaking jobs as chefs and tour guides (information obtained from field trip to Belize, 2008, and the Mesoamerican Barrier Reef Project).

23. That has been a particular issue in Belize, where overproduced, closed-down shrimp farms are termed "scars" on nature.

24. Although the focus here is on land-based aquaculture, it is important to note that sea-based aquaculture is a viable alternative and needs to be explored further, as well as invested in, though also with a focus on small-scale, community-based development.

25. With the assistance of Conservation International and the Inter-American Development Bank.

26. Presentation entitled "Chalalan: A Story of Success," World Bank, n.d., author unknown.

27. For example, homeowners in Buenos Aires have sought to reduce the risk of flooding to their homes by bringing in additional soil to elevate their plots. Now their plots drain into adjacent, lower-lying lands (Barros 2005). Some people invest in retaining walls to prevent soil erosion on steep slopes. The walls stop soil erosion in the short run, but in the long run they increase the risk of mudslides, because a heavy volume of water builds up behind them.

28. Gabion baskets are wire mesh baskets filled with cobble or small boulder material, which is able to adjust to varying conditions. They function much like retention walls, but unlike walls they allow water to trickle through, hence preventing erosion and flooding on stream banks and steep slopes (Freeman and Fischenich 2000).

29. In some places, relocation has had some success. In Belize, attempts to move the entire capital city were made following Hurricane Hatti in the 1960s. Belize City remains the major financial center in Belize, but the seat of government was moved in the early 1970s to Belmopan. An increasing number of people are moving to Belmopan because of the risk of coastal inundation both in the short term during hurricanes and in the long term due to sea level rise.

30. Interview with Howie Prince, director of National Emergency Management Organization, St. Vincent, 2008.

## References

Aguayo, Eva, Pilar Exposito, and Lawelas Nelida. 2001. "Econometric Model of Service Sector Development and Impact of Tourism in Latin American Countries." *Applied Econometrics and International Development* 1, no. 2.

Allison, Edward H., W. Neil Adger, Marie-Caroline Badjeck, Katrina Brown, Decian Conway, Nick K. Dulvy, Ashley Halls, Allison Perry, and John D. Reynolds. 2005. *Effects of Climate Change on the Sustainability of Capture and*

*Enhancement Fisheries Important to the Poor: Analysis of the Vulnerability and Adaptability of Fisherfolk Living in Poverty.* Fisheries Management Science Program: Final Technical Report. London: U.K. Department for International Development.

Allison, Edward H., et al. 2009. "Vulnerability of National Economies to the Impacts of Climate Change on Fisheries." *Fish and Fisheries* 10: 173–96.

Anderson, James J. 2000. "Decadal Climate Cycles and Declining Columbia River Salmon." In *Sustainable Fisheries Management: Pacific Salmon*, ed. E. E. Knudsen and Donald MacDonald, 467–84. New York: CRC Press.

Barros, V. 2005. "Global Climate Change and the Coastal Areas of the Río de la Plata." Final Report for Project No. LA 26. Submitted to Assessments of Impacts and Adaptation to Climate Change. http://www.aiaccproject.org/.

Becken, Suanne, and John Hay. 2007. *Tourism and Climate Change: Risk and Opportunities.* Clevedon, U.K.: Channel View Publications.

Bueno, Ramón, Cornelia Herzfeld, Elizabeth A. Stanton, and Frank Ackerman. 2008. "The Caribbean and Climate Change: The Costs of Inaction." Tufts University. Available at http://ase.tufts.edu/gdae/Pubs/rp/Caribbean-full-Eng.pdf.

Burke, L., and J. Maidens. 2004. *Reefs at Risk.* Washington, DC: World Resources Institute.

Carrol, Rory. 2008. "Tourism Curbed in Bid to Save Galapagos Haven," *Mail and Guardian*, October 12. http://www.mg.co.za/article/2008-10-12-tourism-curbed-in-bid-to-save-galapagos-haven.

Cavallos, Diego. 2005. "Untreated Wastewater Making the Sea Sick." Interpress Service News Agency.

CHA-CTO (Caribbean Hotel Association and Caribbean Tourism Organization). 2007. "CHA-CTO Position Paper on Global Climate Change and the Caribbean Tourism Industry." http://caribbeanhotelassociation.com/downloads/Pubs_ClimateChange0307.pdf.

CReSIS (Center for Remote Sensing of Ice Sheets). 2008. Sea Level Rise Maps and GIS Data–Southeast USA. https://www.cresis.Ku.edu.

De la Torre, Augusto, Pablo Fajnzylber, and John Nash. 2009. *Low Carbon High Growth: Latin American Responses to Climate Change. An Overview.* World Bank Latin American and Caribbean Studies. Washington, DC: World Bank.

DFID (Department for International Development, U.K.). 2004. *Climate Change in Latin America.* Key Fact Sheet No. 12. http://www.dfid.gov.uk/pubs/files/climatechange/keysheetsindex.asp.

Drosdoff, Daniel. 2009. "Barbados Acts to Prevent Water Crisis." Newsbeat, *IDB America: Magazine of the Interamerican Development Bank.* http://www.iadb.org/idbamerica/index.cfm?thisid=2793.

Ehmer, Philippe, and Eric Heymann. 2008. *Climate Change and Tourism: Where Will the Journey Lead?* Deutsche Bank Research, Frankfurt am Main.

Fay, Marianna, ed. 2005. *The Urban Poor in Latin America. A World Bank Report.* Washington, DC: World Bank.

Freeman, Gary E., and Craig Fischenich. 2000. "Gabions for Streambank Erosion Control." U.S. Army Corps of Engineers, Environmental Laboratory. http://el .erdc.usace.army.mil/elpubs/pdf/sr22.pdf.

FAO (Food and Agriculture Organization). 2007. Fisheries and Aquaculture Information and Statistics Service.

——. 2008. Fisheries and Aquaculture Country Profiles. http://www.fao.org/ fishery/countryprofiles/search/en.

Hall, Michael C., and James Higham, eds. 2005. *Tourism, Recreation, and Climate Change.* Channel View Publications.

Hamilton, Jacqueline M., D. J. Maddison., and R. S. J. Tol. 2005. "Climate Change and International Tourism: A Simulation Study." *Global Environmental Change* 15: 253–66.

Hendershot, Rick. 2006. "Cancun Beach Being Restored at Record Pace." Health Guidance for Better Health. http://www.healthguidance.org.

Hinrichson, Don. 2008. "Ocean Planet in Decline." *People and Planet: People and Coasts and Oceans.* http://www.peopleandplanet.net/doc.php?id=429& section=6.

Hoeg-Guldberg, O., et al. 2007. "Coral Reefs under Rapid Climate Change and Ocean Acidification." *Science* 318 (December 14).

ILO (International Labor Organization). 2008. LABORSTA—Database of Labor Statistics. http://laborsta.ilo.org.

IUCN. 2007. "Latin American Park Congress: 2008–2018 to Be 'Decade of MPAs.'" http://www.iucn.org/.

Jackson, Ivor. n.d. "Potential Impacts of Climate Change on Tourism." Issues paper prepared for OAS, Mainstreaming Adaptation to Climate Change (MACC) Project.

Kawasaki, T. 2001. "Global Warming Could Have a Tremendous Effect on World Fisheries Production." In *Microbehavior and Macroresults: Proceedings of the Tenth Biennial Conference of the International Institute of Fishery Economics and Trade,* ed. R. S. Johnston and A. L. Shriver. July 10–15, 2000, Oregon, U.S.

Kelly, Kathryn, and Ronald Sanabria. 2008. "Mainstreaming Biodiversity Conservation into Tourism through the Development and Dissemination of Best Practices." World Bank presentation, September 18, World Bank, Washington, DC.

LCSSD (Latin American and Caribbean Region Social Development). 2008. "Tourism in LCSSD: Building a New Beam." A World Bank presentation. Author unknown. World Bank, Washington, DC.

Lydersen, Kari. 2008. "Risk of Diseases Rises with Water Temperatures." *Washington Post*, October 20, A08.

Martinez, Julia. 2008. "Impacts of Climate Change in the Tourism Sector in Mexico." World Bank presentation, April 15, World Bank, Washington, DC.

McField, Melanie, and Patricia Kramer. 2007. "Healthy Reefs for Healthy People: A Guide to Indicators of Reef Health and Social Well-Being in the Mesoamerican Reef Region." Washington DC: Smithsonian Institution.

McGoodwin, James R. 2001. "Understanding the Cultures of Fishing Communities: A Key to Fisheries Management and Food Security." Rome: FAO. http://www.fao.org/docrep/004/y1290e/y1290e00.HTM.

McGranahan, Gordon, Deborah Balk, and Bridget Anderson. 2007. "The Rising Tide: Assessing the Risks of Climate Change and Human Settlements in Low-Elevation Coastal Zones." *Environment and Urbanization* 19 (1): 17–37.

McLeod, Elizabeth, and Rodney V. Salm. 2006. *Managing Mangroves for Resilience to Climate Change*. Gland, Switzerland: World Conservation Union.

Morrison, Jason, Mari Morikawa, Michael Murphy, and Peter Schulte. 2009. "Water Scarcity and Climate Change: Growing Risks for Business and Investors." A CERES and Pacific Institute Report, February.

Murray, Peter A. Undated. *Climate Change, Marine Ecosystems, and Fisheries: Some Possible Interactions in the Eastern Caribbean*. Environment and Sustainable Development Unit, Organization of Eastern Caribbean States.

Nellemann, C., and E. Corcoran, eds. 2006. *Our Precious Coasts—Marine Pollution, Climate Change, and the Resilience of Coastal Ecosystems*. Grid-Arendal, Norway: UNEP.

OECD (Organization for Economic Cooperation and Development). 2007. *Ranking Port Cities with High Exposure and Vulnerability to Climate Extremes: Exposure Estimates*. OECD Environment Directorate. Paris: OECD.

Perry, Allison L., et al. 2005."Climate Change and Distribution Shifts in Marine Fishes." *Science* 308, June 24.

Ravallion, Martin, Shaohua Chen, and Prem Sangraula. 2008. "New Evidence on the Urbanization of Global Poverty." Background Paper for the 2008 World Development Repor, Development Research Group, World Bank, Washington, DC.

Sale, P. F., M. J. Butler IV, A. J. Hooten, J. P. Kritzer, K. C. Lindeman, Y. J. Sadovy de Metcheson, R. S. Steneck, and H. van Lavieren. 2008. *Stemming Decline of the Coastal Ocean: Rethinking Environmental Management*. United Nations University, UNU-UNWEH, Hamilton, Canada.

Sumaila, Ussif Rashid, Sylvie Guenetta, Jackie Alder, David Pollard, and Ratana Chuenpagdee. 1999. *Marine Protected Areas and Managing Fished Ecosystems.* CMI Reports. Bergen: Chr. Michelsens Institute.

Tropical Re-Leaf Foundation. 2008. "Nariva Swamp Restoration Initiative." http://www.ema.co.tt/main.htm.

UNEP (United Nations Environment Program). 2007. "Climate Change Hits Hard on Latin America and the Caribbean." UNEP Press Release. http://www .unep.org.

UN Habitat (United Nations Human Settlement Program). 2003. *The Challenge of Slums. Global Report on Human Settlements.* London: Earthscan Publications.

United Nations Population Division. 2006. *World Urbanization Prospects: The 2005 Revision.* Department of Economic and Social Affairs, Urban and Rural Areas Dataset (POP/DB/WUP/Rev.2005/1/Table A.6).

UNWTO (United Nations World Tourism Organization). 2008. "Climate Change and Tourism: Responding to Global Challenges." Advance Summary. http://www.unwto.org/index.php.

Vergara, Walter. 2005. *Adapting to Climate Change: Lessons Learned, Work in Progress, and Proposed Next Steps for the World Bank in Latin America.* Sustainable Development Working Paper 25, World Bank, Washington, DC.

Vergara, Walter, Natsuko Toba, Daniel Mira-Salawa and Alejandro Deeb. 2009. "The Consequences of climate-induced coral loss in the Caribbean by 2050–2080." In *Assessing the Potential Consequences of Climate Destablization in Latin America.* Sustainable Development Working Paper, World Bank, Washington DC.

Verner, Dorte, and Willy Egset. 2008. *Social Resilience and State of Fragility in Haiti.* A World Bank Report. Washington, DC: World Bank.

WHO (World Health Organization). 2008. *Global Epidemics and Impacts on Cholera.* http://www.who.int/topics/cholera/impact/en/index.html.

Wilkinson, C. R., ed. 2008. *Status of Coral Reefs of the World: 2008.* Townsville, Australia: Global Coral Reef Monitoring Network and Rainforest Research Center.

Wilkinson, Clive, and David Souter, eds. 2008. *Status of the Caribbean Coral Reefs After Bleaching and Hurricanes in 2005.* Townsville, Australia: Global Coral Reef Monitoring Network.

World Atlas of Coral Reefs. 2001, rev. 2007. UNEP/WCM. http://www .unep-wcmc.org/marine/coralatlas/index.htm.

World Bank. 2001. "Attacking Brazil's Poverty: A Poverty Report with a Focus on Urban Poverty Reduction Policies." Report No. 20475-BR, World Bank, Washington, DC.

————. 2007. "Sustainable Fisheries through Improved Management and Policies." Chapter 6 in *Republic of Peru Environmental Sustainability: A Key to Poverty Reduction in Peru*. Environmentally and Socially Sustainable Development in Latin America and the Caribbean. Washington, DC: World Bank.

————. 2008. "At a Glance: Poverty Numbers in LAC." http://www.worldbank .org/lacpoverty.

WRM (World Rainforest Movement). 2002. "Latin America: Mangroves: Local Livelihoods vs. Corporate Profits." December.

WTTC (World Travel and Tourism Council). 2005. "Global Travel and Tourism Poised for Continued Growth in 2005: Tsunami Impact on Travel and Tourism is Significant but Limited." *Tourism News*. Web site article August 4.

# CHAPTER 6

# Human Health and Climate Change

## Lykke Andersen, John Geary, Claus Pörtner, and Dorte Verner

Climate change and variability pose a serious health risk to many developing countries, most of which are poorly prepared to face it. By 2000, according to estimates from the World Health Organization (WHO), climate change and variability had caused the loss of more than 150,000 lives around the world and 5.5 million disability-adjusted life years (DALY).[1] For Latin America and the Caribbean, the human cost of climate change and variability by 2000 was an estimated 92,000 DALY, placing the region among those with the most negative health effects.[2] These data show that climate change and variability have already had a significant impact on health, as the number of cases of illness and mortality from disease and extreme weather events has increased.

Malaria has increased in several countries in Latin America and the Caribbean (LAC) over the last two decades, particularly in Bolivia, Brazil, Colombia, Guyana, Peru, and Suriname. The increase has occurred despite improvements in public health infrastructure that brought a decrease in child mortality in every country of the region over the same period (table 6.1). However, while malaria is likely to extend its reach into new areas as a result of climate change, other areas can expect a decline in the incidence of the disease, so that much uncertainty exists about the net effect on malaria rates in the region as a whole.

**Table 6.1    Selected Health Indicators**

| | Under-5 mortality rate (per 1,000 births) | | | Malaria cases reported (total cases) | | |
|---|---|---|---|---|---|---|
| | 1985 | 1995 | 2005 | 1982 | 1989 | 1997 |
| Antigua and Barbuda | n.a. | 21.0 | 12.0 | n.a. | n.a. | n.a. |
| Argentina | 33.3 | 23.5 | 16.3 | 567 | 1,620 | 592 |
| The Bahamas | n.a. | 24.0 | 15.0 | n.a. | n.a. | n.a. |
| Barbados | n.a. | 15.0 | 12.0 | n.a. | n.a. | n.a. |
| Belize | 55.3 | 30.4 | 17.2 | 3,868 | 3,285 | 4,014 |
| Bolivia | 147.0 | 105.0 | 65.0 | 6,699 | 25,367 | 51,478 |
| Brazil | 74.0 | 41.2 | 21.3 | 221,939 | 577,520 | 392,976 |
| Chile | 30.0 | 14.0 | 9.5 | n.a. | n.a. | n.a. |
| Colombia | 42.7 | 31.4 | 21.4 | 78,601 | 100,286 | 180,898 |
| Costa Rica | 23.0 | 16.0 | 12.2 | 110 | 699 | 4,712 |
| Cuba | 19.7 | 12.5 | 7.0 | n.a. | n.a. | n.a. |
| Dominica | 19.9 | 17.9 | 15.2 | n.a. | n.a. | n.a. |
| Dominican Republic | 77.0 | 53.0 | 31.0 | 4,654 | 1,275 | 816 |
| Ecuador | 72.0 | 43.0 | 25.0 | 14,633 | 23,274 | 16,356 |
| El Salvador | 83.0 | 46.0 | 27.0 | 86,202 | 9,605 | 2,719 |
| Grenada | n.a. | 33.0 | 21.0 | n.a. | n.a. | n.a. |
| Guatemala | 107.0 | 64.0 | 43.0 | 77,375 | 42,453 | 32,099 |
| Guyana | 96.0 | 79.0 | 63.0 | 1,700 | 20,822 | 32,103 |
| Haiti | 170.8 | 141.6 | 84.3 | 65,354 | 23,231 | n.a. |
| Honduras | 75.3 | 49.3 | 28.7 | 57,482 | 45,922 | 65,863 |
| Jamaica | 38.9 | 32.6 | 31.3 | n.a. | n.a. | n.a. |
| Mexico | 56.4 | 45.0 | 35.8 | 49,993 | 101,241 | 5,046 |
| Nicaragua | 85.0 | 53.0 | 37.0 | 15,601 | 45,982 | 42,819 |
| Panama | 39.0 | 29.5 | 23.5 | 334 | 427 | 505 |
| Paraguay | 51.0 | 33.0 | 23.0 | 66 | 5,247 | 567 |
| Peru | 100.7 | 62.5 | 27.3 | 20,483 | 32,114 | 193,740 |
| St. Kitts and Nevis | n.a. | 30.0 | 20.0 | n.a. | n.a. | n.a. |
| St. Lucia | 26.3 | 20.5 | 14.4 | n.a. | n.a. | n.a. |
| St. Vincent and the Grenadines | 31.5 | 21.9 | 20.4 | n.a. | n.a | n.a. |
| Suriname | 52.0 | 44.0 | 39.0 | 2,805 | 1,704 | 11,323 |
| Trinidad and Tobago | 36.1 | 32.5 | 37.0 | n.a. | n.a. | n.a. |
| Uruguay | 29.1 | 22.4 | 13.4 | n.a. | n.a. | n.a. |
| Venezuela, R.B. de | 38.1 | 28.3 | 21.3 | 4,269 | 43,374 | 22,400 |

*Source:* Authors' tabulations based on World Bank indicators.
*Note:* n.a. indicates no data available.

As climate change and variability cause changes in temperature, precipitation patterns, and extreme weather events, both direct and indirect effects on health are projected to intensify. The direct effects include an increase in morbidity from disease and malnutrition. Malnutrition, which may increase as declining crop yields cause more food insecurity, will be an underlying cause of many deaths from vector- and waterborne diseases. At the same time, vector- and waterborne diseases may also inhibit the uptake of nutrients, causing malnutrition. Indirect impacts include the effects of poor health on social and human development; for example, poor health can reduce educational performance as well as household productivity and income. Decreasing human capital increases the risk that asset-poor households will fall into poverty, while making it more difficult for others to escape poverty.

The first section of this chapter analyzes the major health risks associated with climate change and variability, specifically the risks of vector- and waterborne diseases and of malnutrition. The role that the lack of assets such as clean water plays in exacerbating the risk of disease and the role that El Niño plays in disease rates are examined. In the second section, the social livelihoods framework (SLF; introduced in chapter 1) is applied to analyze the wider social and human development effects arising from the deterioration of health as a human capital asset. Subsequently the chapter reports (a) an analysis of the effects of different natural disasters on the health status of children and (b) a simulation of the overall effects of gradual climate change on health, as proxied by life expectancy. The chapter concludes with policy recommendations, highlighting some of the most important areas for adaptation to reduce health risks from climate change and variability.

## Main Effects of Climate Change on Human Health

Historically, variations in climate and extreme weather have increased the rates of illness and morbidity in the areas affected. As the Intergovernmental Panel on Climate Change (IPCC 2007) notes, "highly unusual extreme weather events" have been affecting various parts of Latin America in recent years. One outcome of these changes is an increase in the health risk from climatic factors—a trend that is to some degree supported by data on malaria and dengue fever from the region. Much research has already been done on the relationship between weather extremes produced by El Niño and epidemic diseases, which is relevant for understanding some of the likely health impacts of climate change and variability (IPCC 2007; Moreno 2006; WHO 2003). El Niño patterns have been found to have a strong impact on the incidence of both vector- and waterborne diseases.

Figure 6.1 illustrates the principal ways in which climate change and variability affect health through gradual changes in temperature and precipitation, as well as through weather-related disasters. Temperature and

**Figure 6.1    Main Effects of Climate Change on Health**

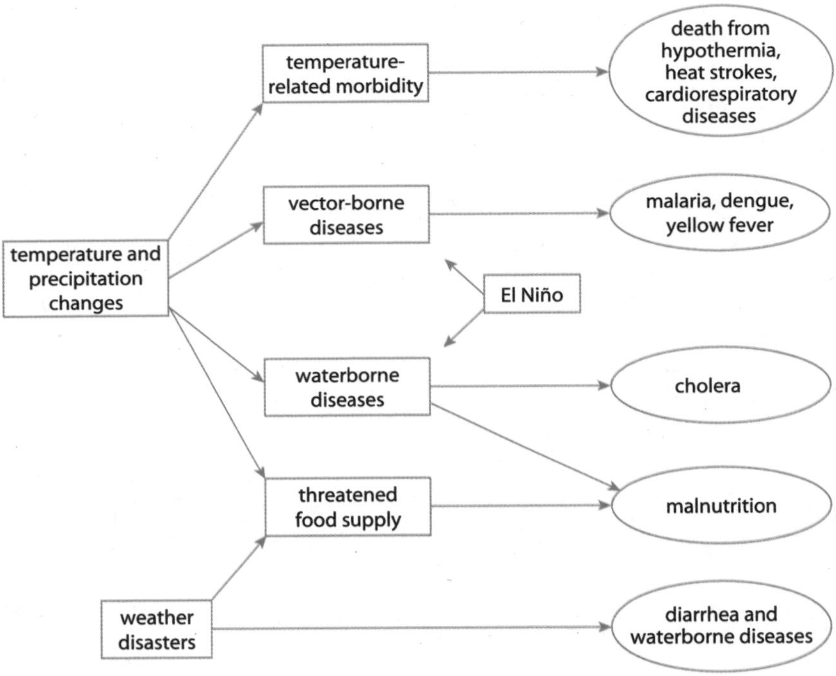

*Source:* Adapted from Patz and Balbus 1996.

precipitation changes increase the risk of both vector- and waterborne
diseases, but they also affect health indirectly because they increase the
risk of malnutrition. Both disease and nutritional status are affected by El
Niño weather changes. Droughts, floods, and other weather-related disasters increase the risk of mortality in the areas where they occur, but their
indirect impacts can present an even greater health risk. Such natural disasters can affect livelihoods and threaten food supplies (WHO 2003) and
can also reduce access to safe water, thereby increasing the risk of diarrhea and other waterborne diseases.

In Latin America and the Caribbean, the main health concerns associated with climate change and variability are the vector-borne diseases
malaria and dengue fever, diarrheal and other waterborne diseases, and
malnutrition. Each will be discussed in turn. Although the discussion here
is limited to these major effects, other health effects should also be recognized and addressed. They include cardiorespiratory diseases caused by

changes in air quality brought on by increased temperatures, which can increase particulate matter, pollen, and ozone in the atmosphere. Increased temperatures, particularly in cities, where heat-island effects can push temperatures very high, can be expected to lead to increased mortality and morbidity, especially among the elderly and among poor and homeless people who lack the means to take evasive action.[3] Other health effects include increases in parasitic diseases and food poisoning, as the parasites and bacteria that cause them thrive at higher temperatures.

According to Patz and Kovats (2002), the health impacts of climate change and variability, especially those due to changes in precipitation and temperature, will vary greatly across the region, but the areas most at risk appear to be the following:

- Populations that border regions with endemic diseases that are sensitive to climate conditions
- Regions for which a correlation has been observed between epidemic diseases and weather extremes such as El Niño episodes
- Areas where several climate impacts are projected to affect health (for example, stress on food and water supplies and risk of coastal flooding)
- Areas at risk because of environmental and socioeconomic stresses, whose populations have little capacity to adapt (for example, areas with poor land use or underdeveloped health infrastructure)

In assessing the health risks from climate change it is important to pay close attention to socioeconomic factors that can exacerbate the risks and to appreciate that the risk to health increases when multiple stresses are at work simultaneously, such as when droughts or floods result in malnutrition and thereby increase susceptibility to disease.

### Vector-Borne Diseases

Vector-borne diseases, especially malaria and dengue fever, constitute a major health risk from climate change and variability, particularly because these diseases are highly sensitive to even small changes in temperature and precipitation. Notably, the number of cases of both diseases has spiked following strong El Niño years, which are characterized by changes in temperature and precipitation. The future impact of climate change will depend on local conditions, as each locality presents a different threshold for the vector. One of the major risks is that malaria will spread into highland areas as mean temperatures rise. In Africa, malaria has begun to occur in highland areas with cooler climates where it had not previously been observed. The reason is most likely that temperatures in those areas have risen (WHO 2003). In general, the expansion of a disease to new areas presents a much greater health risk than does an extension of the transmission season in an area already affected, as regions formerly without malaria have low levels of immunity. Dengue fever has become a serious health problem in much of the LAC region, and new strains are being spread. Dengue is mainly found in urban areas. It is exacerbated by unplanned urbanization, with poor health infrastructure and lack of clean water and sanitation, and accordingly it presents a major health risk for some of the largest cities in the LAC region (WHO 2008).

Because the mosquitoes that spread most vector-borne diseases cannot regulate their own body temperature, they are very sensitive to changes in moisture levels and temperature. Thus, to the degree that climate change and variability alter temperatures, precipitation, and humidity, they will also alter the distribution and incidence of vector-borne diseases. Changes in temperature can alter the length of the transmission season or change the range of the disease, as the vectors that carry diseases respond to changes in temperature. Thus, a temperature increase in an area with low average temperatures can enable mosquitoes to extend their transmission season and range, whereas a temperature increase in a warm area where temperatures are already near their limits of tolerance could reduce the incidence of malaria by killing the mosquito and the parasite (WHO 2003).

An increase in precipitation is in general likely to raise the incidence of vector-borne diseases because it will create new habitats for larvae, although flooding caused by severe rainfall can destroy the same habitats. What happens in a given area will partly depend on the specific weather conditions: dry and hot weather following El Niño has induced disease outbreaks in the humid coastal regions of Colombia and Venezuela, while flooding has generated epidemics in the northern coastal areas of Peru that are normally very dry (IPCC 2007).

The disease pattern following El Niño years suggests that climate change carries a significant risk of an increase in the incidence of malaria and dengue fever (figure 6.2). After the severe El Niño of 1997–98, a great spike occurred in cases of both malaria and dengue fever, and during the 2002–03 El Niño, the number of dengue cases reached a record level. As mentioned previously, the effects of climate change and variability on vector-borne diseases will be highly localized, and small changes in temperature or precipitation can have significant positive or negative impacts on the incidence of vector-borne diseases.

**Figure 6.2    Incidence of Malaria and Dengue Fever in Latin America and the Caribbean, 1996–2006**

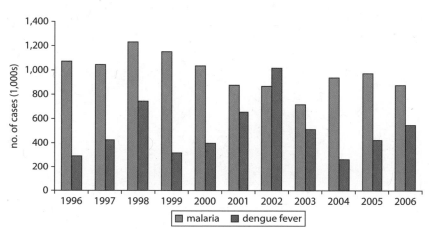

*Source:* PAHO 1995–2006.
*Note:* Both 1997–98 and 2002–03 were designated as El Niño years; 1998–99 and 2000–01 were La Niña years.

***Malaria.*** Malaria has the greatest death toll of any vector-borne disease in the world, annually claiming more than a million lives worldwide (WHO 2005). Even small changes in temperature and precipitation could significantly alter the distribution of this disease. The prospects include the spread of malaria to regions not previously plagued by it, where the populations enjoy little immunity.

There is great uncertainty about the impact that climate change and variability will have on malaria incidence. According to the IPCC (2007), "there is no clear evidence that malaria has been affected by climate change in South America," whereas the World Health Organization (WHO) has argued that malaria is probably the vector-borne disease that will be most affected by climate change (WHO 2003). It is agreed, however, that climate change and variability will likely reduce the incidence of malaria in some areas and increase it in others, depending on local factors.

In the Peruvian Amazon near Iquitos, small temperature variations of 1–2°C have been found to determine the seasonality of malaria transmission (WHO 2003). As shown in figure 6.3, an inverse relationship exists between malaria incidence and temperature. That relationship can be explained by the fact that temperatures in this particular area are near the upper limits of the vector's tolerance.

**Figure 6.3    Malaria Incidence and Temperature in Iquitos, Peru**

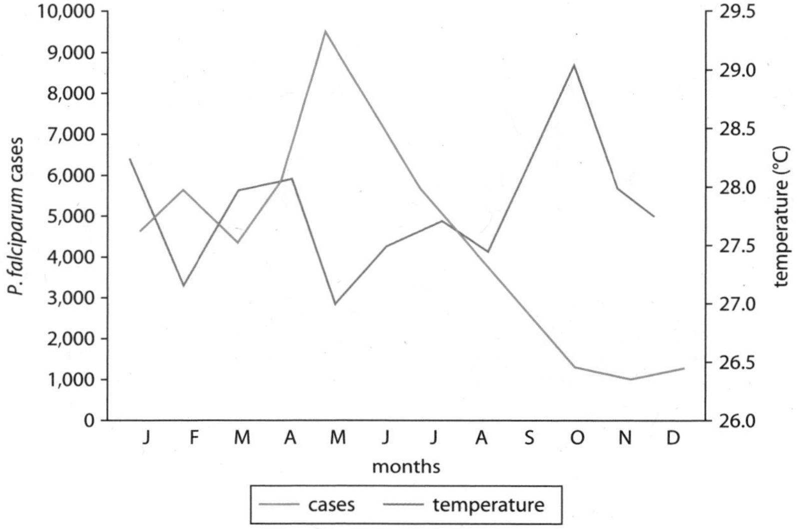

*Source:* Aramburú, Ramal, and Witzig 1999.

Historically, the levels of malaria in 715 municipalities in Colombia show a statistically significant relationship with precipitation but not with temperature. The threshold level for ideal vector breeding conditions was found to be annual precipitation in excess of 2,300 millimeters (Blanco and Hernandez 2009). Since the 1970s, when the rate was about 400 cases per 100,000 people, an increase in the incidence of malaria has been observed; by the 1990s cases were about 800 per 100,000 (De la Torre, Fajnzylber, and Nash 2008).

As noted above, one of the greatest health risks that climate change and variability present is the risk of temperature increases bringing vectors into areas where they did not previously thrive. Due to the low immunity of the population in areas where malaria is not endemic, the spread of the disease to new areas would have much more serious consequences than any increases in transmission rates within areas already affected (Iwanciw, Giles, and Effen 2004). In Colombia, rises in minimum nocturnal temperatures have led to greater incidence of malaria in the Andean piedmont (WHO 2003). A small temperature increase could potentially extend the malaria risk to major urban areas such as Quito and Mexico City (Moreno 2006). Puerto Perez, a hyperendemic region in Bolivia located 4,000 meters above sea level, shows that altitude is not necessarily an obstacle to malaria. Regions with hyperendemic malaria have high humidity, providing sources of stagnant water for the mosquitoes to breed in, even if temperatures are not necessarily high (Iwanciw, Giles, and Effen 2004). A decrease in precipitation, on the other hand, can reduce the incidence of malaria, as has been seen in the Brazilian state of Roraima (IPCC 2001b).

Migration heightens the risk of malaria transmission. To the degree that climate change and variability affect migration rates, they will affect the incidence of malaria, since rapid and unplanned urbanization can make it more difficult to control the disease and the migration of people from malaria-endemic areas to urban areas increases transmission risks (IPCC 2007). A study of the Guayamerin and Riberalta municipalities in Bolivia (Iwanciw, Giles, and Effen 2004) found that the large number of migrants to the cities was an important risk factor for malaria transmission. Similarly, a study in Oaxaca, Mexico, found that migration toward the main population centers was an important contributor to malaria transmission (PAHO 1995–2006).

As shown in table 6.2, the Andean region is most at risk of increases in transmission rates, as are parts of the Southern Cone, where there is presently little immunity to the disease. In contrast, a significant reduction in malaria is expected in the Amazon, and varying impacts are predicted for Central America.

In Mexico, although the incidence of malaria has fallen significantly in recent years, the disease is still a major health issue, with more than 2,400 registered cases in 2006 (PAHO 1995–2006). The main transmission area thus far has been the coastal state of Oaxaca, especially in connection with extreme weather events. The last major epidemic broke out in the wake of Hurricane Pauline in 1998, with 80 percent of Mexican cases occurring in the area where the hurricane hit. Mexico City is one of the high-altitude cities projected to experience an increase in malaria due to higher temperatures.

**Table 6.2    Expected Changes in Malaria Transmission Rates Due to Climate Change**

| The Amazon | Andean region | Central America | Southern cone |
|---|---|---|---|
| Significant reduction expected as the climate becomes too dry for mosquitoes. | Increase expected for cities such as La Paz and Quito due to temperature increases. | Increase expected for Nicaragua. General decrease expected; already observed in Honduras. | Increased risk along southern limits of range due to temperature increases. |

*Source:* IPCC 2007.

Table 6.3 reports forecast increases in the numbers of malaria and dengue cases due to climate change based on statistical models of the incidence of both diseases and projections of change in precipitation and temperatures (derived from eight global circulation models used in the IPCC's fourth assessment report). The incidence of malaria is projected to increase, mainly in rural areas, by 8 percent over the next 50 years and 23 percent over the next 100 years. The incidence of dengue, mostly affecting urban areas, is projected to increase by 21 percent by 2050 and by 64 percent by 2100.

*Dengue fever.* As table 6.3 indicates, climate change and variability are expected to increase the incidence of dengue fever significantly; the disease is already spreading in Mexico and central South America (Warren and others 2006). With transmission rates likely to grow by 21 percent by the 2050s in most parts of South America,[4] dengue fever could become one of the major health risks resulting from climate change and variability. Although dengue fever epidemics are closely correlated with changes in temperature and precipitation, as has been observed following El Niño years, they are also strongly influenced by socioeconomic factors and water storage practices. As dengue fever is primarily an urban disease, whose risk is heightened by unplanned urbanization, poor housing, and poor public health infrastructure, it presents a major health risk for some of the largest cities in the LAC region.

Worldwide, 3 billion people are currently at risk from dengue fever and tens of millions are affected, resulting in at least 653,000 DALY worldwide (WHO 2003).[5] In the last few decades, the disease has extended its reach through the LAC region (figure 6.2), particularly affecting urban areas (WHO 2008). The rates of both dengue and dengue hemorrhagic fever have risen in the region during the last two decades

**Table 6.3    Additional Cases of Malaria and Dengue Fever under 50- and 100-Year Future Scenarios**

| Vector-borne disease | Total cases recorded 2000–05 | Additional number of cases predicted for a 6-year period | |
| --- | --- | --- | --- |
| | | 50-year scenario | 100-year scenario |
| P. falciparum malaria | 184,350 | 19,098 | 56,901 |
| P. vivax malaria | 274,513 | 16,247 | 48,207 |
| Dengue fever | 194,330 | 41,296 | 123,445 |
| Total | 653,193 | 76,641 | 228,553 |

*Source:* Blanco and Hernandez 2009.

(Torres and Castro 2007). Compared with the 6-year period from 2000 until 2005, when 194,330 cases were recorded, the first quarter of 2008 alone saw more than 100,000 registered cases and 40 fatalities, most of which occurred in Brazil (PAHO 2008).

Dengue fever is spread by mosquitoes, primarily the *Aedes aegypti* mosquito, which breeds in urban environments in water-holding containers.[6] The disease is seasonal and predominates during warmer, more humid weather. Rainfall also appears to affect its transmission potential. For the correlation between rainfall and temperature, on one hand, and dengue epidemics on the other, a statistically significant lag has been found of 3–7 months for temperature and 1–3 months for rainfall (Chen 2006).

As shown in figure 6.4, the *Aedes aegypti* mosquito had been eradicated in most of Latin America and the Caribbean by 1970, but by 2001 it had made a significant comeback. The key reason was the massive, uncontrolled urbanization that took place in tropical America during the 1970s and 1980s, coupled with poor maintenance of public health infrastructure. The mosquito's spread through much of Central and South America has put many countries at risk of dengue epidemics. Moreover, new virus strains have been detected in new areas. That increases the potential of secondary dengue infections, which are one factor leading to the more serious

**Figure 6.4    *Aedes Aegypti* Distribution in the Americas**

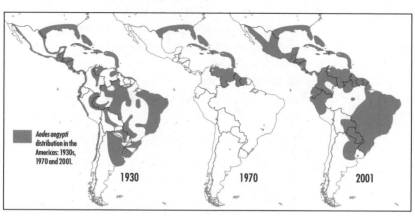

Aedes aegypti distribution in the Americas: 1930s, 1970 and 2001.

1930    1970    2001

IBRD 37786
MAY 2010

*Source:* Gubler 2002.

dengue hemorrhagic fever, which causes both internal and external bleeding (Gubler 2002; pers. com. 2009).

Mexico is increasingly plagued by dengue fever—the number of cases there has increased by 600 percent since 2001—and by dengue hemorrhagic fever (AP 2007). Increased precipitation has been cited as a factor. According to Joel Navarette, of the Mexican Social Security Institute, dengue fever has changed from being a seasonal to a year-round phenomenon in parts of the country. This corresponds with projections by the IPCC that climate change and variability will cause an increase in the number of dengue cases.

As with malaria, migration and other population movements have played an important role in spreading different strains of dengue fever across Mexico. A person can develop immunity to one strain, but subsequent exposure to a different strain increases his or her risk of developing the more dangerous hemorrhagic type (AP 2007).

In urban areas, change in precipitation might be a less important causative factor than changes in water storage practices. Droughts lead households to store water in tanks and other containers, which can serve as breeding sites.[7] As mentioned previously, dengue fever is primarily an urban disease that is greatly influenced by socioeconomic factors such as housing, public health infrastructure, lack of basic sanitation, and water storage practices (Torres and Castro 2007). The importance of socioeconomic factors is suggested by comparing the incidence of dengue fever on the Texan and Mexican sides of the U.S.-Mexican border: between 1980 and 1996, 50,333 cases were reported in the three Mexican states bordering Texas, compared to only 43 cases in Texas (WHO 2003).

Population growth and unplanned urbanization are thought to be important contributors to dengue fever, as is increased travel, which can bring the disease to areas previously free from it (Torres and Castro 2007). In the LAC region, São Paulo, Caracas, Mexico City, and Buenos Aires are just some of the cities with unplanned growth that are at risk from epidemics (WHO 2003).

## Waterborne Diseases

Waterborne diseases present another serious health risk in Latin America, and climate change and variability are likely to increase their incidence. Evidence from El Niño years suggests that cholera and other waterborne diseases are highly sensitive to changes in climate. Diarrheal diseases are both the most common type of waterborne illness and by far the deadliest group of diseases associated with the environment. In Latin America and the Caribbean in 2002 they killed almost 20 times as many people as malaria and dengue fever combined (Warren and others 2006). Extreme weather events, such as floods and heavy rains, increase the incidence of waterborne diseases, but other important causes are the risky water storage and water use practices of people who lack access to clean water and sanitation. Accordingly, the risk of waterborne diseases is expected to increase as a result of climate change and variability—an expectation that is supported by the evidence of El Niño's influence on diarrheal diseases.

Evidence is ample of the link between extreme weather events and waterborne diseases is ample. Following Hurricane Mitch, the number of cholera cases was four times higher than normal in Guatemala and six times higher in Nicaragua (IPCC 2001b). On the Pacific Coast of South America, outbreaks of cholera have been tied to extreme weather caused by El Niño; about 11,700 people died from cholera in Peru between 1991 and 1996. Similarly, when temperatures were higher than normal during El Niño in 1997–98, an unusually large number of people suffering from diarrhea and dehydration were admitted to hospitals for rehydration treatment in Lima. Indeed, a time-series analysis showed an 8 percent increase in admissions due to diarrhea for each 1°C increase in temperature (WHO 2003).

In tropical climates diarrheal diseases normally peak during the rainy season, and studies from the United States have shown that about half of all waterborne diseases can be attributed to extreme rainfall (WHO 2003). Most of the waterborne microorganisms that cause disease derive from human or animal feces, which heavy rains and flooding can transport into groundwater as well as surface and well water. Sources of drinking water can thus become contaminated with *E. coli* bacteria and other microorganisms. Communities that have combined systems for storm water drainage and sewage run the risk of accidental releases of raw sewage. Heavy rainfall can lead to surface runoff that allows pathogens to enter aquifers, wells, and drinking water. Because one of the projected aspects of climate change is more frequent heavy rains, such contamination is likely to escalate, causing a surge in the incidence of waterborne diseases. However, underlining the complexity of forecasting the health impact of climate change, the seasonal pattern of cholera outbreaks in the Amazon basin has been associated with lower river flow in the dry season (Gerolomo and Penna 1999).

Rising temperatures may lengthen the season, or alter the geographical distribution, of waterborne diseases, inasmuch as most of the protozoa, viruses, and bacteria that cause them thrive in warm water and weather.

Besides changes in temperature and precipitation, socioeconomic conditions and human behavior play a major role in spreading water-borne disease. Water storage practices, as well as lack of clean water and sanitation, are important determinants of disease incidence and can themselves be affected by weather extremes. There is evidence that water shortage causes an increase in diarrheal disease due to unsanitary water practices. Lack of access to natural capital assets, such as clean water and basic sanitation, makes the poor more vulnerable to waterborne diseases. For adaptation purposes, it is important to keep in mind that waterborne diseases are driven not only by natural events but also by socioeconomic conditions and human behavior.

## Malnutrition

The World Health Organization considers malnutrition the single most important risk to global health, making up 15 percent of the total disease burden in DALY (WHO 2003). According to the IPCC, the additional number of people in Latin America at risk of suffering hunger in 2020 will be 5 million under the IPCC's A2 emission scenario (IPCC 2001a).[8] However that projection does not take account of the fertilization effect of the additional carbon dioxide ($CO_2$) in the atmosphere, which will increase yields of some crops. That effect has been estimated to reduce the additional number of people at risk from hunger by 2020 to 1 million (Parry and others 2004). Malnutrition in Latin America and the Caribbean will be affected by droughts as well as by floods, which not only destroy croplands but also cause vector- and waterborne diseases that increase the risk of malnutrition by inhibiting the body's uptake of nutrients (World Bank 2008). Evidence from drought periods suggests that malnutrition will increase the number of children suffering from stunted growth and irreversibly impaired brain development; indeed young children are the ones most at risk from malnutrition. For them the effects can be lifelong because they are related to development, whereas adults can recover. Malnutrition also raises the risk of mortality from other diseases, especially malaria and diarrhea; it is the main source of mortality from both diseases (Parry and others 2004). Malnutrition in pregnant mothers increases the risk of subsequent infant mortality.

More frequent droughts will diminish crop production in some parts of the LAC region, as discussed in chapter 4 (UNDP 2007). In Mexico, crop reductions of 30 percent, the upper end of the range projected by the IPCC by 2020 (IPCC 2007), could threaten livelihoods and food security and be exacerbated by loss of livestock. In northeast

Brazil, agricultural gross domestic product (GDP) has declined by up to 25 percent in past years of severe drought, causing the displacement of several million rural poor (Mata and Nobre 2006). Empirical studies from Africa have found a close correlation between being born in a drought year and malnutrition and stunted growth. In Kenya, for example, children born in drought years were 50 percent more likely to be malnourished and 72 percent more likely to suffer stunted growth than other children (UNDP 2007).

Other factors also cause malnutrition. They include repeated infections (for example, ones caused by poor sanitation), which can reduce dietary intake. Women suffering malaria or diarrhea during pregnancy give birth to children with low birth weight, who subsequently are at greater risk of infections and chronic disease (Acharya and Paunio 2008). Both malaria and diarrhea affect nutritional development and increase the risk of malnutrition, which increases the body's susceptibility to disease. Malnutrition is a strong contributor to mortality from other diseases, being the source of 61 percent of the mortality from diarrhea and 57 percent of that from malaria (ECLAC, UN, and UNICEF 2006). Increased fatalities from malaria have been observed following a drought year; malnutrition among pregnant women can adversely affect fetal health and is the main cause of infant and preschool mortality (WHO 2003).

The number of people at risk of malnutrition could increase significantly as both floods and droughts intensify in parts of Latin America and the Caribbean. The effects on health of an increase in malnutrition would include rises in the incidence of stunted growth and irreversibly impaired brain development, as well as in mortality from other diseases, especially malaria and diarrhea, both of whose severity is affected by malnutrition.

The extent to which conditions early in a child's life affect later outcomes has attracted substantial interest in recent years.[9] In general, the conclusion is that early childhood malnutrition can significantly depress performance in school and hence presumably earnings later in life, although it is difficult to pinpoint an exact period in the child's life that is more critical than others. For example, in a study from Indonesia, Maccini and Yang (2008) showed that the amount of rainfall experienced during early life had statistically significant effects on self-reported adult health for women but not for men. Women born between 1953 and 1974 who as children experienced rainfall 20 percent above average were also slightly taller (0.57 centimeters) and tended to be more educated and to live in households with more assets. Further, Strauss and Thomas (1998) found a substantial return to height in labor markets, even after controlling for education: taller people got better jobs. Hence natural hazards, working through their effects on nutrition, have the potential to affect not only children's current health status but also how they fare in school and perform in the labor market as adults.

## Impacts of Poor Health on Livelihood Assets

Not only will climate change and variability cause higher rates of morbidity and malnutrition, but these direct impacts will extend beyond health to education and poverty reduction. As reflected in figure 6.5, poor health affects educational attainment directly by reducing cognitive ability, thus increasing the risk of failure and dropping out. Sickness calls for increased spending on treatment and prevention, which reduces a household's financial capital and its ability to invest in education. Another effect of poor health is reduced productivity and income, not only for the member of the household who is sick but also for other members who are forced to spend time taking care of him or her. The reduction of a household's income erodes its financial assets, while making it more difficult to invest in education. By reducing financial and human capital, the health effects of climate change and variability can play a role in pushing households into a poverty trap from which their lack of assets makes it impossible to escape (UNDP 2007).

One of the most significant impacts of disease and malnutrition is that on educational attainment. Both malaria and waterborne diseases

**Figure 6.5    Selected Social Impacts of an Increase in Illness and Morbidity at the Household Level**

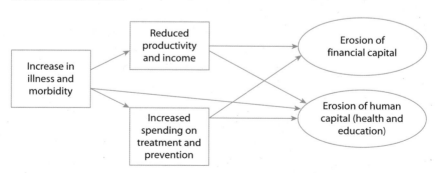

*Source:* Authors.

can cause anemia due to iron deficiency. As seen above, if gross nutritional development is impeded, especially in children under the age of two, brain development is irreversibly impaired, affecting cognitive ability, language, attention, and behavior (Holding and Snow 2001). Thus the consequences of ill health and malnutrition in childhood are often long lasting, so that children (also once they become adult) require special support from family, community, health services, and education services.

Besides impairing brain development and cognitive abilities, malaria can interfere with educational attainment in other ways. School attendance might be interrupted by illness, or children might only be able to attend school if it is located close to home (Holding and Snow 2001). Although few empirical studies examine the effect that malaria can have on educational performance, the limited evidence suggests that a relationship exists (Chima, Goodman, and Mills 2003). Losing many days because of absences increases the likelihood of failing and having to repeat a grade, as well as of dropping out of school altogether (Sachs and Malaney 2002). As for the effects on educational performance, a study in Kenya of 87 case patients found that 3–4 years after being discharged from treatment for malaria, 14 percent were performing at least two standard deviations below the mean in school tests, compared to only 5 percent among the control group (Holding and Snow 2001). When treatment costs are high, a family with limited assets may decide to stop supporting the education of some or all of its children (Holding and Kitsao-Wekulo 2003). In addition, if parents believe that the long-term effects of the disease will compromise their child's educational prospects or health status, they may decide not to invest in the child's education.

Malaria's impacts on educational attainment and on labor productivity are likely to be more significant in areas located on the edge of endemicity because in those areas a smaller percentage of children will have developed immunity by the time they reach school age. (Currently, areas with endemic malaria have low rates of school absence because school-age children have already developed immunity [Holding and Snow 2001]). Productivity losses from malaria are smaller in areas where the disease is widespread, compared to areas where transmission is less common and immunity is rare (Malaney, Spielman, and Sachs 2004).

Household income is affected by disease in several ways. Health is a human capital asset, and its value at the household level is determined both by the number of individuals who are able to work and by the quality of the work they are able to do (IISD 2003). Because malaria and malnutrition can reduce both the quantity and the quality of household labor, they can significantly reduce income. Children who become ill or weak might not be able to take up chores for the family, such as working in the fields, thus reducing their contribution to family productivity and income. Similarly, as the primary caregivers, women may have to spend more time on child care when a child is sick, reducing their availability for activities that contribute to household food production or income generation.

Besides lowering labor productivity, diseases such as malaria also force households to spend on prevention and treatment, the cost of which may erode their financial assets if they are poor (Malaney, Spielman, and Sachs 2004). Sometimes a rural household must pay hired workers to offset the loss in productivity of its own family members (Sachs and Malaney 2002).

## Natural Disasters and Child Health

Little work has been done on how shocks from natural disasters affect the health of children.[10] Natural hazards such as hurricanes, floods, and droughts are potentially very destructive, and climate change and variability are likely to put more and more households at their mercy. If these events damage children's health or raise child mortality, the effects on long-term growth and social development will be large. This is made more likely by the fact that children often bear a disproportionately large share of the burden when a shock such as a natural disaster forces households to focus resources on their most productive members.

The authors examined how natural disasters affect the health status of children, using data from Guatemala (appendix C). The regression results are summarized in table 6.4. Looking in more detail at the effect of shocks from natural disasters on children's height for age and weight for height, and on three symptoms of illness (diarrhea, fever, and cough) revealed that girls, on average, are more resilient than boys and that children with educated parents are more resilient than others. Land ownership by the family had little effect on the three measures of illness—which may indicate that the variable does not adequately capture whether or not the household owns land. Indigenous children were substantially shorter than nonindigenous children of the same age, even when controlling for the literacy and education level of the mother.[11] It appears, however, that indigenous children do well in terms of current health conditions (weight for height).

**Table 6.4    Effects of Shocks on Child Anthropometrics and Illness Symptoms, by Type of Shock**

|  | Height for age | Weight for height | Diarrhea | Cough | Fever |
|---|---|---|---|---|---|
| Frost | −0.094** | −0.080* | 0.003 | 0.051*** | −0.011 |
|  | −0.045 | −0.044 | −0.013 | −0.017 | −0.016 |
| Hurricane | −0.051 | −0.024 | 0.022 | 0.042 | 0.067 |
|  | −0.121 | −0.118 | −0.033 | −0.073 | −0.071 |
| Strong wind | −0.115*** | 0.008 | 0.010 | 0.040** | −0.038** |
|  | −0.035 | −0.034 | −0.010 | −0.016 | −0.015 |
| Drought | −0.066 | 0.112** | −0.020 | −0.030 | −0.022 |
|  | −0.056 | −0.057 | −0.017 | −0.021 | −0.021 |
| Heavy rain | −0.125*** | −0.020 | −0.004 | 0.081*** | 0.250*** |
|  | −0.044 | −0.043 | −0.012 | −0.031 | −0.030 |
| Flooding | −0.064*** | 0.038* | −0.003 | −0.025** | −0.010 |
|  | −0.020 | −0.020 | −0.006 | −0.012 | −0.011 |

*Source:* Authors' calculations. Each result is from a separate regression; complete results on request.
*Note:* All shocks 1–6 months prior to survey.
* significant at 10%; ** significant at 5%; *** significant at 1%.

## Simulations of the Effect of Climate Change on Health

The authors also undertook a regression analysis of the relationship between climate and health indicators in five countries in Latin America, all of which have sufficient internal climate variation to allow a cross-section estimation of the general relationships between average climate and health (appendix C). Using municipality-level data, the analysis focused on the net effect of climate change (proxied by changes in average temperature and precipitation patterns, that is, not including climatic variability or weather-related disasters) on health (proxied by life expectancy at birth and child mortality).[12]

The findings suggest that the relationships between climate and health vary in strength from country to country. In Peru and Chile, moderate temperatures are associated with significantly better health indicators than either cold or hot temperatures (all other things equal), but in Bolivia, Brazil, and Mexico no significant relationships were found. Rainfall shows a mixed relationship with health indicators. In Brazil, health indicators were found to be significantly better for intermediate amounts of rain (all other things equal), but in Chile and Peru they seem to be better for extreme amounts of rain (either very little or a lot), while in Bolivia and Mexico no significant relationships between health and precipitation were found.

Although it is always risky to use differences found in a cross section to analyze possible changes over time, the authors used the estimated relationships from the municipality-level regressions to simulate the effects of expected future climate change on health indicators as a way to get some idea about the direction and magnitude of the effects that might be expected if climate changes as suggested by the IPCC model materialize.[13] The simulated impacts of expected climate change on life expectancy were found to be weak, ambiguous, and relatively small. Contrary to expectations, the poorest country of the five investigated, Bolivia, appeared to be the least affected by expected climate change, and Mexico also appeared quite insensitive.

Peru is the country where the most dramatic effects, both positive and negative, were found. Future climate change is estimated to cause a small reduction in average life expectancy of about 0.2 years over the next 50 years (table 6.5). Other things equal, Peru's currently poor and cold highland regions will benefit, while the hotter and richer coastal regions will lose, so that overall future climate change in Peru is expected to reduce health inequalities between the country's districts.

According to the simulations, the average loss of life expectancy due to climate change over the next 50 years is about one year for Brazil. A strong positive relationship ($\rho = 0.65$) appears between current life expectancy and the estimated impact of future climate change on life expectancy—implying that the municipalities that are currently worse off will suffer larger-than-average deteriorations in life expectancy. That means that the climate change expected in Brazil is likely to contribute to a worsening in the distribution of health outcomes across Brazilian municipalities.

**Table 6.5    Estimated Impacts of Future Climate Change on Life Expectancy, 2008–58**

|  | Overall impact on life expectancy | Most adverse effect found | Most beneficial effect found | Impact on the distribution of health outcomes |
|---|---|---|---|---|
| Bolivia | No effect | No effect | No effect | No effect |
| Peru | −2 months | −35 months | +29 months | Less inequality |
| Chile | −6 months | −20 months | +4 months | Less inequality |
| Brazil | −12 months | −19 months | +6 months | More inequality |

*Source:* Authors' estimations.

*Note:* The data available for Mexico are on infant mortality rather than life expectancy; therefore Mexico is not included. − = reduction in life expectancy. + = increase in life expectancy.

In Chile, expected climate change is estimated to reduce overall life expectancy by about 6 months. This level of reduction is representative for the Metropolitan Region in the middle of the country, whereas northern states are expected to lose more—up to 19 months in Tarapacá—and southern states are expected to benefit slightly from a warmer climate. A weak but statistically significant negative relationship ($\rho = -0.22$) appears between current life expectancy and the estimated future impact of climate change, indicating that although climate change in Chile is likely to have a negative effect on health outcomes, it will have a positive effect on the distribution of health outcomes.

It is worth emphasizing that while the relationships found between climate variables and life expectancy were relatively weak, the regressions show strong and unequivocally positive relationships between education and life expectancy. According to these simulations, the estimated adverse health effects of climate change in Brazil over the next 50 years, for example, could be countered by an increase in average education levels of just 1.1 year. Education therefore seems to be a very effective "no-regrets" policy to counter potential adverse health effects of climate change. It should also be kept in mind, however, that even greater health benefits could be achieved through education in the absence of climate change. Climate change weakens the contributions of education to well-being. In interpreting these findings it is important to keep in mind that the analyses considered changes in *mean* temperatures and precipitation only; the effects of increased climate variability and more severe weather-related disasters were not included.

## Building Adaptive Capacity: Policy Perspectives

Climate change and variability present not only an environmental issue but also a development issue involving serious health risks. Indeed, the LAC region has already seen increased incidence of diseases associated with climate change and variability. Whereas small changes in temperature and precipitation can have a large impact on disease rates, socioeconomic factors, such as poor health infrastructure, migration, and lack of assets such as safe water and sanitation, exacerbate the risks.

Many of the anticipated health risks to Latin America and the Caribbean concern diseases that already challenge public health in the region. The increased risks arise from the projected expansion of their geographical range or lengthening of their transmission season. This is particularly the case for the vector-borne diseases malaria and dengue fever, as well as for waterborne diseases. Risks are particularly serious when diseases are brought into new areas, where immunity levels are low, or when new virus strains are created.

Further, some of the health impacts from climate change and variability could have wider social consequences as human capital assets are weakened. That could hold back children's educational performance and reduce financial and other household assets, thereby reducing household resilience and adaptive capacity.

To sum up, the potential health risks to the LAC region are very serious, and as the IPCC noted in its last report, adaptive capacity in the region is low (IPCC 2007). Building up the region's adaptive capacity and reducing health risks resulting from climate change and variability, especially among the groups with the fewest assets, arguably are the most important development issues facing the region today. Recommendations for those purposes are presented below.

1. *Improve the knowledge base to facilitate prediction and reduce health risks.*

In adapting to climate change and variability, an essential goal is to develop models for projecting disease patterns. One of the problems in projecting the health impacts of climate change and variability and designing adaptation strategies is the current lack of adequate models. Many health outcomes affected by climate are also influenced by socioeconomic and environmental factors and occur with a lag between exposure and response. These relationships need to be better understood if robust models for prediction are to be developed. Partly because of lack of funding, epidemiological data that can be used to describe the relationship between climatic factors and health impacts are scarce (Ebi 2008).[14]

Functioning health surveillance systems are needed to monitor the incidence and spread of diseases. Currently, many countries are underreporting diseases. Improving health surveillance systems will allow countries to report diseases more accurately and better measure the effectiveness of health programs.

Assessments are needed at the national and subnational levels of the relative burden of future climate change–related health impacts and vulnerabilities, to enable the most effective adaptation strategies to be designed.

2. *Increase adaptive capacity by strengthening assets.*
Involving communities in assessing and preparing for climate change and variability risks can help to build up adaptive capacity. As discussed in chapter 3, community-centered risk assessment projects have been shown to facilitate adaptation while helping to create social capital— an important asset for adapting to climate change and variability. This is particularly important for communities suffering extreme weather events because of the extra risks to health that they introduce. It is crucial that disaster risk reduction and emergency response plans are prepared involving the most vulnerable communities.

Health considerations should be incorporated in measures to combat water scarcity to prevent water from being contaminated by runoff during floods and heavy rain or by the use of sewage or wastewater for irrigation. Additionally, government agencies and nongovernmental organizations should augment their efforts to widen the access of households and communities to simple mechanisms for purifying water and encourage the use of water storage practices that prevent parasites from thriving. Low-cost programs promoting household hygiene would reduce the incidence of diarrhea and malnutrition.

Part of the reason why the importance of environmental health risks has often been overlooked is the cross-sectoral nature of environmental health (Acharya and Paunio 2008). Public health agencies should increase their focus on the role of environmental factors, such as the degradation of ecosystems and other forms of natural capital, in spreading diseases and heightening health risks.

3. *Reduce the vulnerability of groups facing the greatest health risk.*
Access to medical care for migrant populations should be expanded, with an emphasis on reducing the spread of diseases.

The risk of malnutrition should be reduced by implementing or expanding nutritional programs for asset-poor households, particularly those with young children.

## Notes

1. DALY refers to the sum of years of life lost due to premature death (YLL), and years of life lived with disability (YLD).

2. Floods have caused the majority of climate change–related DALY in Latin America, at 72,000, followed by diarrhea, at 17,000, and malaria at 2,000 (WHO 2003).

3. It has been estimated that the 2003 heat wave in Europe led to between 40,000 and 52,000 excess deaths (Larsen 2006; Sardon 2007).

4. Even greater increases are estimated by Stern (2007), who projects dengue transmission to increase by two to five times by the 2050s.

5. This number has been put as high as 1.8 million DALY.

6. The *Aedes aegypti* mosquito is also known as *Aedes stegomyia*.

7. Anecdotal evidence from St. Lucia suggests that a public education campaign stressing the importance of keeping water containers covered, to deny mosquitoes access to standing water, has successfully reduced the incidence of vector-borne diseases.

8. This scenario envisages a mean temperature change of 3.4°C by 2090–99, relative to 1980–99 (see IPCC 2001a).

9. See, for example, Glewwe and King 2001; Glewwe, Jacoby, and King 2001; Alderman, Hoddinott, and Kinsey 2006; and van den Berg, Lindeboom, and Portrait 2007. In addition there is some evidence that conditions while *in utero* can affect later outcomes, as shown by Behrman and Rosenzweig 2004; and Almond 2006.

10. As discussed in Strauss and Thomas (1995) and Wolpin (1997), the literature on the determinants of child health and mortality in developing countries is large. Despite this large literature, there is surprisingly little that directly deals with the potentially adverse effects of shocks from natural hazards. Foster (1995) showed for Bangladesh that floods lead to substantially lower weight for the children affected and argued that credit market imperfections were the main factor behind the differences in how children's weights responded. More recently, Baez and Santos (2007) examined the effects of Hurricane Mitch in Nicaragua on children's health, schooling, and labor force participation. They found that those affected by the hurricane were more likely to be undernourished and that the distribution of nutritional status worsened, especially for those at the bottom of the distribution.

11. Recall that the WHO does not find any substantial evidence of different growth patterns for children of different races exposed to "optimal" conditions.

12. For each country the relationship between climate (temperature and precipitation) and life expectancy (or child mortality) was estimated, using municipality-level data, controlling for other factors that also might affect life expectancy (education and urbanization rates). These estimated relationships were then used to simulate the effects of past (past 50 years) climate change as well as expected future (future 50 years) climate change, in each municipality in each of the five countries. Estimating the effects at the municipal level has the advantage of permitting an analysis of the distributional consequences of climate change on health. Effects of increased climatic variability,

of extreme events such as floods and hurricanes, or of the disappearance of tropical glaciers were not analyzed.

13. The simulation of expected future climate change (2008–58) relied on central projections from the IPCC's set of 21 coordinated Atmosphere-Ocean General Circulation Models (AOGCMs; Magrin and others 2007).

14. The World Bank is currently implementing an Early Warning System for Malaria and Dengue Surveillance and Control (DMEWS) in pilot municipalities in Colombia. The project aims to develop a framework that allows the local risk of dengue and malaria transmission in the face of climate change to be continuously evaluated, so as to determine the most appropriate actions to prevent epidemics before they begin (World Bank P083075 - Colombia: Integrated National Adaptation Program).

## References

Acharya, A., and M. K. Paunio. 2008. *Environmental Health and Child Survival: Epidemiology, Economics, Experiences.* Washington, DC: World Bank.

Alderman, H., J. Hoddinott, and B. Kinsey. 2006. "Long-Term Consequences of Early Childhood Malnutrition." *Oxford Economic Papers* 58 (3): 450–74.

Almond, D. 2006. "Is the 1918 Influenza Pandemic Over? Long-Term Effects of In Utero Influenza Exposure in the Post–1940 U.S. Population." *Journal of Political Economy* 114 (4): 672–712.

AP (Associated Press). 2007. "Hemorrhagic Dengue Fever Surges in Mexico with Climate Changes, Migration, Urbanization." March 31.

Aramburú, J., C. Ramal, and R. Witzig. 1999. "Malaria Reemergence in the Peruvian Amazon Region." *Emerging Infectious Diseases* 5 (2): 209–15.

Baez, J. E., and I. V. Santos. 2007. "Children's Vulnerability to Weather Shocks: A Natural Disaster as a Natural Experiment." Mimeo. Syracuse University. http://student.maxwell.syr.edu/jebaez/Mitch&Children%27sVulnerability (FinalVersion-Sept2007).pdf.

Behrman, J. R., and M. R. Rosenzweig. 2004. "Returns to Birthweight." *Review of Economics and Statistics* 86 (2): 586–601.

Blanco, J., and D. Hernandez. 2009. "The Costs of Climate Change in Tropical Vector-Borne Diseases: A Case Study of Malaria and Dengue in Colombia." *LCR Sustainable Development Working Paper* 32: 69–87, World Bank, Washington, DC.

Chen, A. 2006. *The Threat of Dengue Fever in the Caribbean: Impacts and Adaptation.* Washington, DC: International START Secretariat.

Chima R. I., C. A. Goodman, and A. Mills. 2003. "The Economic Impact of Malaria in Africa: A Critical Review of the Evidence." *Health Policy* 63 (1): 17–36.

De la Torre, A., P. Fajnzylber, and J. Nash. 2008. *Low Carbon, High Growth. Latin American Responses to Climate Change.* Washington, DC: World Bank.

Ebi, Kristie L. 2008. "Healthy People 2100: Modeling Population Health Impacts of Climatic Change." *Climatic Change* 88: 5–19.

ECLAC, UN, and UNICEF. 2006. "Child Malnutrition in Latin America and the Caribbean." *Challenges 2.* United Nations publication ISSN 1816-7543.

Foster, A. D. 1995. "Prices, Credit Markets, and Child Growth in Low-Income Rural Areas." *Economic Journal* 105 (430): 551–70.

Gerolomo, M., and M. L. F. Penna. 1999. "The Seventh Pandemic of Cholera in Brazil." *Informe Epidemiologico do Sus* 8 (3): 49–58.

Glewwe, P., H. G. Jacoby, and E. M. King. 2001. "Early Childhood Nutrition and Academic Achievement: A Longitudinal Analysis." *Journal of Public Economics* 81 (3): 345–68.

Glewwe, P., and E. M. King. 2001. "The Impact of Early Childhood Nutritional Status on Cognitive Development: Does the Timing of Malnutrition Matter?" *World Bank Economic Review* 15 (1): 81–113.

Gubler, Duane J. 2002. "Epidemic Dengue/DHF as a Public Health, Social, and Economic Problem in the 21st Century." *Trends in Microbiology* 10 (2): 100–03.

Holding, P., and P. Kitsao-Wekulo. 2003. *New Perspectives on the Causes and Potential Costs of Malaria.* Disease Control Priorities Project Working Paper No 7, World Bank, Washington, DC. http://www.dcp2.org/file/23/WP7.pdf.

Holding, P., and R. Snow. 2001. "Impact of Plasmodium Falciparum Malaria on Performance and Learning: Review of the Evidence." *American Journal of Tropical Medicine* 64 (1): 68–75.

IISD (International Institute for Sustainable Development). 2003. *Livelihoods and Climate Change.* Winnipeg, Canada: IISD.

IPCC (Intergovernmental Panel on Climate Change). 2001a. *Special Report on Emission Scenarios.* Geneva: IPCC.

———. 2001b. *Third Assessment Report, Working Group II.* Geneva: IPCC.

———. 2007. *Climate Change 2007: Impacts, Adaptation, and Vulnerability.* Contribution of Working Group II to the Fourth Assessment Report of the IPCC. Geneva: IPCC.

Iwanciw, Javier Gonzales, Jorge Cusicanqui Giles, and Marilyn Aparicio Effen, eds. 2004. *Vulnerabilidad y Adaptación al Cambio Climático en las Regiones del Lago Titicaca y los Valles Cruceños de Bolivia.* La Paz: Government of Bolivia, Ministry of Development Planning.

Larsen, Janet. 2006. "Setting the Record Straight: More than 52,000 Europeans Died from Heat in Summer 2003." Earth Policy Institute. Web article, July 28, 2006. http://www.earth-policy.org/Updates/2006/Update56.htm.

Maccini, S., and D. Yang. 2008. "Under the Weather: Health, Schooling, and Socioeconomic Consequences of Early-Life Rainfall." Mimeo. University of Michigan.

Magrin, G., C. Gay García, D. Cruz Choque, J. C. Giménez, A. R. Moreno, G. J. Nagy, C. Nobre, and A. Villamizar. 2007. *Latin America. Climate Change 2007: Impacts, Adaptation and Vulnerability.* Contribution of Working Group II to the Fourth Assessment Report of the Intergovernmental Panel on Climate Change, ed. M. L. Parry, O. F. Canziani, J. P. Palutikof, P. J. van der Linden and C. E. Hanson. 581–615. Cambridge, U.K.: Cambridge University Press.

Malaney, P., A. Spielman, and J. Sachs. 2004. "The Malaria Gap." *American Journal of Tropical Medicine* 71 (suppl. 2): 141–46.

Mata, L., and C. Nobre. 2006. "Impacts, Vulnerability, and Adaptation to Climate Change in Latin America." Background paper commissioned by UN Framework Convention on Climate Change, Lima, Peru.

Moreno, Ana Rosa. 2006. "Climate Change and Human Health in Latin America: Drivers, Effects, and Policies." *Regional Environmental Change* 6: 157–64.

PAHO (Pan-American Health Organization). 1995–2006. *Numbers of Reported Cases of Dengue and Malaria.* Washington DC: PAHO.

———. 2008. *Numbers of Reported Cases of Dengue and Dengue Hemorrhagic Fever (DHF) in the Americas, by Country: Figures for 2008.* Washington DC: PAHO.

Parry, M. L., C. Rosenzweig, A. Iglesias, M. Livermore, and G. Fischer. 2004. "Effects of Climate Change on Global Food Production under the SRES Emissions and Socioeconomic Scenarios." *Global Environmental Change* 14: 53–67.

Patz, Jonathan A., and John M. Balbus. 1996. "Methods for Assessing Public Health Vulnerability to Global Climate Change." *Climate Research* 6: 113–25.

Patz, Jonathan A., and R. Sari Kovats. 2002. "Hotspots in Climate Change and Human Health." *BMJ* 325: 1094–98.

Sachs, Jeffrey, and Pia Malaney. 2002. "The Economic and Social Burden of Malaria." *Nature* 415: 113–25.

Sardon, J. P. 2007. "The 2003 Heat Wave." *Eurosurveillance* 12, no. 3 (March): 694. 01 European Centre for Disease Prevention and Control, Stockholm. http://www.eurosurveillance.org/ViewArticle.aspx?ArticleId=694.

Stern, Nicholas. 2007. *Stern Review on the Economics of Climate Change.* Cambridge, U.K.: Cambridge University Press.

Strauss, J., and D. Thomas. 1995. "Human Resources: Empirical Modeling of Household and Family Decisions." In *Handbook of Development Economics*, vol. 3A, ed. J. Behrman and T. N. Srinivasan, 1883–2023. Amsterdam, New York, and Oxford: Elsevier Science.

———. 1998. "Health, Nutrition, and Economic Development." *Journal of Economic Literature* 36 (2): 766–817.

Torres, Jaime, and Julio Castro. 2007. "The Health and Economic Impact of Dengue in Latin America." *Cadernos de Saúde Pública* 23 (suppl. 1): 23–31. Rio de Janeiro. doi: 10.1590.

UNDP (United Nations Development Program). 2007. *Fighting Climate Change: Human Solidarity in a Divided World, Human Development Report 2007/08.* New York: Oxford University Press.

Van den Berg, G. J., M. Lindeboom, and F. Portrait. 2007. *Long-Run Longevity Effects of a Nutritional Shock Early in Life: The Dutch Potato Famine of 1846–1847.* Discussion Paper 3123, Institute for the Study of Labor (IZA), Bonn, Germany.

Warren, R., N. Arnell, R. Nicholls, P. Levy, and J. Price. 2006. *Understanding the Regional Impacts of Climate Change.* Tyndall Center for Climate Change Research Working Paper 90, Norwich, U.K.

Wolpin, K. I. 1997. "Determinants and Consequences of the Mortality and Health of Infants and Children." In *Handbook of Population and Family Economics*, vol. 1A, ed. M. R. Rosenzweig and O. Stark, 483–557. Amsterdam: Elsevier Science.

WHO (World Health Organization). 2003. *Climate Change and Human Health.* Geneva: WHO.

———. 2005. *Using Climate to Predict Infectious Disease Epidemics.* Geneva: WHO.

———. 2008. *Dengue and Dengue Hemorrhagic Fever.* Fact Sheet No. 117. Geneva: WHO.

World Bank. 2008. *Environmental Health and Child Survival: Epidemiology, Economics, Experiences.* Washington, DC: World Bank.

# CHAPTER 7

# Migration and Climate Change

## Lykke Andersen, Lotte Lund, and Dorte Verner

Estimates of the number of global climate change migrants by the year 2050 range widely, from 25 million to 1 billion people. The range of estimates illustrates the difficulty of forecasting migration in response to climate change in Latin America and the Caribbean, or indeed anywhere (Brown 2008). The decision to migrate is a complex one involving both push and pull factors and many different trade-offs. And even if a family wants to migrate, it may not be able to, because of lack of resources, poverty, or legal constraints. This complexity makes it difficult to assess how many migrants can be attributed to a single cause, such as climate change.

Nonetheless, by using climate change projections it is possible to indicate which areas in Latin America and the Caribbean (LAC) are most prone to extreme events—such as droughts, floods, landslides, and windstorms—that will create a strong potential for migration. Clearly the livelihoods of many communities across the region—from the poor rural areas to the city slums, from the mountains to the coasts—are under threat, making migration the only option for finding alternative livelihoods (table 7.1).

"Human migration" denotes any movement by humans from one locality to another, sometimes over long distances or in large groups. Push and pull factors are those factors that either push people forcefully into migration or attract them. Climate change is typically considered a push

**Table 7.1  Hot Spots for Selected Climate Change Events with Potential to Spur Migration**

| Climate event | Hot spots with intensified climate events | Type of migration | Communities or distinct groups at risk | Migration potential |
|---|---|---|---|---|
| Drought | • Argentina, Brazil, Chile, Paraguay, Uruguay | • Labor migration.<br>• Typically temporary, but also permanent. | • Landless rural households. | • High for landless rural households— mostly young men, often also whole families.<br>• Low for poorest of poor.<br>• Examples of planned population movement to more fertile areas in Mexico. |
| Floods and landslides | • Bolivia, Ecuador, Peru, and Uruguay<br>• Many Latin American cities<br>• Small island states and low-lying deltaic areas | • Distress migration.<br>• Typically temporary. | • City slum communities and households.<br>• Rural communities and households.<br>• Coastal communities and households. | • High for poor city dwellers, rural households, coastal households; typically whole families and communities going to relief sites. |
| Hurricanes and tropical wind storms | • Central America, Northwestern Latin America, the Caribbean | • Distress migration.<br>• Typically temporary. | • Coastal communities and households. | • High for coastal households; typically whole families and communities going to relief sites. |
| Sea level rise | • South American and Caribbean coastal areas<br>• Large coastal cities<br>• Caribbean islands, especially the Bahamas | • Labor migration.<br>• Managed retreat. | • Coastal communities and households. | • In the long term, high potential for "managed retreat."<br>• Typically whole communities will be forced to move. |
| Extreme temperatures and glacial melt | • Andes | • Labor migration.<br>• Permanent. | • Primarily poor indigenous communities and households. | • High potential for indigenous communities and households. |

*Sources:* Christensen 2008; WWF 2008.

factor, through causing increased numbers of extreme events (droughts, floods, hurricanes) that negatively affect people's livelihoods. But a person's decision to leave home is rarely based solely on climate change (IOM 2008). Except in cases of sudden-onset climate events, which cause people to flee for their lives, migration also requires a pull factor, be it environmental, social, or economic. From an economic perspective, migration—including climate-induced migration—can be viewed as a response to economic opportunity: people migrate seeking higher returns on their individual assets (Fiess and Verner 2003).

Migration is also an important adaptation mechanism. Assunção and Feres (2008a) undertook simulations of climate change effects on agricultural productivity and hence on poverty levels in Brazil. They found an average effect of climate change on Brazilian poverty levels of 3.2 percentage points for the period 2030–49, reduced to 2 percentage points if sectoral and geographic labor mobility is allowed for. In other words, migration can substantially reduce the potential impacts of climate change on poverty. However, the poorest of the poor, who lack the resources and social ties enabling migration, may be left in a downward spiral of worsening poverty, a pattern that is replicated across the world.

As climate change and variability threaten an unknown number of mostly poor people in developing countries, migration is becoming a top priority issue in the discourse on climate change adaptation. At the same time, the debate is changing. The Intergovernmental Panel on Climate Change (IPCC) now recognizes that *physical* vulnerability constitutes only one aspect of a person's vulnerability to environmental hazards, and that economic, political, and social vulnerability at the individual, community, and national levels must feature in the assessment. The discussion is, however, made more difficult by the lack of an internationally accepted definition of "forced climate migration" or "environmental refugee."[1]

This chapter first reviews existing migration and urbanization patterns in Latin America and the Caribbean, in general and in relation to climate change and variability. The second section introduces a typology of potential climate-related migration, which is then used as the basis for an analysis of social implications of climate change–induced migration using the sustainable livelihoods framework (SLF). Then two estimations of climate change–induced migration are undertaken—one for internal migration within Bolivia and the other for international migration between Mexico and the United States—to show how it is possible to quantify the amount of migration that can be attributed to past, present, or future climate change.

## Migration Patterns

Migration, internal as well as intraregional and international, plays an important role in Latin America and the Caribbean, which had the largest number of migrants of any developing region in 2000–05 (World Bank 2008). As of 2005, some 25 million Latin American and Caribbean migrants were abroad, representing more than 13 percent of migrants worldwide (ECLAC 2006) and about 4 percent of the region's total population of 559 million.

Since the 1950s the net emigration rate from the LAC region has steadily risen, but it is expected to peak at approximately 1.1 million per year between 2000 and 2010. Subsequently it is projected to fall to close to 600,000 per year by 2020 and to stabilize at that level until 2050, which is the horizon for the current forecasts (figure 7.1).

Migrant remittances to LAC countries have been growing steadily, in absolute terms (figure 7.2) and as a share of the region's gross domestic product (GDP), reaching 2 percent in 2006 (ECLAC 2006). By 2006, remittances to LAC countries, at US$68 billion, were surpassed only by those into Asian countries, at US$114 billion (IADB 2007). However, the economic crisis that gripped the world in 2008 led remittances to Latin America and the Caribbean to remain stagnant in the third quarter of 2008, and the fourth quarter saw a decline for the first time since tracking of them began in 2000 (IADB 2009).

**Figure 7.1    Net Emigration, Latin America and the Caribbean, 1950–2050**

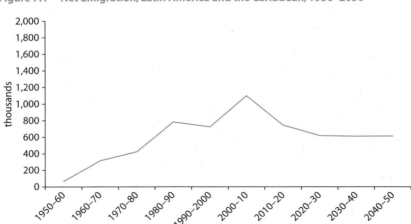

*Source:* United Nations Population Division 2009.

**Figure 7.2    Growth in Remittances Received in Latin America and the Caribbean, 1997–2006**

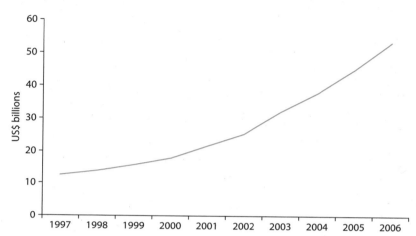

*Source:* Authors' elaborations based on data from World Bank, *World Development Indicators,* various years.

Internal rural-to-urban migration, related to industrialization, has been a driver of the high level of urbanization in the LAC region. Whereas in 1950 only around 40 percent of the region's population was urban, more than three-quarters was urban by the turn of the millennium, and the figure is now nearing 80 percent (Cerutti and Bertoncello 2003; United Nations Population Division 2000 and 2007).

The rate of internal migration is now unexpectedly falling, however (ECLAC 2007). A possible reason, which will require further research, is the replacement of internal migration by international migration, as seems to have happened in countries such as Haiti, Dominican Republic, Ecuador, or Bolivia (Morales 2008). In all these countries, degradation of natural resources, desertification, and extreme climatic events have caused further impoverishment and increasing pressure on land still capable of sustaining livelihoods. As a result, the benefits from internal migration are likely to diminish and make the economic opportunities of international migration all the more alluring. However, the economic crisis that spread across the world in late 2008–09 will likely cause at least a temporary decline in international migration.

Temporary migration as an adaptive response to climate stress is already happening in many areas of Latin America and the Caribbean. Long-term or repeated droughts and other climate-related events have already made many of the region's rural emigrants environmental, as well as economic, refugees. No official statistics are available to show how many migrants in the region are moving in response to climate-related or environmental factors, but the link seems to be obvious and is indicated in the apparently strong interconnection between loss from natural disasters and remittances received (figure 7.3 and figure 7.4).

Extreme climate events may trigger migration, as suggested by Murrugarra and Herrera (2008). Their study shows that the migration decision is connected to the migrant's asset base. The more social and financial assets are available (access to money, family, networks, and contacts at the destination), the farther the migrant is likely to travel, and the higher the person's level of education, the more destinations and economic opportunities are available to him or her. The exception is land.

**Figure 7.3    Remittances Received during Natural Disasters, 1970–2007**

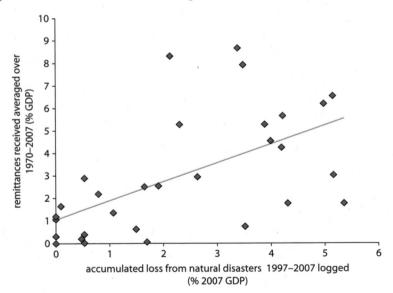

*Source:* Authors' elaboration.

*Note:* A strong correlation appears between loss from natural disasters and remittances received. In this graph, the authors have accounted for the size of the economy in both variables. When a disaster strikes, a high level of remittances received, in relation to GDP, could be caused by an increase in remittances or by an economic contraction caused by the natural disaster, turning remittances into a higher *proportion* of GDP even if *nominal* amount remains unchanged. Authors cannot say anything about the relative contribution of these two causes.

**Figure 7.4    Nicaragua: Natural Disaster Impact on Remittances Received as Percentage of GDP (Indexed)**

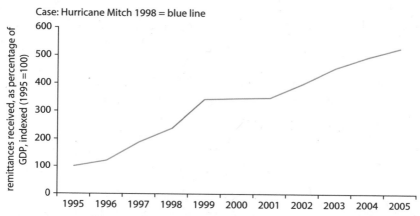

Case: Hurricane Mitch 1998 = blue line

*Source:* Authors' elaboration.

*Note:* It has been difficult to find countries where there is a discernible effect on remittances. In the case of Nicaragua, one could say that an increase can be observed at the time of Hurricane Mitch in 1998. The remittances (as percentage of GDP) shown on the y-axis have been indexed (1995 = 100). The explosive increase in remittances covers up small variations caused by natural disasters, many of which are local. Thus even if local populations are very dependent on remittances it might not show up in aggregate statistics.

Because land is not a mobile asset, landowners are to a large extent tied down by their land, especially if they do not have a title and therefore cannot easily sell the land at an acceptable price. Landowners and their families seem to be willing to stay in place longer than households with limited access to land and other physical assets.

Most of the region's migrants are young men, but the region now also has one of the world's highest migration rates for women (Deshingkar and Grimm 2004). In some Latin American countries, access to land is so limited that nearly all the poor young people in rural areas view migration as their main, and perhaps only, livelihood option. Across the region during 1988–2003, 15.5 million rural young people (aged 15–29) became part of the urban population. This translates into a reduction of one-third of the rural population of this age group (table 7.2).

## A Typology of Climate Change–Induced Migration

Climate-driven migration can broadly be divided into three categories: labor migration, distress migration, and planned population movement. It may be temporary or permanent.

*Climate-induced labor migration* is an adaptive response, typically undertaken as a last resort, by households confronted by climate-related stress

**Table 7.2    Youth and Rural-Urban Migration, 1985, 2000, 2015**

| Rural population | | | | | |
|---|---|---|---|---|---|
| Age groups | 1985 | | 2000 | | 2015 |
| 0 to 14 years | 53,991,606 | | 47,444,897 | | 40,576,184 |
| 15 to 29 years | 32,367,343 | −39.7% | 32,574,098 | −36.1% | 30,317,944 |
| *Urban population* | | | | | |
| Age groups | 1985 | | 2000 | | 2015 |
| 0 to 14 years | 94,498,537 | | 113,051,382 | | 122,264,802 |
| 15 to 29 years | 78,484,724 | +16.5% | 110,067,008 | +11.1% | 125,625,565 |
| *Total population* | | | | | |
| Age groups | 1985 | | 2000 | | 2015 |
| 0 to 14 years | 148,490,143 | | 160,496,279 | | 162,840,986 |
| 15 to 29 years | 110,852,067 | −3.9% | 142,641,106 | −2.8% | 155,943,509 |

*Source:* Cerutti and Bertoncello 2003.

such as gradual or chronic drought. Other types of gradual climate-related stress include sea level rise, salinization of agricultural land, desertification, growing water scarcity, and food insecurity. Gradual environmental change will usually, at least at first, result in mainly internal rather than international migration flows.

Some examples include the flow of migrants pushed by drought and the resulting desertification from Brazil's and Argentina's northeast regions to the state capitals and the south-central regions of these countries. In Bolivia, Chile, the Dominican Republic, Ecuador, Haiti, Mexico, and Peru, most migrants move from degraded areas to the main cities, to provincial or state and national capitals, or abroad (Morales 2008). In Bolivia, rural-urban migrants, driven by desertification, have moved in large numbers to the city of El Alto. Lima, Peru, has also grown exponentially because of the flow of environmental migrants and is joining the ranks of the world's megacities. Other countries, such as El Salvador, see a larger proportion of their migratory flows going to neighboring countries and the United States.

*Climate-induced distress migration* is a result of natural disasters. These migrant flows comprise large numbers of distressed people seeking aid until they may be able to return to their homes. The characteristics of distress migration differ within and across countries, as they are shaped by the severity and geography of a crisis, the ability of individual households to respond, evacuation opportunities, existing and perpetuating vulnerabilities, available relief, and intervening government policies (Raleigh, Jordan, and Salehyan 2008). But distress migration has two common and important features. First, the distress migrants are often seeking internal rather than international relocation, and second, their migration is typically temporary, not permanent. If permanent migration is the result, that is seen to reflect the state's deficient response rather than the natural disaster itself (Oliver-Smith 2004). It should also be noted, that distress migrants are typically disempowered and poorly positioned to negotiate the terms of their displacement.

A typical example of distress migration took place when Hurricane Mitch struck Central America in 1998. It was the deadliest Atlantic hurricane since the Great Hurricane of 1870. Honduras evacuated 45,000 citizens on the Bay Island. The government of Belize issued a red alert and asked citizens on offshore islands to leave for the mainland. Much of Belize City was evacuated. Guatamala issued a red alert as well. By the time Mitch made landfall those evacuated along the western Caribbean coastline included 100,000 in Honduras, 10,000 in Guatemala, and 20,000 in the Mexican state of Quintana Roo. Still, nearly 11,000 people were killed and more than 11,000 were still missing by the end of 1998. In all, 2.7 million were left homeless or missing. The flooding caused extreme damage, estimated at over US$5 billion (1998 dollars; US$ $6.5 billion in 2008 dollars).[2]

*Climate-induced planned population movement*, also called "managed retreat," or the progressive abandonment of land and structures in highly vulnerable areas and resettlement of inhabitants, is frequently mentioned in reference to erosion and sea level rise, but it can also be used as a policy instrument in areas with increasing and repeated climate events such as droughts. One of the few examples is the Plan Puebla Panamá, discussed in the following section of this chapter.

## Social Implications of Climate Change–Induced Migration

As noted above, studies show that to migrate in response to climate processes typically requires access to money, family networks, and contacts at the destination (IOM 2008). Even in the most extreme unanticipated natural disasters, migrants who have any choice tend to travel along preexisting paths, to places where they have family, support networks, and historical ties (Murrugarra and Herrera 2008). Like traditional migrants, climate migrants are generally positively selected by education and skills. Migration benefits them and their families by allowing the redistribution of labor toward more productive occupations (Bravo 2008). But migration can also have a strong negative social impact. The poorest of the poor may not have the resources to migrate, and thus they risk being trapped in a downward spiral of poverty.

An indication of the positive and negative social implications of the different categories of climate change–induced migration can be achieved by using the SLF (introduced in chapter 1) to look at the impact on people's assets, interpreting migration as a livelihood strategy. That is done in general terms below; an applied analysis of this kind should be adapted to the circumstances at hand.

*Labor migration.* Climate change–induced labor migration has great social implications, comparable to those of traditional types of labor migration, for both the migrant and the family left behind. Climate-induced migration is expected only to take place as a last resort, when all other adaptation measures have proved insufficient.

Migration can provide several benefits for migrants and their families. The most important is normally an increase in financial assets, as work in the new location often pays a higher wage. A large share of households in poor regions receive income from relatives abroad; for example, 31 percent of all households in drought-ridden northeast Brazil frequently receive some form of remittances (Lemos and others 1997). Remittances to the family left at home will raise the household's income, perhaps improving food security and access to health care, or allowing children to go to school instead of working. Physical assets such as housing or livestock may also increase. With a higher income, the household may obtain access to financial services. In Catamayo, Ecuador, for example, three credit and savings cooperatives and one bank offer basic financial services to recipients of remittances (Orozco 2006). Migrants' new environment may also offer them an opportunity to increase their knowledge through new work experiences and training, and even new language skills, and to improve their health as the result of better food and access to health services. Migrants may also be able to gain access to new personal networks and formalized groups, which can also act as safety nets.

But climate migrants also face a series of risks. The cost of migration (travel, food, housing, and so forth) can be so unexpectedly high that it ends up worsening the family's financial situation. Migrants' working and housing standards can be so poor as to damage their health. Migrants without networks in place face the risk of social isolation. They may be discriminated against and even be the cause of conflict. And migrants' ties to their family, networks, and formalized groups at home may deteriorate while they are away. Children may suffer if one or both parents are away from the family for a long time.

Women left behind in charge of the household often face stigmatization and breakdown in their social networks (Orozco 2006). And losing a valuable member of the household can lead to poor maintenance of the household's physical assets.

*Distress migration.* The social implications of climate-related distress migration are by nature almost all negative. In their community of origin, people displaced by natural disasters suffer serious damage to their physical assets, such as housing and livestock, and usually to natural resources on which they depend. The natural disaster may also damage their health, for example, as the result of lack of access to clean water and sanitation in the aftermath of the event (see chapters 3 and 6 for further discussion). Financially, a disastrous climate event can cost a household all its worldly possessions—especially in the case of poor households with no insurance. At the new destination, there is risk of conflict with the receiving population.

Evidence shows that distress migrants return to their home areas at a remarkably high rate (Suhrke 1993). Following earthquakes in Nicaragua in 1972 and Guatemala in 1976, researchers found a 90 percent population retention rate in both damaged and undamaged areas, indicating that the permanent emigration rate from disaster-affected communities may be similar to overall emigration rates and not driven by the natural hazard (Belcher and Bates 1983; Perch-Nielsen 2004). Studies of post-disaster migrants and nonmigrants in Guatemala and the Dominican Republic found that people's intent to stay in their villages was not related to the damage they had experienced but rather to the type of work they used to do. Specifically, those working in coffee plantations decided to move, as their economic future looked bleak. Further, people who had invested more in their home area were less likely to move (Belcher and Bates 1983; Morrow-Jones and Morrow-Jones 1991; Perch-Nielsen 2004).

*Planned population movement.* Planned population movement is a policy instrument intended to benefit particular communities that are victims of, and often at the same time causes of, environmental degradation. Mexico's Plan Puebla Panamá is designed to assist small and isolated communities by relocating them to rural towns. The social argument is that establishing rural towns will facilitate the provision of public services to a higher proportion of the population in the targeted areas. The plan makes it a condition for the farmers who participate that they abandon their traditional crops, such as corn and beans, and replace them with crops that are in high demand in international markets, such as oranges and palm oil. The productive conversion should increase the households' income, as well as their knowledge regarding sustainable land use. However, abandoning staple crops completely in favor of cash crops can leave the farmers vulnerable to volatility in world markets. Families are expected to benefit from improved health services, electricity, information technology, schools, child care, and other services, which were much less available in their old settlements. More general experience with resettlement schemes, for example, those associated with big dam projects, shows that when done properly and inclusively, planned population movement can avoid the danger of sparking potential conflict with and within the communities. Still, a risk exists that the households and their members may lose their traditions and spiritual connection to their land. This is a risk especially for indigenous peoples (box 7.1).

---

**Box 7.1**

## Indigenous Peoples and Induced Migration

Estimates are sketchy of the amount of climate change–related migration taking place among the indigenous peoples of Latin America and the Caribbean. Very few empirical studies exist on the nexus between climate change and migration among these peoples. As with other vulnerable groups, the conditions that drive them to migrate are diverse and are linked to access to economic, cultural, and social capital (Perch-Nielsen, Bättig, and Imboden 2008). Indigenous groups such as the Misquito Indians of Nicaragua's Caribbean coast have access to natural resources, such as lobster and shrimp, that command high international prices, and so only as a last resort do young Misquito Indians migrate to help sustain their families and communities by taking temporary jobs on Caribbean cruise ships. But for many indigenous communities across the region, migration has gradually become essential to subsistence (Altamirano, Hirabayashi, and Albó 1997). In impoverished indigenous communities, men migrate seasonally to work in construction or other nonskilled jobs, leaving women in charge of children and domestic work. The nonfarm income that men earn is invested in crops, domestic animals, and occasionally in children's education.

*(continued)*

**Box 7.1** *(continued)*

Traditionally young males have migrated, domestically or abroad, with their home communities' approval, and over the past two decades rural indigenous women have joined the flow, typically also supported by social and family networks (Herrera and Ramirez 2008). But migrant indigenous women typically have less-positive migration experiences than men: in the receiving communities they are vulnerable and disadvantaged by discrimination, lack of previous cross-cultural experiences, illiteracy, and language barriers (Torres and Carrasco 2008). Because of these barriers, low-wage employment in the informal sector is typically their only option. Kniveton and others (2008) found that indigenous communities' migratory behavior, choices of destination, social networks, and strategies show that migration involves power inequalities within families. It can therefore be expected that climate change will affect indigenous women disproportionately.

Indigenous people are extremely attached to their communities and territories. If forced to leave their land by extreme climate events, they tend to return as soon as basic conditions are reestablished. In many cases, they are so reluctant to abandon their homes when hit by climate events that the result is an increased number of fatalities. Evidence shows that when indigenous people do migrate, they are more likely to thrive if they can maintain or reestablish links to their ancestral communities and territories, preserving their social bonds and cultural identity (CEPAL 2006). The recent rapid spread of communications technology, even into remote rural areas, has made these links much easier to maintain.

*Source:* Author's compilation, Kronik and Verner 2010.

## Quantitative Analysis of Climate-Induced Migration in Bolivia and Mexico

Given the complexity of decisions about migration, even the order of magnitude of the migration flows that climate change may induce in the LAC region as a whole is not known. But to show how it is possible to quantify the amount of migration that can be attributed to past, current, or future climate change, this section presents two sample estimations of climate-induced migration: one for internal migration within Bolivia and the other for international migration between Mexico and the United States. Both estimations use municipality-level data to estimate the relationships between climate and migration patterns and subsequently use the estimated relationships to assess how climate change would affect those patterns.

Since climate change is rarely the only reason for migrating, or even the main one, it is difficult to identify who is a climate change migrant and who is not, and thus it is difficult to actually count the number of climate change migrants. However, econometric methods allow researchers to quantify the marginal contribution of climate change to migration flows, and so get a rough idea about how migration patterns are affected by climate change. The models used for the analysis are set out in appendix C.[3]

### Climate Change and Internal Migration in Bolivia

Most climate-induced migration is likely to take place within countries, so this analysis estimates how climate affects migration patterns within Bolivia, which contains almost every conceivable climate, from glaciers and deserts to tropical rainforest. The data used to estimate the migration models come from the country's 2001 population census, which asked everybody about their current place of residence, as well as their residence five years earlier. These data were used to estimate a complete, municipality-to-municipality migration matrix.

Because Bolivia had 314 municipalities at the time of the census, a complete municipality-to-municipality migration matrix comprises $314 \times 313 = 98,282$ entries. For each entry, the migration rate is calculated. Three different models are estimated. The first is a model of out-migration and includes only explanatory variables from the municipality of origin (municipality $i$). The second is a model of destination choice and includes variables from the destination municipality (municipality $j$), as well as the distance between origin and destination (Distance$_{ij}$). The third model explains specific migration rates and includes variables about both origin and destination. All variables that were not statistically significant at the 95 percent level were automatically deleted.[4]

***Determinants of out-migration.*** The estimations found that out-migration rates vary substantially with both temperature and rainfall, in a nonlinear fashion (table 7.3). Since the average out-migration rate is 83 per 1,000 inhabitants over 5 years of age (with a standard deviation of 37), temperature alone explains a substantial part of the variation in rates (54 percent of the total explained variation, according to Fields' decomposition).

The relationship with temperature implies that if all other factors were held constant, out-migration rates would be high in cold and hot municipalities, but low in municipalities with average temperatures around

**Table 7.3    Bolivia: Regression Findings for the Three Migration Models**

| Explanatory variables | Model 1: Out-migration | Model 2: Destination choice | Model 3: Migration rates |
|---|---|---|---|
| Constant | 105.385 | 12.845 | 1.484 |
| | (6.01) | (13.65) | (13.76) |
| Distance$_{ij}$ | | −14.771 | −1.687 |
| | | (−47.73) | (−51.42) |
| *Characteristics of sending municipality* | | | |
| Average temperature$_i$ | −10.163 | | −0.107 |
| | (−4.21) | | (−12.48) |
| (Average temperature$_i$)$^2$ | 0.401 | | 0.004 |
| | (4.73) | | (14.26) |
| Average rainfall$_i$ | −51.864 | | −0.175 |
| | (−3.15) | | (−7.62) |
| (Average rainfall$_i$)$^2$ | 12.720 | | |
| | (2.59) | | |
| Per capita consumption$_i$ | 0.268 | | 0.002 |
| | (2.97) | | (5.34) |
| (Per capita consumption$_i$)$^2$ | −0.0004 | | −0.000003 |
| | (−2.21) | | (−4.39) |
| Share in secondary and tertiary sectors$_i$ | | | −0.370 |
| | | | (−2.42) |
| (Share in secondary and tertiary sectors$_i$)$^2$ | 55.2970 | | 0.756 |
| | (3.67) | | (4.59) |
| Urbanization rate$_i$ | | | −0.197 |
| | | | (−4.12) |
| Education level$_i$ | | | |
| Municipal spending per capita$_i$ | −0.171 | | 0.001 |
| | (−2.24) | | (4.98) |
| Population density$_i$ | −0.410 | | −0.003 |
| | (−2.92) | | (−5.45) |
| Share of municipality with steep slope$_i$ | 0.512 | | 0.002 |
| | (5.09) | | (6.76) |
| Density of secondary roads$_i$ | | | |
| Oil concession$_i$ | | | −0.066 |
| | | | (−3.12) |
| Forestry concession$_i$ | | | 0.110 |
| | | | (4.66) |
| *Characteristics of receiving municipality* | | | |
| Average temperature$_j$ | | 11.105 | −0.119 |
| | | (−11.00) | (−12.98) |
| (Average temperature$_j$)$^2$ | | 0.045 | 0.005 |
| | | (12.76) | (15.11) |

*(continued)*

**Table 7.3    Bolivia: Regression Findings for the Three Migration Models** (continued)

| Explanatory variables | Model 1: Out-migration | Model 2: Destination choice | Model 3: Migration rates |
|---|---|---|---|
| Average rainfall$_j$ | | −1.9608 | −0.1497 |
| | | (−2.99) | (−6.32) |
| (Average rainfall$_j$)$^2$ | | 0.3831 | |
| | | (2.00) | |
| Population size$_j$ | | 0.9955 | 0.0870 |
| | | (84.26) | (78.59) |
| Per capita consumption$_j$ | | -0.0255 | −0.0021 |
| | | (−12.17) | (−11.45) |
| Education level$_j$ | | 0.4781 | 0.0472 |
| | | (6.04) | (6.57) |
| Municipal spending per capita$_j$ | | | 0.0022 |
| | | | (8.09) |
| Share in secondary and tertiary sectors$_j$ | | 6.7452 | 0.5533 |
| | | (10.42) | (9.21) |
| Pressure$_j^a$ | | −0.9509 | −0.0751 |
| | | (−5.31) | (−4.53) |
| Municipality area$_j$ | | −0.0382 | −0.0028 |
| | | (−3.17) | (−2.47) |
| Population density$_j$ | | 0.1710 | 0.0107 |
| | | (28.12) | (19.20) |
| Urbanization rate$_j$ | | | |
| Density of primary roads$_j$ | | | |
| Density of secondary roads$_j$ | | −0.0072 | −0.0005 |
| | | (−5.33) | (−4.23) |
| Share of municipality with steep slope$_j$ | | 0.0130 | 0.0013 |
| | | (3.41) | (4.03) |
| Oil concession$_j$ | | | |
| Forest concession$_j$ | | 0.8824 | 0.0877 |
| | | (3.43) | (3.68) |
| Number of observations | 312 | 98282 | 97032 |
| $R^2$ | 0.2648 | 0.2172 | 0.1959 |

*Source:* Authors' estimation.

*Notes:* The subscript *i*'s in the stub column refer to sending municipality information; the subscript *j*'s refer to receiving municipality information; *t*-values in parentheses.

a. Pressure is the ratio of population to total employment: an indication of the level of unemployment or lack of economic opportunities. It is expected to be positively related to out-migration rates and negatively related to in-migration rates.

11–14°C (figure 7.5). The difference in predicted out-migration rates between municipalities with the least desired temperature and the most desired is 82 points, corresponding to 2.2 times the standard deviation of out-migration rates.

**Figure 7.5    Bolivia: Temperature Sensitivity of Out-Migration Rates**

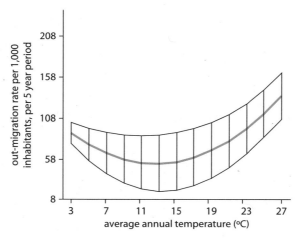

*Source:* Authors' elaboration.
*Note:* The green central line represents the point estimate from Model 1 of table 7.3. The thinner black lines delimit the 95 percent confidence interval as estimated by Stata's lincom command.

Out-migration rates are also sensitive to the amount of rainfall at the place of origin. People are more likely to leave municipalities that have very little or excessive rainfall. The optimal level of rainfall (to avoid losing population to migration) appears to be about 2 meters per year (figure 7.6), which is found in lowland rainforest areas.

Apart from the climate variables, the most important variable explaining out-migration is the share of the economically active population that is occupied in the secondary or tertiary sectors. Out-migration is low in municipalities dominated by farmers, but it rises exponentially as the share in the manufacturing and service sectors increases. The predicted difference in out-migration rates between the lowest and the highest observed share in the secondary and tertiary sectors is 55 points. This variable was originally meant to capture the level of development of the municipality; the results are contrary to expectations because they show that people are more likely to leave the most developed municipalities. The likely explanation for this result is that people working in the manufacturing and service sectors are much more mobile than farmers because farmers are tied to their land. This is particularly true if farmers do not have title to their land (as is the case for most subsistence farmers) and thus find it difficult to sell it at a reasonable price.

Out-migration rates also vary nonlinearly with per capita consumption levels. They are low in the richest municipalities, as expected, but they are even lower in the poorest. The reason is probably that people there are

**Figure 7.6    Bolivia: Rainfall Sensitivity of Out-Migration Rates**

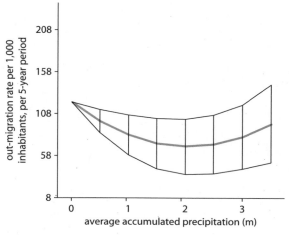

*Source:* Authors' elaboration.
*Note:* The green central line represents the point estimate from Model 1 of table 7.3. The thinner black lines
delimit the 95 percent confidence interval as estimated by Stata's lincom command.

simply too poor to shoulder the costs of migrating. Finally, the remaining
coefficients indicate that people are abandoning the most sparsely populated municipalities and the most rugged terrain, but municipal spending
can help to keep them in place.

***Determinants of the choice of location.*** Migrants clearly prefer destinations that are not too far from their place of origin. They also tend to
choose small municipalities with large populations and high population
densities, probably because these offer more economic opportunities
through economies of scale. They also prefer municipalities with relatively developed secondary and tertiary sectors and high levels of public
investment, and they avoid municipalities that offer few economic opportunities (as proxied by the variable *pressure*). Strangely, they avoid municipalities with high per capita consumption levels, possibly because these
have higher costs of living.

Migrants seem to be ambiguous about climatic conditions at the destination. There is a slight preference for high or low temperatures
instead of temperate climates, and for little rain, but the differences,
although statistically highly significant, are very small. The predicted in-
migration rate only varies by 10 points between the most preferred
temperature and the least preferred, and by 2 points between most and

least preferred amount of rainfall. That is not much compared to the standard deviation of 24 points.

*Determinants of migration rates from a specific municipality to another.* When the factors in both the sending and receiving municipalities are combined, so as to analyze specific migration rates, we find that all the estimated signs are consistent with the findings of the two previous models and that the interpretations are basically the same, though the relative importance changes.

The single most important factor explaining migration patterns (as judged by the Fields' decomposition) is the size of the population at the destination. Then follow the population density at the destination, the migration distance, and the education level at the destination. This finding suggests that people are moving to the nearest big city they can find. All the remaining variables, while statistically highly significant, add very little (less than 0.02) to the explanatory power of the model.

Thus, while there is some evidence that people are moving away from places with extreme climates, their adaptation strategy is less to find a place with a better climate than to find a place with economic opportunities that are less sensitive to climatic conditions (big cities).

*The impact of recent climate change (1958–2008) on migration in Bolivia.* According to data from meteorological stations across Bolivia, the country's highlands have experienced reductions in average temperatures of about 1°C over the last 50 years,[5] and the lowlands have experienced increases in average temperatures of about 0.25°C. Using the estimated cross-section relationship illustrated in figure 7.5, the authors simulated the effects of this recent climate change and found that the effect at the national level is about one extra migrant per 1,000 inhabitants per 5-year period (table 7.4). This implies that recent climate change is causing almost 2,000 people who would not otherwise have migrated to do so every year. This number should be compared with the total of about 135,000 people who migrate within Bolivia every year for a multitude of reasons.

The climate change migrants come mainly from the cold states of Potosí and Oruro, while the states of Beni and Tarija have become better at retaining their populations due to favorable changes in their local climates.

Figure 7.7 shows estimated changes in out-migration rates due to recent climate change, by municipality. Clearly, the poorest municipalities are experiencing the largest increases in out-migration rates due to climate

**Table 7.4    Bolivia: Effects of Recent Climate Change on Out-Migration Rates, by State (per 1,000 inhabitants)**

| State | Estimated effect of recent climate change |
| --- | --- |
| Beni | −5.35 |
| Chuquisaca | 0.81 |
| Cochabamba | 0.00 |
| La Paz | 1.07 |
| Oruro | 4.38 |
| Pando | 1.05 |
| Potosí | 5.63 |
| Santa Cruz | 0.00 |
| Tarija | −4.75 |
| Total | 1.05 |

*Source:* Authors' elaboration.

**Figure 7.7    Estimated Effects of Recent Climate Change on Out-Migration Rates, by Municipality in Bolivia**

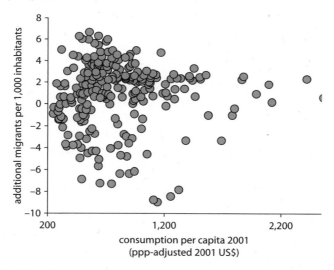

*Source:* Authors' elaboration.

change; in some of the very poorest, the rate is 6–8 persons per 1,000 inhabitants every 5 years.[6]

***The impact of future (2008–58) climate change on migration from Mexico to the United States.*** Climate change may also cause international migration, but the shortage of detailed data on international migration in Latin America and the Caribbean makes reliable estimation difficult. This section takes advantage of a migration intensity index that has been estimated for each of 2,350 municipalities in Mexico by the National Population Council of Mexico, using information from the 2000 population census (Consejo Nacional de Población 2002). The index integrates information about current migrants in the United States, migrants recently returned from the United States, and remittances received. The explanatory variables of principal interest are average annual temperature and annual rainfall, and as control variables, the literacy rate and urbanization rate in each municipality have been included. The calculations allow for nonlinear effects for all four variables, by including the square of each.

The regression results indicate that temperature is one of the most important variables explaining migration intensities in Mexico (table 7.5). It appears to be even more important than education, and much more important than income.

Figure 7.8 illustrates the estimated nonlinear relationships between climate and migration intensities. Contrary to expectations, migration intensities are maximized at moderate temperatures (around 20°C), even when controlling for other factors that also might affect migration rates.

Migration rates are also significantly related to average annual rainfall, being highest in municipalities with very little rain. This implies that a reduction in rainfall would tend to increase migration to the United States, whereas a general temperature increase would have ambiguous effects. It is therefore necessary to do a detailed, municipality-level simulation to see which effects dominate.

**Table 7.5    Mexico–United States: Municipal-Level Regression Explaining Migration Intensity, 2000**

| Explanatory variable | Coefficient estimate | t-value |
|---|---|---|
| Temperature | 0.2912 | 11.15 |
| Temperature$^2$ | −0.0073 | −11.21 |
| Rainfall | −0.2038 | −3.14 |
| Rainfall$^2$ | −0.0279 | −1.42 |
| Per capita income | 0.3438 | 0.95 |
| Per capita income$^2$ | −0.0257 | −1.25 |
| Literacy rate | 0.1135 | 7.33 |
| Literacy rate$^2$ | −0.0007 | −6.80 |
| Urbanization rate | 0.0034 | 2.07 |
| Urbanization rate$^2$ | −0.0001 | −5.66 |
| Constant | −7.3700 | −11.89 |
| $N = 2,350$ | $R^2 = 0.2714$ | |

*Source:* Authors' elaboration.

**Figure 7.8 Estimated Relationships between Temperature and Rainfall and Migration Intensity**

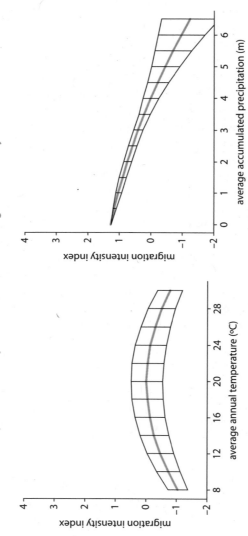

*Source:* Authors' elaboration.

*Note:* Green central lines represent the point estimates from table 7.5. The thinner black lines delimit the 95 percent confidence interval as estimated by Stata's lincom command.

According to the latest generation of IPCC models, during the next 50 years the coastal areas of Mexico are expected to warm less than the center, and especially less than the northern part of the central region (IPCC 2007). Here were simulated the effects of a 1.25°C warming in all coastal states, a 1.5°C increase in the central states to the south of Zacatecas, and a 1.75°C increase in the central states to the north of Zacatecas. Also simulated was a 5 percent reduction in rainfall across all of Mexico, as that is what the majority of IPCC models suggest.

Applying these expected climate changes over the next 50 years to the estimated relationships shown in figure 7.8 for each municipality in Mexico indicates, perhaps somewhat unexpectedly, that future climate change is likely to reduce migration to the United States slightly. Since this is a composite index, it is difficult to quantify the reduction exactly.[7] Nonetheless, these simulations suggest that no dramatic increase in Mexico-United States migration should be expected as the result of the gradual climate change that is expected over the next 50 years.

No reason is seen to assume that the simulation of Mexico-United States migration patterns is representative of future international migration patterns due to climate change for the rest of the LAC region: Great differences exist among countries and subregions in existing climatic conditions, levels of vulnerability, and projected climate changes. Also, this analysis completely ignores possible changes in extreme events, since researchers do not yet have sufficient information to be able to quantify such changes at the municipal level.

This chapter has shown how it is possible to quantify the amount of internal or international migration that can be attributed to past, current, or future climate change. The two examples indicate quite limited effects of climate change on migration, but to obtain an overall sense of future climate migration it would clearly be necessary to repeat this type of analysis for many more countries.

## Notes

1. The definition used here is that put forward by the International Organization for Migration (IOM): "Persons or groups of persons who, for compelling reasons of sudden or progressive changes in the environment that adversely affect their lives or living conditions, are obliged to leave their habitual homes, or choose to do so, either temporarily or permanently, and who move either within their country or abroad."

2. http://en.wikipedia.org/wiki/Hurricane_Mitch.

3. Note that the climate data used relate to average monthly temperatures and precipitation levels; increases in climate variability and severe weather events are not included, except to the extent that they are related to average temperature and precipitation. Specific effects of changes in glacier melt are not included either.

4. This was done by using the stepwise function in Stata.

5. This is driven by relatively large drops in nighttime temperatures, as summer- and daytime temperatures have risen.

6. In another municipality-based study of migration, in this case for Brazil, Assunção and Feres (2008b) found that high temperatures and low precipitation levels at the origin act as push factors, while lower temperatures and more precipitation at the destination act as pull factors. They also found the greatest effect when the distanced traveled is greater than 100 kilometers; for migrations of less than 100 kilometers from the origin hardly any climate effect is detectable.

7. For this reason it would be preferable to work with raw migration rates instead of composite indexes.

## References

Altamirano, T., R. Hirabayashi, and X. Albó. 1997. *Migrants, Regional Identities, and Latin American Cities.* Society for Latin American Anthropology Publication Series 13. Arlington, VA: American Anthropological Association.

Assunção, J., and F. C. Feres. 2008a. "Climate Change, Agricultural Productivity, and Poverty." Background paper for De la Torre, A., P. Fajnzylber, and J. Nash (2009), *Low Carbon, High Growth—Latin American Responses to Climate Change: An Overview.*" Washington, DC: World Bank.

———. 2008b. "Climate Migration." Background paper for De la Torre, A., P. Fajnzylber, and J. Nash (2009), *Low Carbon, High Growth—Latin American Responses to Climate Change: An Overview.* Washington, DC: World Bank.

Belcher, J. C., and F. L. Bates. 1983. "Aftermath of Natural Disasters: Coping through Residential Mobility." *Demography* 7.

Bravo, J. 2008. "Migration, Population Distribution, and Development." Presentation at Expert Group Meeting on Urbanization, Internal Migration, Population Distribution, and Development, New York, January 21–23, United Nations Population Division.

Brown, Oli. 2008. *Migration and Climate Change.* IOM Migration Research Series 31. Geneva: International Organization for Migration.

CEPAL (Comisión Económica para América Latina y el Caribe). 2006. *Panorama Social de América Latina 2006.* Publicaciones de las Naciones Unidas.

Cerutti, M., and R. Bertoncello. 2003. "Urbanization and Internal Migration Patterns in Latin America." Paper prepared for Conference on African Migration in Comparative Perspective, Johannesburg, South Africa, 4–7 June.

Christensen, J. H. 2008. "Physically Based Climate and Climate Change Related Issues on a Country Basis for Latin America." Manuscript. Denmark, Dansk Meterologisk Institut.

Consejo Nacional de Población. 2002. "Índice de Intensidad Migratoria México–Estados Unidos, 2002." December. Mexico City.

Deshingkar, P., and S. Grimm, S. 2004. *Voluntary Internal Migration—An Update.* London: Overseas Development Institute.

ECLAC (Economic Commission for Latin America and the Caribbean). 2006. "International Migration, Human Rights, and Development in Latin America and the Caribbean, Summary and Conclusions." Thirty-first Session of ECLAC, Montevideo, Uruguay, March 20–24.

———. 2007. "Social Panorama of Latin America 2007 on the Basis of Special Processing of Census Microdatabases." Santiago, Chile: Latin American and Caribbean Center (CELADE)–Population Division of ECLAC.

Fiess, N., and D. Verner. 2003. *Migration and Human Capital in Brazil during the 1990s.* World Bank Policy Research Working Paper 3093, World Bank, Washington, DC.

Herrera, G., and J. Ramirez. 2008. *America Latina Migrante: Estado, Familias, Identidades.* Quito, Ecuador: FLACSO–Ministerio de Cultura.

IADB (Inter-American Development Bank). 2007. "Migrant Workers Worldwide Sent Home More than US$300 billion in 2006, New Study Finds." News Release, Oct 17.

———. 2009. "IDB Sees Remittances to Latin America and the Caribbean Declining in 2009." News Release, March 16.

IPCC (Intergovernmental Panel on Climate Change). 2007. *Climate Change 2007: The Physical Science Basis.* Contribution of Working Group I to the Fourth Assessment Report of the IPCC. Geneva: IPCC.

Kniveton, D., K. Schmidt-Verkerk, C. Smith, and R. Black. 2008. *Climate Change and Migration: Improving Methodologies to Estimate Flows.* IOM Research Series Paper No. 33. Geneva: International Organization for Migration.

Lemos, M. C., D. Nelson, F. Finan, R. Fox, D. Mayorga, and I. Mayorga. 1997. "The Social and Policy Implications of Seasonal Forecasting: A Case Study of Ceará, Northeast Brazil." NOAA Report, Award # NA76GPO385.

Morales, C. 2008. "Desertification, Degradation, and Migration in Latin America and the Caribbean." Position paper presented at International Organization for Migration and UN Research Workshop on Migration and the Environment: Developing a Global Research Agenda, Munich, April.

Morrow-Jones, H. A., and C. R. Morrow-Jones. 1991. "Mobility Due to Natural Disaster: Theoretical Considerations and Preliminary Analyses." *Disasters*.

Murrugarra, E., and C. Herrera. 2008. "Natural Disasters and Migration Responses: Recent Evidence from Nicaragua." Manuscript. World Bank, Washington, DC.

Oliver-Smith, A. 2004. "Theorizing Vulnerability in a Globalized World: A Political Ecological Perspective." In *Mapping Vulnerability: Disasters, Development, and People*, ed. G. Bankoff and D. Hillhorst. London: Earthscan.

Orozco, M. 2006. *Between Hardship and Hope: Remittances and the Local Economy in Latin America*. Washington, DC: Multilateral Investment Fund, Inter-American Development Bank

Perch-Nielsen, S. 2004. "Understanding the Effect of Climate Change on Human Migration: The Contribution of Mathematical and Conceptual Models." Department of Environmental Studies, Swizz Federal Institute of Technology, Zürich.

Perch-Nielsen, S., M. B. Bättig, and D. Imboden. 2008. "Exploring the Link between Climate Change and Migration." *Climatic Change* 91: 375–93.

Raleigh, C., L. Jordan, and I. Salehyan. 2008. "Assessing the Impact of Climate Change on Migration and Conflict." Paper presented at Workshop on Social Dimensions of Climate Change, Social Development Department, World Bank, Washington, DC.

Suhrke, A. 1993. "Pressure Points: Environmental Degradation, Migration and Conflict." Cambridge, MA: American Academy of Art and Science.

Torres, A., and J. Carrasco, J. 2008. *Al filo del identidad. La migración indigena en America Latina*. Quito, Ecuador: FLACSO/UNICEF/AECID.

United Nations. 2007. Climate Change—An overview." Paper prepared by the Secretariat of the UN Permanent Forum on Indigenous Issues, UN Department of Economic and Social Affairs, New York.

United Nations Population Division. 2000 and 2007. *Revised Urbanization Prospects for LAC*. New York: United Nations.

———. 2009. *World Population Prospects: The 2008 Revision. Highlights*. New York: United Nations.

World Bank. 2008. *World Development Indicators, 2008*. Washington, DC: World Bank. http://go.worldbank.org/6HAYAHG8H0.

WWF (World Wildlife Fund). 2008. *Up in Smoke? Latin America and the Caribbean. The threat from Climate Change to the Environment and Human Development*. Third Report from the Working Group on Climate Change and Development.

# CHAPTER 8

# Conflict and Climate Change

## Olivier Rubin

This chapter investigates the potential effect of climate change on conflict in Latin America and the Caribbean (LAC). Conflicts profoundly
affect people's livelihoods. Physical assets may be eroded as infrastructure
becomes unsafe or is destroyed, as state capacity is undermined, and as
homes and basic services are damaged. Natural assets may deteriorate
through direct destruction, expropriation for military purposes, or confis-
cation by armed groups. Financial assets are also likely to decline, eroded
by job losses, high inflation, forced migration, and the potential collapse
of legal and economic structures. Collier and Hoeffler (2006) estimate
that it takes a country involved in a civil war an average of 21 years
to reach the gross domestic product (GDP) it would otherwise have
achieved. Human assets suffer from violent conflict as people are killed
and injured, schools are attacked, random violence abounds, people
engage in fighting rather than productive occupations, children suffer
mental traumas, and the numbers of orphans and widows increase. Social
assets may be adversely affected by conflict-induced mistrust, forced
migration, exacerbation of existing divisions, rape, stigmatization, and
exclusion.

Existing studies, reviewed in the first section below, find little evidence
of links between climate change and armed conflict. Many qualitative

and quantitative studies on both conflict and climate change, however, are lacking from a social perspective, so the subsequent section explores ways of improving such analyses. Those ways include using disaggregated rather than aggregated data; looking at dynamic effects of climate change and variability, as opposed to level effects; focusing on changes in opportunities rather than scarcity itself as a driver of conflict; considering cumulative and threshold effects of climate change and variability; and considering how climate change could be used as a cause to mobilize people for conflict, by laying blame at the door of the wealthy, whether domestically or internationally, rather than that of Mother Nature. Then, because people's livelihoods also depend crucially on the outcome of nonviolent conflicts, the third section of the chapter presents cases of nonviolent conflicts associated with environmental degradation. The fourth section focuses on socioeconomic factors that might channel environmental changes into conflict in the region. The chapter concludes by drawing implications for policy; it recommends addressing the more direct adverse socioeconomic outcomes of climate change and variability to reduce the scope for conflict.

Compared to other areas, the LAC region has a low risk of armed conflict,[1] and during the last 15 years the incidence of armed conflicts there has declined substantially. With the downfall of many military regimes in the 1980s, and continued economic integration (for example, through the Southern Cone Common Market [Mercosur] and the North American Free Trade Agreement [NAFTA]), the region has transformed itself from a zone of conflict formation to a security community with Brazil as the dominant, stabilizing force. Despite the region's overall stability, however, many small-scale conflicts continue. Internal armed conflicts (as opposed to interstate or international ones) account for 97 percent of the conflicts in the region, compared to the worldwide average of 78 percent. They mainly take place between government and paramilitary groups (Colombia), insurgents fighting for autonomy (Bolivia), or criminal groups and organized crime (Brazil and Mexico).

Table 8.1 identifies the most conflict-prone countries in Latin America and the Caribbean. Colombia, Guatemala, Peru, El Salvador, and Nicaragua account for more than 75 percent of the total years of conflict in the region, and the risk of climate change and variability playing a role in armed conflict in these countries might be greater than elsewhere.

**Table 8.1    Number of Years in Conflict during 1946–2006 in LAC Countries**

|  | Years | % | 1946–90 | 1991–2006 |
|---|---|---|---|---|
| Colombia | 41 | 25.2 | 25 | 16 |
| Guatemala | 33 | 20.2 | 28 | 5 |
| Peru | 22 | 13.5 | 12 | 10 |
| El Salvador | 15 | 9.2 | 14 | 1 |
| Nicaragua | 12 | 7.4 | 12 | 0 |
| Argentina | 7 | 4.3 | 7 | 0 |
| Cuba | 5 | 3.1 | 5 | 0 |
| Bolivia | 3 | 1.8 | 3 | 0 |
| Haiti | 3 | 1.8 | 1 | 2 |
| Paraguay | 3 | 1.8 | 3 | 0 |
| Suriname | 3 | 1.8 | 3 | 0 |
| Ecuador | 2 | 1.2 | 1 | 1 |
| Honduras | 2 | 1.2 | 2 | 0 |
| Mexico | 2 | 1.2 | 0 | 2 |
| Panama | 2 | 1.2 | 2 | 0 |
| Venezuela, R.B. de | 2 | 1.2 | 1 | 1 |
| Chile | 1 | 0.6 | 1 | 0 |
| Costa Rica | 1 | 0.6 | 1 | 0 |
| Dominican Republic | 1 | 0.6 | 1 | 0 |
| Grenada | 1 | 0.6 | 1 | 0 |
| Trinidad and Tobago | 1 | 0.6 | 1 | 0 |
| Uruguay | 1 | 0.6 | 1 | 0 |

*Source:* Author's compilation based on Uppsala/PRIO (2009) armed conflict dataset; http://www.prio.no/CSCW/Datasets/Armed-Conflict/UCDP-PRIO/.

## Links between Climate Change and Conflict

Climate change may increase the risk of conflict through several different channels:

- *Resource scarcities.* Climate change is likely to exacerbate resource scarcities in Latin America and the Caribbean, as described in previous chapters. Greater scarcity of food, water, forests, energy, and land could increase competition for the remaining resources, spurring internal unrest or even border conflicts.

- *Enhancing the likelihood of migration.* Households facing increased scarcities, rising sea level, and natural disasters may decide to migrate. With climate change and variability likely to perpetuate current trends of environmental degradation, ecological migration can be expected to become more prevalent (chapter 7). Studies suggest that international migration might destabilize the immigrants' destination countries, making them more prone to conflict.

- *Undermining state capacity.* Climate change and variability may undermine the capacity of the state in numerous ways: They may (a) raise the costs of infrastructure in remote rural areas, thus limiting the reach of the state. They may also (b) erode fiscal resources. They do that by requiring funds for the (unproductive) purpose of disaster management; by reducing the tax base, as natural degradation undermines production; by raising the amount of agricultural subsidies needed to maintain adequate food production; and by generally requiring increased spending for adaptation. If natural disasters inhibit the state's ability to provide security and a minimum of basic needs for affected populations, popular support and the legitimacy of the state may be weakened. This weakening of state power has the dual adverse consequence of limiting the state's developmental efforts (in turn providing incentives for challenging the state) and eroding the state's monopoly over the legitimate use of physical force (providing opportunities for challenging the state). Or, rather than actually undermining state capacity, climate change may simply expose existing state fragilities—which would also affect the incentives and opportunities for challenging the state or taking advantage of its exposed weaknesses.

- *Increasing the frequency and intensity of natural disasters.* The chaotic conditions that follow in the wake of natural disasters may provide opportunities to rebel groups to challenge the government's authority— an authority that may already have suffered because of inadequate disaster response. A power vacuum in the affected regions would provide recruitment opportunities for militant groups if people find that the opportunity costs of refraining from engaging in conflict become negligible.

Empirical studies provide only limited support for a link between climate change and an increasing risk of conflict. Whereas most research on climate change and variability is published in respected, peer-reviewed journals, the research that relates climate change to conflict consists largely of speculative scenarios published on the Web sites of political organizations and think tanks.[2] That is cause for concern.

Two main strands of academic research linking climate change and variability to conflict can be identified:

- Qualitative studies that analyze existing cases of environmental degradation and conflict to make projections
- Econometric studies based on quantitative data that examine the robustness and magnitude of key causal correlations to predict future developments

Under certain circumstances environmental degradation may increase the risk of conflict. But whereas environmental scarcity does not by itself increase the risk of conflict in isolation from other socioeconomic factors, it often acts as a catalyst, amplifying traditional causes of conflict such as historical grievances, ethnic divisions, and political differences.[3] Both qualitative and quantitative studies conclude that climatic factors can play a role in conflicts but that they do not constitute the main drivers of conflict. The qualitative analyses in the form of case studies find that most of the conflicts involving environmental factors have been intrastate and small in scale, and that although environmental scarcities could perpetuate tensions in society, other factors need to be present for violent conflict to break out.[4] Even if one deliberately selects cases that involve both conflict and environmental degradation (and thus disregards cases where ecological scarcity does not generate conflict, as well as cases in which it spurs increased cooperation), one finds no case that unequivocally points to environmental degradation as the main factor driving conflict.[5]

Systematic quantitative research yields even less evidence of a link between climate change and conflict. The pioneers of multivariate analyses of environmental factors and conflict, Hauge and Ellingsen (1998), concluded that land degradation, freshwater scarcity, population density, and deforestation all increased the number of years of civil war. However, their result was countered the very same year by Esty and others (1998), whose similar multivariate analyses found that environmental change did not appear to be directly linked to state failure. Urdal (2005) focused on scarcities in his cross-country logistical regressions covering the period 1950–2000 and found no robust evidence that demographic/environmental factors affected the risk of armed conflict. Scarcity considerations were also the focal point of the analysis by Binningsbø, Soysa, and Gleditsch (2006), which used the World Wildlife Fund's (WWF) ecological footprint measure in assessing the impact of environmental sustainability on the risk of violent conflicts. These authors report that the strongest "result is that the ecological footprint, perhaps the most widely accepted measure of man's use (or abuse) of ecoservices, is positively correlated with peace, a result that is stubborn and substantively large." A recent assessment of the empirical literature on climate change and conflict (Buhaug, Gleditsch, and Theisen 2008) highlights the lack of robustness by concluding that several of the earlier findings are either not replicable or do not hold with improved data.

Evidence indicates that climate change will have only a negligible impact on the risk of armed or violent conflict. The conclusions from the literature, combined with the LAC region's limited propensity for violent conflict, suggest that climate change will not significantly influence the risk of conflict. Even so, such an inference seems problematic. The effects of climate change and variability work through multiple and often inter-related channels, some of which have not yet received enough academic scrutiny. Whereas the effects of climate change are local, dynamic, and multidimensional, the existing studies rely strongly on aggregate-level effects of environmental scarcity on conflict. And whereas the focus has traditionally been on violent conflicts, a need also exists for analysis that encompasses nonviolent conflicts. The latter are more numerous, and the way in which they are resolved (inclusively and equitably, or by means that perpetuate tendencies of coercion, exclusion, inequality, and stigma-tization) affects social development outcomes.

In short, it seems premature to dismiss the potential threats that climate change and variability pose to national security and development. It is noteworthy that from a social development perspective many of the existing studies rest on misspecifications of the two variables of interest—climate change (the explanatory variable) and conflict (the dependent variable)—and thus risk seriously underestimating the impact of climate change on conflict in Latin America and the Caribbean. In the following section these two variables are recalibrated from a social perspective to obtain better approximations of the extent and nature of climate change–induced conflicts in the LAC region.

## Social Perspective on Climate Change in Conflict Analysis

Five aspects of climate change and variability are examined from a social development perspective: (a) the disaggregated impacts of environmental scarcities on conflict; (b) the dynamic effects of climate change on livelihoods; (c) the changes in opportunities as well as scarcities caused by climate change; (d) the cumulative and threshold effects of climate change; and (e) the prospects for using climate change as a cause for political mobilization.

### Disaggregated Analysis of Environmental Scarcities and Violent Conflicts

A need exists for disaggregated analysis of climate change and variability. Part of the lack of strong, unequivocal findings in quantitative studies of environmental scarcities and conflict can be attributed to the use of aggregate proxies based on data with limited reliability. The problem of working with aggregative research designs is that they do not reflect the fact that both environmental degradation and conflict often take place locally; the most common type of violent conflict, internal conflict, usually engulfs a few provinces or municipalities rather than entire states. A related problem with much of the quantitative evidence is a lack of a regional focus. Regional dummies are rarely included in the models, despite the fact that variations in climate change and variability (or other environmental factors) will have different impacts across different regions. Accordingly, policy recommendations might also differ across regions.

Disaggregated studies appear to support a more robust relationship between climate change and variability and conflict. In a recent article, Urdal (2008) reported a subnational analysis based on a time series of data for 27 Indian states, with the significant finding that scarcity of productive land is associated with higher risks of violent conflict, in particular when it interacts with high rural population growth and low agricultural yields. Another way of addressing the problem with aggregate data while still adhering to the principles of a cross-country design has been used by Raleigh and Urdal (2007). These authors include newly available geospatial disaggregated global data in logistic models regressing freshwater scarcity and land degradation on violent conflicts. Considering various socioeconomic control variables and interaction terms, they conclude that demographic and environmental variables do indeed have a moderate effect on the risk of civil conflict.

### Dynamic and Broad-Based Effects of Climate Change and Variability

A major weakness of the disaggregated analyses described above is their reliance on level data. Though rarely treated in the empirical literature, the dynamic effects of climate change on livelihoods are important.

From a conflict perspective, the change in key variables (such as economic decline or increasing inequality) might be more important than a stable state of deprivation (Nafziger and Auvinen 2002). For example, studies indicate that transient poverty is a stronger determinant of war and conflict than is chronic poverty (Goodhand 2003; Nafziger and Auvinen 2002). Discrepancies between what social groups expect and what they actually get are the most important drivers of conflict. Deprivation itself seldom produces strong grievances, but where people perceive a gap between the situation they believe they deserve and the situation they have actually achieved, they are more likely to feel aggrieved (Gleditsch and Urdal 2002). Increasing scarcity is therefore more dangerous than a situation of chronic scarcity, and fluctuating scarcity patterns that upset existing power relations could be even more conducive to conflict. This point is important because global warming changes environmental variables and increases their variability—which in turn cause power relations to shift and perceptions of fairness and envy to be recalibrated according to the new situation.

In addition, climate change leads to other impacts that alone or in combination with scarcities could produce disastrous outcomes that lead to conflict. One example of such a broader impact is the link between environmental migration and conflict. Theoretically, migrants can increase the risk of conflict at their destination by

- changing the religious and ethnic composition,
- increasing the competition for scarce resources,
- disrupting social networks in the receiving community,
- bringing in arms and enemy combatants,
- introducing new ideologies and ideas,
- providing a source of recruitment for military groups (refugee camps),
- bringing with them an antiauthoritarian and conflict-prone mindset,
- attempting to mobilize opposition directed at their country of origin, and
- taking advantage of the sheer scale of migration, generating disorder that both the refugees and residents can exploit.

The adverse effect of migration on the risk of conflict has been well documented. Although Urdal (2005) did not find evidence of a link between migration and conflict, both Salehyan and Gleditsch (2006) and Reuveny (2007, 2008) report that migration appears to increase the risk of conflict.[6]

## Beyond Ecological Scarcities: Changes in Opportunities

Future impacts of global warming on conflicts are likely to be triggered as much by changes in opportunities as by scarcities and grievances. Some of the most elaborate quantitative studies of conflict conclude that proxies for greed (or opportunity) all do well at explaining variation in civil wars between 1960 and 1990, while most proxies for grievances turn out to be insignificant (Collier and Hoeffler 1998, 2004; Collier, Hoeffler, and Rohner 2006).[7] For instance, the studies have shown that having a dispersed population or mountainous terrain appears to increase the risk of conflict, since these two factors work to the advantage of rebel groups that are capable of operating beyond the government's reach. The effects of climate change and variability could mimic these variables by undermining state logistics, making the terrain impassable, or dispersing vulnerable populations.

Relatively little research has been done on two other important questions. The first of these is the impact of climate change and variability on state institutions, even though many traditional, quantitative studies have found that those institutions have had significant effects on conflict. The second is the consequences to society of a large exogenous shock in the form of a natural disaster. Exogenous shocks to the economy in general have a destabilizing effect on societies, and little reason exists to think that shocks induced by climate change and variability should have any less effect. A quantitative analysis by Nel and Righarts (2008) produced the quite robust conclusion, despite problems with data, that natural disasters, in particular, geological ones, had a significant impact on violent civil conflict.

## Unknown Cumulative and Threshold Effects of Climate Change and Variability

It could be hypothesized that the natural events hitherto observed have not affected conflict because they have stayed within a moderate range. It may be that equilibria could become unstable if key environmental variables cross a tipping point, in terms of either direct consequences or indirect socioeconomic impacts of climate change.[8] To be prudent, such worst-case scenarios should be considered when devising policies. Most studies have only focused on one or two dimensions of climate change and variability, and historical data have not yet been able to mimic the cumulative consequences of simultaneous changes on several environmental fronts—ecological scarcity, rising sea levels, more floods, extreme climate, environmental migration, use of climate change to mobilize groups for conflict, and so forth.

Hence, although the evidence thus far indicates that environmental scarcities do not have huge implications for the risk of conflict, one cannot rule out the possibility that the cumulative impacts of climate change and variability would affect it. Research on large exogenous shocks to society (in the form of large-scale reforms or financial instability), as well as research on changes in key variables (poverty and inequality), suggests that climate change and variability might perpetuate conflict in much the same way.[9]

### Climate Change as a Tool for Political Mobilization

Existing quantitative studies do not address the fact that the perception of the explanatory variable, climate change, has been altered. Natural hazards that used to be blamed on nature are now increasingly blamed on humankind. It is difficult to mobilize against Mother Nature—and easier to mobilize against someone who can be held accountable for the disasters. Droughts and famines might increasingly be blamed on the effects of global warming by a government eager to avoid being held culpable; and climate change could broaden the cleavage between the poor (who bear the brunt of climate change and climate variability) and the wealthy (who are mainly responsible for global warming) locally, nationally, and globally. Such factors could lay the groundwork for climate change to be used politically as a tool for conflictual mobilization. What often matters in conflicts, according to Martin (2005), is "the actors' perceptions of conflict situations, rather than simple causal links between contributory factors and outcomes." Such perceptions can easily be formed by the prevalent discourses.

Haldén (2007) introduces the idea of regarding climate change as a "conditioning factor—a permissive cause—of human action, not a deterministic, effective cause." Thus, rather than arguing that environmental factors might have causal effects on the risk of conflict, it might be fruitful to look at climate change as a potential tool for mobilization—in much the same way that other major societal transformations (political reforms, financial crises, and so on) might mobilize different groups for violent purposes.

### Social Perspective on Conflict in Conflict Analysis

This section will emphasize the importance of nonviolent conflicts and of the daily violence in nonconflict situations. In the context of social development, conflict resolution does not take the form of a dichoto-mous outcome that is either violent (bad) or peaceful (good). Peaceful resolutions are not always conducive to social development because they could be forced through by powerful stakeholders even if they adversely affect many social development objectives. Power and coer-cion can be exercised in a multidimensional way that might or might not include violence. Three case studies of nonviolent conflicts associ-ated with ecological degradation illustrate some adverse social dynam-ics that could result from climate change and variability.

### Nonviolent Conflict over Declining Fish Stocks in Brazil

In Brazil, declining fish stocks have led to tensions among different occu-pational groups—fishermen and farmers—as well as between artisanal and commercial fishermen. Fisheries in the São Francisco River are becoming unsustainable as a result of overfishing, dam construction, pollution, and other forms of environmental degradation. In the absence of functioning institutions for resource management, the rapid depletion of the resource has spurred conflict among those with an important stake in it.

Conflicts between fishermen and farmers have become more common because of the construction of more irrigation dams for agriculture, which adversely affects fish spawning; widespread water contamination by agri-cultural toxins and animal manure; and the closing of access points on the rivers that fishermen have traditionally used. The fishermen also increas-ingly clash with the understaffed government agencies managing the river. The Brazilian Agency for the Environment and Natural Resources should enforce the regulation of the fisheries and watercourses but has left this task to the military police. Fishermen have complained to the state of undue violence and disrespectful treatment by the military police and feel that they are "being punished unduly as resource predators, while the high-impact resource users, such as large-scale farmers, hydroelectric compa-nies, and sports fishermen, manage to avoid the penalties" (Gutberlet and others 2007). Animosity has also been directed at the private hydroelec-tric companies and industries along the river. Dams change water flows and reduce fish populations, and polluting industries (a zinc refinery close to the river was highly unwelcome) contaminate the water, with adverse consequences for fish and humans alike.

Changes in the river have also led to clashes among fishermen. Those from one municipality were forced to intrude on fishing grounds in other municipalities because dams, pollution, and overfishing had exhausted their own fishing grounds. Moreover, artisanal fishermen complained about commercial fishermen's use of gill nets, while commercial fishermen complained about artisanal fishermen's illegal fishing. Artisanal fishermen were rarely invited to contribute to resource management decisions. This type of nonviolent conflict over environmental resources has the potential to turn violent, but even if it does not, it might lead to a suboptimal situation of distrust, lack of cooperation, marginalization of some groups, and unproductive, rent-seeking behavior.

## Water Tensions in Mexico

Water shortage in Mexico has spawned both internal conflicts (between vulnerable groups on one side and the state and water-intensive industries on the other) and international tensions between the United States and Mexico. Although the LAC region as a whole has abundant freshwater resources, Mexico is a relatively parched country that faces water shortages in many of its 31 states (see chapter 2 for further discussion of the paradox of regional water abundance and local water scarcity).

Pressure on water resources is highest around the Mexico City Metropolitan Area, where between 1985 and 1992, close to 2,000 events of mobilization over water issues occurred (Castro 2006). They ranged from bureaucratic complaints and nonpayment of water bills in protest against price hikes, to mass demonstrations and civil disobedience, destruction of infrastructure, and violent conflict with the police or the water operators. Although most of the conflicts were triggered by political reforms (many of which were privatization or decentralization schemes), the chronic shortage of water was the underlying condition. With climate change and variability, the shortage of water is predicted to increase around the important Lerma-Chapala basin as the result of increases in temperature coupled with a decrease in rainfall in the region. Thus, growing social unrest around water and water services might be one of the consequences of climate change and variability in Mexico. Chapter 2 discusses the attempts to establish water basin councils in Mexico, to manage water and hence reduce the potential for water-related conflicts, but points to the problems that have arisen over effective representation of all stakeholders and over decentralization of decision-making powers, investment funds, and necessary technical expertise.

Water shortages have spurred interstate conflicts between Mexico and the United States over water rights along the Mexico–U.S. border (Rio Grande and the Colorado River). Because California is the upstream user, that state's increasing consumption effectively determines the amount of water available for many Mexican farmers in drought situations. The latest water conflict has been triggered by the U.S. plan to reline part of the All-American Canal. Such a project would minimize leakage for California users but in the process would impose water shortages on villagers in Mexico. The plan has spurred organized protests from Mexico, with the president noting that it would be "terrible for Mexico."

### Destabilization of Traditional Indigenous Structures

Indigenous peoples are particularly vulnerable to the increased variability in nature's cycles. Climate change and variability could lead to internal conflicts both within and among indigenous communities, as well as to conflicts with nonindigenous communities.[10] With regard to the latter, some of the main conflicts seen are those between mestizo (nonindigenous) and indigenous people over access to farmland and forest resources, with the advance of the agricultural frontier. Aside from the well-described adverse effects on agricultural productivity caused by climate change, the increasing severity of hurricanes, for example, can lead to large tracts of forest being devastated. That together with more droughts increases the risk of forest fires. This condition provides opportunities for farmers to expand the agricultural frontier further into forests inhabited by indigenous people, displacing them from their traditional territories and hence potentially causing conflict.

These three cases illustrate a few of the adverse consequences that nonviolent conflicts induced by climate change and variability can generate. They emphasize that we should be concerned with the socioeconomic outcome of both violent and nonviolent conflicts. They imply that to minimize adverse impacts and expand the capabilities of vulnerable groups, social policies should attempt to influence the processes that translate climate change impacts into increased tensions and disputes.

### Day-to-Day Violence and Climate Change

Climate change might also have adverse impacts on violence in small communal or household settings. This type of unstructured violence without any clear link to conflicts appears widespread. Gender-based violence, for instance, despite being the least visible manifestation of violence (precisely because it is unstructured and exercised in the private domain), is by far the most widespread type of non–conflict-related violence (Kelly and Radford 1998). It is beyond the scope of this chapter to explore in any depth the consequences of climate change and variability for day-to-day violence in general. However, there appears to be a real risk that climate change (by transforming livelihoods and social structures) could spur social violence in nonconflict situations. Moser and Rodgers (2005) show how rapid socioeconomic changes might have destabilizing effects not only on societies but also within families, leading to an increased risk of domestic violence.

## Vulnerability to Climate Change–Induced Conflict

Environmental scarcities are only one of the channels through which climate change is likely to affect violent and nonviolent conflict. When analyzing the scope for climate change–induced conflict in Latin America and the Caribbean, therefore, the focus should ideally be not only on the traditional environmental scarcity hypothesis but also on broader socioeconomic factors that could channel environmental changes into conflicts, as well as on the prospects that climate change will be used as an argument to mobilize groups for conflict. This approach closely follows the suggestion of Peluso and Watts (2003) that "rather than presuming or starting with scarcity (or abundance), analysis of these cases of violence should begin with the precise and changing relations between political economy and mechanisms of access, control, and struggle over environmental resources."

Thus the discussion below focuses on aspects of the socioeconomic context that might catalyze scarcities (or other adverse impacts of climate change and variability) into conflict. Most studies of conflict have found that history matters (past conflicts increase the risk of present conflicts), as do demography (a large young male population can be destabilizing), regime type and state strength, and level of development (the level, growth, and structure of the economy). It is within this socioeconomic context that climate change and variability—like any other significant exogenous shocks—might destabilize societies and increase the risk of conflict.

### Socioeconomic Context

Gaining an overall view of the socioeconomic conditions in Latin America and the Caribbean that could exacerbate the conflictual dimension of climate change is essential. For example, the region's high income inequality and crime rates provide a backdrop of tension to which the adverse impacts of climate change and variability are likely to add further stress that could increase the risk of violent and nonviolent conflicts.

Compared to African countries, countries in the LAC region in general are relatively strong, middle-income societies with accountable governments. Rural populations, which are likely to feel the impacts of climate change and variability most directly, constitute 22 percent of the region's total population. The relatively sparse settlement in rural areas is reflected in the small contribution of agriculture to the region's GDP—only 7 percent, compared to 19 percent in South Asia and 16 percent in Africa. Latin America and the Caribbean comprise several countries with high levels of human development, a majority of countries with medium human development, and only Haiti categorized as having low human development.[11] The region's secondary education enrollment rate is high—89 percent in 2005—compared to those of other developing regions and to the global average of 65 percent. The male youth bulge, another significant factor increasing the risk of violent conflicts, does not appear to be a major problem in the region. Young people aged 15–24 made up 18.6 percent of the population in 2005, very close to the global average of 17.9 percent, and the share of 15- to 24-year-olds in the population is projected to fall below the global average to 15 percent by 2030 and 12.6 percent in 2050 (United Nations Population Division 2006).

A robust cross-country finding is that weak states have a greater risk of conflict, even after accounting for reverse causality (having many conflicts makes states weak). Barnett and Adger (2007) suggest that the impact of climate change on conflict is greater in weak states, as indicated by the Failed States Index. Table 8.2 compares Latin America and the Caribbean with the world, based on the Failed States Indexes compiled by the Fund for Peace (2008). The region has only two countries among the top 50 (Haiti at number 11 and Colombia at 33), and the its average failed state score of 70.9 is very close to the global average of 70.5. However, the LAC region appears to fare considerably worse than the global average in terms of brain drain and income inequality.

Several quantitative studies have found that both pure democracies and repressive autocracies are relatively stable but regimes in transition are more prone to violent conflict. Table 8.3 compares Freedom House's Political Rights Index for Latin America and the Caribbean with the rest of the world.[12]

As shown in the table, the political rights score for the average LAC region country (1973–2006) is 2.9, against the world average of 3.9, indicating that countries in Latin America and the Caribbean possess more democratic traits than is average. Within Latin America, Paraguay, Nicaragua, and Guatemala are the countries that have spent most years in the limbo between a functioning democratic regime and authoritarian

Table 8.2 Failed States Index, LAC and the World

| | (1) | (2) | (3) | (4) | (5) | (6) | (7) | (8) | (9) | (10) | (11) | (12) | Total |
|---|---|---|---|---|---|---|---|---|---|---|---|---|---|
| LAC | 6.5 | 4.1 | 5.8 | 6.3 | 7.5 | 5.6 | 6.3 | 5.9 | 6 | 5.8 | 5.8 | 5.3 | 70.9 |
| Global average | 6.4 | 5.1 | 5.8 | 5.6 | 6.7 | 5.6 | 6.4 | 5.7 | 5.9 | 5.5 | 6 | 5.7 | 70.5 |
| Difference | 0.1 | −1 | 0 | 0.7 | 0.8 | 0 | −0.1 | 0.2 | 0.1 | 0.3 | −0.2 | −0.4 | 0.4 |

Source: Calculations based on Fund for Peace, Failed States Index.
Note: Numbers heading the table columns represent the factors making up the index, as follows:
1. Mounting demographic pressures.
2. Massive movement of refugees and internally displaced people, creating complex humanitarian emergencies.
3. Legacy of vengeance-seeking group grievance.
4. Chronic and sustained human flight.
5. Uneven economic development along group lines.
6. Sharp and/or severe economic decline.
7. Criminalization and/or delegitimization of the state.
8. Progressive deterioration of public services.
9. Suspension or arbitrary application of the rule of law and widespread violation of human rights.
10. Security apparatus as "state within a state."
11. Rise of factionalized elites.
13. Intervention of other states or external factors.
Values between 1 (best) and 10 (worst) along twelve dimensions. Unweighted averages.

Table 8.3 Share of Countries in Different Regime Categories and Regime Stability across Categories—Comparison of LAC Region and the World

| Political rights regime score | 1 | 2 | 3 | 4 | 5 | 6 | 7 | Average | Median | Variance | SD |
|---|---|---|---|---|---|---|---|---|---|---|---|
| LAC | 27% | 29% | 13% | 10% | 8% | 7% | 6% | 2.9 | 2.0 | 1.5 | 1.0 |
| World | 23% | 15% | 7% | 9% | 11% | 19% | 16% | 3.9 | 4.0 | 1.2 | 0.9 |
| Bolivia | – | – | – | – | – | – | – | 2.9 | – | 3.1 | – |
| Colombia | – | – | – | – | – | – | – | 2.7 | – | 0.8 | – |
| Mexico | – | – | – | – | – | – | – | 3.4 | – | 0.7 | – |
| Nicaragua | – | – | – | – | – | – | – | 4.2 | – | 1.1 | – |
| Peru | – | – | – | – | – | – | – | 3.8 | – | 3.4 | – |

Source: Author's calculations based on Freedom House data.
Note: The numbers heading the columns indicate the seven Freedom House regime categories, ranging from 1 = most democratic, through 7 = least democratic. Region and selected LAC countries 1973–2006; unweighted averages. Regime stability is proxied by regime category variance and standard deviation within countries and averaged across countries. — = not available.

rule (32, 30, and 28 years respectively). Chile, not surprisingly, is the country with the greatest change in rule, from military rule to the present democratically elected president.

Haldén (2007) notes that although countries in the LAC region are not as poorly integrated economically as those in Africa, "they are nonetheless fraught with tensions and huge disparities in standards of living and income." This, he argues, is worrisome for three reasons: (a) disparities and divisions might by themselves impede growth and undermine adaptation strategies; (b) the substantial inequality might also destabilize societies and increase the risk of conflict in the light of climate change and variability; and (c) the differences between large segments of the populations also imply that climate change will have very unequal impacts on the population, further exacerbating tensions.

Large inequalities between groups (along ethnic, religious, political, or geographical divisions) increase the risk of violent conflict, whereas high individual income inequality is a driver of crime (Dahlberg and Gustavsson 2005; Fajnzylber, Lederman, and Loayza 2002; Kelly 2000; Østby 2007). Violent conflicts entail coordinated group action; as Østby argues, "Even though an individual may feel frustrated if he is poor compared to other individuals in society, he will not start a rebellion on his own."[13] Based on quantitative analysis of survey data from 55 countries, Østby (2007) found that all measures of horizontal inequality—ethnic, religious, political, and geographic—and in particular geographic horizontal inequality, are associated with higher risks of conflict outbreaks. In Bolivia, highly unequal distributions of natural resources create tensions between regions. In response to the nationalization of gas and oil reserves, the resource-rich regions of Santa Cruz, Tarija, Beni, and Pando (comprising 35 percent of the Bolivian population) declared autonomy. Violent disputes erupted between the government and the regions seeking autonomy, and also within those regions between supporters of the president and separatists.

The impact of income inequality on crime rates has also been analyzed. The high crime rates in Latin America and the Caribbean are likely rooted partly rooted in the region's high level of income inequality. Gutierrez and others (2004) concluded, based on their econometric analysis of homicide rates in Brazil, that individual inequality (as measured by the Gini coefficient), but not poverty, played a substantial role in determining criminality. LAC countries in general, and countries such as Brazil, Colombia, and Haiti in particular, are among the most unequal in the world (table 8.4). Climate change and variability might raise crime rates by causing events (natural disasters, sea level rises, and ecological degradation) that worsen inequality. Adverse exogenous shocks to societies are rarely absorbed equally by different income strata.

Higher crime rates could lead to a weakening of state capacity. Crime of most types has been rising in Latin America and the Caribbean since 1980, mirroring global trends. In 1980, the LAC region had a little more

**Table 8.4    Income Inequality across Regions and within LAC Countries (Gini coefficients)**

| | |
|---|---|
| LAC | 50.7 |
|    Haiti | 59.2 |
|    Brazil | 59.0 |
|    Colombia | 56.4 |
|    Bolivia | 54.0 |
|    Mexico | 49.7 |
|    Argentina | 49.5 |
|    Peru | 48.8 |
|    Guyana | 47.4 |
|    Costa Rica | 46.4 |
|    Nicaragua | 46.3 |
|    Trinidad and Tobago | 40.7 |
| Sub-Saharan Africa | 50.1 |
| East Asia | 39.0 |
| Middle East | 38.7 |
| North America | 36.7 |
| South Asia | 34.9 |
| Europe and Central Asia | 31.9 |

*Source: World Development Indicators* (World Bank 2009).
*Note:* Unweighted averages 1950–2006; based on data from 131 countries where data were available.

than 2,000 incidents of reported crime per 100,000 inhabitants, a figure that had risen to more than 3,000 incidents by 2002–03 (Shaw, Dijik, and Rhomberg 2003). Table 8.5 compares homicide rates in 2000–07 with those in the rest of the world. Most of the homicides in the LAC region are drug related, and do not appear to be caused directly by natural scarcities. Climate change could, however, weaken state capacity and shift the balance between drug cartels and the police and military. Financial resources will need to be devoted to disaster management, police and military forces will need to be deployed to uphold states of emergency, and the state's infrastructure might erode as the result of natural disasters.

### Prospects for Conflictual Mobilization in Latin America and the Caribbean

Environmental changes themselves do not engender conflict, but they can be used to mobilize people for conflict. The results of climate change might cause discontent and distress, but to engender violent conflict requires leaders that mobilize groups and create cleavages between them. The essence of conflicts is a struggle among elite factions for control of political power and wealth, in which scarce resources are used as the prize that goes to the winning coalition. In the words of Haldén (2007),

**Table 8.5   Homicide Rates, 2000–07
(per 100,000 population)**

| | |
|---|---|
| World | 7.4 |
| LAC | 22.0 |
| Colombia | 52.3 |
| Venezuela, R.B. de | 45.9 |
| Jamaica | 45.0 |
| Honduras | 43.2 |
| El Salvador | 41.0 |
| Guatemala | 34.3 |
| Belize | 30.8 |
| Brazil | 27.8 |
| Ecuador | 16.7 |
| Mexico | 13.7 |
| Suriname | 12.7 |
| Paraguay | 12.1 |
| Haiti | 11.5 |
| Argentina | 8.4 |
| Costa Rica | 6.6 |
| Uruguay | 5.7 |
| Peru | 4.9 |
| Bolivia | 3.3 |
| Chile | 1.7 |

*Source:* UNODC 2008.
*Note:* Unweighted averages.

"the most important proximate causes of conflicts may be the presence of politicians who want to exploit the situation, or the absence of politicians who make conscious choices to avoid conflict by addressing socioeconomic hardships resulting from climate change." The factor that determines whether violent conflicts arise, Oldstone (2002) asserts, is the extent to which elites succeed in channeling group-based grievances into open conflict with other groups or the state.

Several African leaders are using climate change as a cause for mobilization and attack. For example, Uganda's president has called climate change "an act of aggression from the Western World" and a Namibian UN diplomat has called climate change "low-intensity biological and chemical warfare" (Brown, Hammill, and Mcleman 2007; UN General Assembly 2008).

The rhetoric used in the LAC region has not been as conflictual. Natural disasters, according to the Bolivian president, Evo Morales, are getting steadily worse and are brought on "by a system, the capitalist system, the unbridled industrialization of the resources of the planet Earth"

(Sheridan 2008). Brazilian president Lula da Silva has criticized the wealthy countries for their contributions to global warming and told them to stay out of Brazil's business when it comes to the fate of the Amazon rainforest: "The wealthy countries are very smart, approving protocols, holding big speeches on the need to avoid deforestation, but they already deforested everything" (Yardley 2007). At the 2008 UN General Assembly climate change debate, the consumer angle was emphasized by the Bolivian speaker who posed the rhetorical question of whether climate change was "more dangerous to human beings than the insensitivity of people who had everything." The UN speaker from St. Vincent and the Grenadines paired debt relief with climate change by suggesting that debt obligations of developing countries should be reviewed through the prism of climate change and variability. The Salvadoran speaker was the only speaker from Latin America and the Caribbean to make a direct link between climate change and variability and conflict. She found climate change to be a serious threat to international security that could "no longer be perceived in the traditional context of war and peace, as they had been since the United Nations founding" (UN General Assembly 2008).

Somewhat contrary to what one might expect, the incidence of political activism in the LAC region does not appear to be higher than in other regions.[14] However, the use of climate change as an idea to mobilize groups for conflict seems likely to increase, and given the inequality in many LAC countries, that mobilization may be targeted not only against the wealthy developed nations but also against wealthy segments of domestic populations in times of crisis.

As noted in chapter 1, for their study the authors chose to consider only the social impacts of climate change and appropriate adaptation responses. It is beyond the scope of the book to look also at mitigation of climate change in any detail. However, that is not to say that mitigation measures will not have social impacts, some of which have the potential to cause conflict. For example, throughout Latin America and the Caribbean a drive is occurring toward hydropower in an effort to diminish reliance on fossil fuels, whether for fiscal and national security reasons or to reduce greenhouse gas emissions. However, unless the planning and construction of hydropower plants are carried out sensitively and inclusively, and offer acceptable relocation, including guarantees of income-earning opportunities for people displaced by dam building, the potential for conflict is there as they lose their homes and livelihoods. Similarly, the use of carbon trading facilities in forest areas could create conflict if it displaces indigenous peoples from their traditional territories.

It would therefore be important to include these groups in any such arrangement; indeed experience has shown that indigenous peoples offer better forest protection than government-run programs (Nepstad and others 2006). The rush to produce biofuels as an alternative to fossil fuels also has the potential to cause conflict. If the cultivation of biofuel crops displaces food crops, that could lead to food price hikes, perhaps food insecurity, and potentially disgruntled subsistence farmers who have been displaced from their land. Beyond mentioning these few examples, however, we do not address the conflictual potential of climate change mitigation.

To sum up, climate change and variability may undermine human security for certain groups, but whether it leads to violent conflict depends on other key socioeconomic and political conditions—most importantly on how it is perceived and communicated by key political actors. Most studies of climate-related factors have focused on physical consequences, such as soil degradation, deforestation, and water scarcity. Inasmuch as these physical consequences seem inevitable, it would be beneficial to expand the focus to encompass the political economy conditions that might generate conflict. Instead of stressing the environmental scarcity aspect, a fruitful social perspective would place more emphasis on opportunities that environmental change presents, as well as on the sociopolitical channels that translate environmental consequences into different outcomes across various groups.

## Policy Perspectives

Climate change does not seem to call for an immediate change in policies or for specific programs in the area of conflict prevention for the time being. With respect to conflict, the focus should be on considering the political economy dimension because the political economy that (a) dictates the policies that can be implemented to address the adverse socioeconomic consequences of climate change; (b) produces the structures that reproduce and perpetuate adverse conflictual or nonconflictual outcomes from climate change; and (c) influences the risk that existing socioeconomic factors become conduits for violent conflict.

Because multiple channels exist through which climate change might affect conflict, policies need to address the consequences of climate change on several fronts. Doing that will reduce the risk of violent conflict and ensure that conflicts are resolved in ways that do not jeopardize key development objectives. Policies can either seek to prevent the adverse socioeconomic consequences of climate change from turning into violent conflict, or seek to prevent the adverse socioeconomic consequences from materializing in the first place. However, only the latter objective directly affects the outcome of nonviolent conflicts.

Interventions to address broader socioeconomic consequences constitute "no-regrets" policies: regardless of the future implications of climate change, and regardless of how strong the link is between those implications and violent conflict, addressing broader socioeconomic deficiencies in societies will benefit human well-being. From a social development perspective, such policies should be the policies of focus, and the social development discipline provides a very useful, albeit somewhat overlooked, perspective on how they can be pursued. In a changing climate, vulnerable groups with a weak political voice will lose out to more powerful groups. Social development analysis has proved adept at addressing such issues of power relations and political institutions.

Social development tools can provide indispensable contributions to adaptive measures because they have the advantage of being context specific, multidimensional, and disaggregated across groups and are often conducted upstream. These are advantages when dealing with the impact of climate change on conflict, which is likely to (a) be indirect, weak, and strongly dependent on the country-specific political economy context; (b) work through multiple channels with different impacts on different groups; and (c) be subject to many adaptation and mitigation measures that themselves could generate cleavages and conflict. Poverty and social impact analysis (PSIA) would be a suitable framework in which to analyze the potential implications of climate change for violent and nonviolent conflict across social groups, in different sectors, and at different levels of analysis. As argued above, analyzing the socioeconomic impacts of climate change need not be different from analyzing any other exogenous change or shock to society.

## Notes

1. Based on conflict data from the Peace Research Institute, Oslo (1946–2006), the LAC average of one conflict per 23,714 inhabitants is higher than that in Asia (one conflict per 57,382 inhabitants) and about the same as in Europe (one conflict per 22,250 inhabitants), but significantly lower than in the Middle East (one conflict per 12,459 inhabitants) or Africa (one conflict per 10,161 inhabitants). Most LAC countries are middle income countries, and while they account for 29 percent of the middle-income countries globally, they are only involved in 19 percent of the middle-income-country conflicts (PRIO 2009).

2. With the award of the 2007 Nobel Peace Prize to Al Gore and the Intergovernmental Panel on Climate Change (IPCC), however, the conflictual consequences of climate change and variability are likely to receive increased academic attention.

3. Despite their clear conviction that population growth and environmental degradation do influence conflict, Homer-Dixon and Percival (1995) acknowledge that "environmental scarcity is always enmeshed in a web of social, political, and economic factors, and its contribution to violence is exceedingly difficult to disentangle from contributions by these other factors."

4. See *Environment, Population, and Security* (1994–96), a joint effort with the American Association for the Advancement of Science, http://www .library.utoronto.ca/pcs/eps.htm; and *Environmental Scarcity, State Capacity, and Civil Violence* (1994–98), undertaken with the American Academy of Arts and Sciences, http://www.library.utoronto.ca/pcs/state .htm, both under the supervision of Homer-Dixon. Under the *Environment and Conflicts Project* (ENCOP, http://www.isn.ethz.ch/isn/Digital-Library/ Publications/Detail/?lng=en&id=235), 60 academic experts analyzed some 40 environmental conflicts. The *Inventory of Conflict and Environment* (ICE), http://american.edu/TED/ice/ice.htm, under the supervision of Prof. Jim Lee at the American University, contains more than 200 case studies based on 16 categories where some kind of interaction has been identified between conflict on one side and environmental factors on the other.

5. One of the few systematic studies that examines the link between environmental scarcity on one hand and both conflict and cooperation on the other is Aaron Wolf's (2007) historical dataset of countries' responses, whether cooperative or conflictual, to water scarcity. Of the 1,831 identified interactions based on water scarcity, 67 percent relied on cooperation, 28 percent displayed a conflictual nature, and only 2 percent involved some type of military intervention.

6. These studies focus on the conflict impact of migration in the recipient country. To estimate the overall impact of migration on the risk of conflict, it would be prudent to also include the effect of migration in the country of origin.

7. Despite the clear conclusion that opportunities matter in conflict, it is the grievance angle that has been pursued in most of the empirical analyses. And it is somewhat ironic that some of the studies most relevant for the debate about climate change and variability and conflict are devoid of environmental indicators.

8. Diamond (2005) provides anecdotal historical evidence of societies that collapsed because ecological degradation apparently crossed a tipping point for civilization survival—Easter Island, Pitcairn Island, Henderson Island, the Native American societies of Anasazi and the Maya, and Norse Greenland. In very isolated island states the response was more-or-less apathetic; in other cases, as with the Vikings, the response was expansionist.

9. It is beyond the scope of this chapter to explore the consequences of climate change and variability for day-to-day violence in general. But based on Moser and Rodgers's (2005) study, which convincingly documents a link between rapid socioeconomic changes and an increasing risk of social violence in non-conflict settings, there appears to be a risk that climate change (by transforming livelihoods and social structures) might have adverse secondary effects on this type of violence, particularly in states with weak political institutions.

10. In a changing climate, from the perspective of the indigenous communities the elders appear to have lost their ability to read the signs provided by nature and translate them into actionable recommendations for planting, harvesting, and so on. Consequently they have lost much of their authority. Younger people, with better language skills and access to state institutions, are able to develop linking social capital and attempt to move quickly up the hierarchy, eroding traditional structures of power and bonding social capital, destabilizing tribal or community political and cultural order, and increasing the risk of violent or nonviolent conflicts.

11. UNDP Human Development Reports; http://hdr.undp.org/en/statistics.

12. Freedom House has surveyed political rights since 1973 based on a broad range of sources of information: foreign and domestic news reports, academic analyses, nongovernmental organizations, think tanks, individual professional contacts, and visits to the region. Ranging from 1 (*best*) to 7 (*worst*), the index measures the degree to which a country can be classified as free (average 1.0–2.5), partly free (3.0–5.0), or not free (5.5–7.0).

13. Oldstone (2002) similarly argues that "degradation often brings misery, yet such misery does not generally trigger the elite alienation and opposition to the government necessary for large-scale violence to occur."

14. Political activism and mobilization are represented by the percentage of people reported to have been involved in five different types of political action (based on tabulations of the World Value Surveys). The average percentage of people involved in these five activities was 10 percent in the LAC region, compared to the worldwide average of 13.6 percent. In the most recent World Value Surveys from 85 countries (11 from the LAC region), the five political actions are signing a petition; joining in boycotts; attending lawful demonstrations; political action: joining unofficial strikes; and political action: occupying buildings or factories. This result also holds up when measured only by the more unlawful, latter two activities.

## References

Barnett, J., and Adger, W. Neil. 2007. "Climate Change, Human Security, and Violent Conflict." *Political Geography* 26 (6): 639–55.

Binningsbø, H., I. Soysa, and N. Gleditsch. 2006. "Green Giant or Straw Man? Environmental Pressure and Civil Conflict, 1961–99." Working Paper presented at the 47th Annual Convention of the International Studies Association, San Diego, CA, March 22–25.

Brown, O., A. Hammill, and R. Mcleman. 2007. "Climate Change as the 'New' Security Threat: Implications for Africa." *International Affairs* 83 (6): 1141–54.

Buhaug, H., N. Gleditsch, and O. Theisen. 2008. "Implications of Climate Change for Armed Conflict." Paper presented at Workshop on Social Dimensions of Climate Change, March 5, Social Development Department, World Bank, Washington, DC.

Castro, José. 2006. *Social Struggle in the Basin of Mexico.* New York: Palgrave Macmillan.

Collier, Paul, and Anke Hoeffler. 1998. "On Economic Causes of Civil War." *Oxford Economic Papers* 50: 563–73.

————. 2004. "Greed and Grievance in Civil War." *Oxford Economic Papers* 56: 563–95.

————. 2006. "The Challenge of Reducing the Global Incidence of Civil War." In *How to Spend $50 Billion to Make the World a Better Place,* ed. Bjørn Lomborg. Cambridge, U.K.: Cambridge University Press.

Collier, Paul, Anke Hoeffler, and Dominic Rohner. 2006. *Beyond Greed and Grievance: Feasibility and Civil War.* Center for the Study of African Economies Working Paper 254, August 7. http://www.bepress.com/csae/paper254.

Dahlberg, M., and M. Gustavsson. 2005. *Inequality and Crime: Separating the Effects of Permanent and Transitory Income.* IFAU Working Paper 2005/19. http://www.ifau.se/upload/pdf/se/2005/wp05-19.pdf.

Diamond, Jared. 2005. *Collapse—How Societies Choose to Fail or Succeed.* New York: Viking Press.

Esty, D., J. Goldstone, T. Gurr, B. Harff, M. Levy, G. Dabelko, P. Surko, and A. Unger. 1998. *State Failure Task Force Report: Phase II Findings.* McLean, VA: Science Applications International Corporation.

Fajnzylber, P., D. Lederman, and N. Loayza. 2002. "Inequality and Violent Crime." *Journal of Law and Economics* 45 (1): 1–40.

Fund for Peace. 2008. Failed States Index. http://www.fundforpeace.org/web/index.php?option=com_content&task=view&id=229&Itemid=366.

Gleditsch, N., and Henrik Urdal. 2002. "Ecoviolence? Links between Population Growth, Environmental Scarcity, and Violent Conflict in Thomas Homer-Dixon's Work." *Journal of International Affairs* 56 (1): 283–302.

Goodhand, J. 2003. "Enduring Disorder and Persistent Poverty: A Review of the Linkages between War and Chronic Poverty." *World Development* 31 (3): 629–46.

Gutberlet, Jutta, Cristiana Simão Seixas, Ana Paula Glinfskoi Thé, and Joachim Carolsfeld. 2007. "Resource Conflicts: Challenges to Fisheries Management at the São Francisco River, Brazil." *Human Ecology* 35: 623–38.

Gutierrez, Maria Bernadete Sarmiento, Mario Jorge Cardoso de Mendonça, Adolfo Sachsida, and Paulo Roberto Amorim Loureiro. 2004. *Inequality and Criminality Revisited: Further Evidence from Brazil.* ANPEC - Associação Nacional dos Centros de Pósgraduação em Economia, Ingá Niterói. http:// www.anpec.org.br/encontro2004/artigos/A04A149.pdf.

Haldén, P. 2007. *The Geopolitics of Climate Change.* Stockholm: Swedish Defense Research Agency. http://www.foi.se/upload/projekt/Climatools/Rapporter/ FOI-R—2377—SE.pdf.

Hauge, W., and T. Ellingsen. 1998. "Beyond Environmental Scarcity: Causal Pathways to Conflict." *Journal of Peace Research* 35 (3): 299–317.

Homer-Dixon, T., and V. Percival. 1995. *Environmental Scarcity and Violent Conflict: The Case of South Africa.* Occasional Paper. Project on Environment, Population and Security. Washington, DC: American Association for the Advancement of Science and the University of Toronto.

Kelly, M. 2000. "Inequality and Crime." *Review of Economics and Statistics* 82 (4): 530–39.

Kelly, L., and J. Radford. 1998. "Sexual Violence against Women and Girls: An Approach to an International Overview." In *Rethinking Violence against Women,* ed. R. Dobash and R. Dobash. Thousand Oaks, CA: Sage.

Martin, Adrian. 2005. "Environmental Conflict between Refugee and Host Communities." *Journal of Peace Research* 40 (3): 329–49.

Moser, Caroline, and D. Rodgers. 2005. *Change, Violence and Insecurity in Non-Conflict Situations.* Working Paper 245. London: Overseas Development Institute. http://www.odi.org.uk/publications/working_ papers/wp245.pdf.

Nafziger, E., and J. Auvinen. 2002. "Economic Development, Inequality, War, and State Violence." *World Development* (30) 2: 153–63.

Nel, P., and M. Righarts. 2008. "Natural Disasters and the Risk of Violent Civil Conflict." *International Studies Quarterly* 52 (1): 159–85.

Nepstad, D., S. Schwartzmann, B. Bamberger, M. Santilli, D. Ray, D. Schlesinger, P. Lefebvre, A. Alencar, E. Prinz, G. Fiske, and A. Rolla. 2006. "Inhibition of Amazon Deforestation and Fire by Parks and Indigenous Lands," *Conservation Biology* 20 (1): 65–73.

Oldstone, J. 2002. "Population and Security: How Demographic Change Can Lead to Violent Conflict." *Journal of International Affairs* 56 (1): 3–22.

Østby, G. 2007. *Horizontal Inequalities, Political Environment, and Civil Conflict— Evidence from 55 Developing Countries, 1986–2003.* World Bank Policy Research Working Paper 4193, World Bank, Washington, DC.

Peluso, N., and M. Watts. 2003. "Violent Environments: Responses." *ECSP Report* 9: 89–96.

PRIO (Peace Research Institute, Oslo) Armed Conflict Dataset. 2009. Version 4. http://www.prio.no/CSCW/Datasets/Armed-Conflict/UCDP-PRIO/.

Raleigh, C., and H. Urdal. 2007. "Climate Change, Environmental Degradation, and Armed Conflict: Disaggregated Dataset on Conflict and Environment." *Political Geography* 26 (6): 674–94.

Reuveny, R. 2007. "Climate Change–Induced Migration and Violent Conflict." *Political Geography* 26 (6): 656–73.

———. 2008. "Ecomigration and Violent Conflict: Case Studies and Public Policy Implications." *Human Ecology* 36 (1): 1–13.

Salehyan, Idean, and Kristian Skrede Gleditsch. 2006. "Refugees and the Spread of Civil War." *International Organization* 60 (2): 335–66.

Shaw, M., J. Dijik, and W. Rhomberg. 2003. "Determining Trends in Global Crime and Justice: An Overview of Results from the United Nations Survey of Crime Trends and Operations of Criminal Justice Systems." *Forum on Crime and Society* 3 (1): 35–62.

Sheridan, Barrett. 2008. "The People's Pugilist." *Newsweek*, April 22. http://www.newsweek.com/id/133279?tid=relatedcl.

United Nations General Assembly. 2008. "General Assembly's debate testament to the importance of taking immediate practical steps to address climate change threat, says Assembly President. . . ." GA/10690. http://www.un.org/News/Press/docs/2008/ga10690.doc.htm.

United Nations Population Division. 2006. *Data on Youth in Latin America.* http://esa.un.org/unpp/index.asp.

UNODC (United Nations Office of Drugs and Crime). 2008. *United Nations Surveys on Crime Trends and the Operations of Criminal Justice Systems.* http://www.unodc.org/unodc/en/data-and-analysis/United-Nations-Surveys-on-Crime-Trends-and-the-Operations-of-Criminal-Justice-Systems.html.

Urdal, H. 2005. "People vs. Malthus: Population Pressure, Environmental Degradation, and Armed Conflict Revisited." *Journal of Peace Research* 42 (4): 417–34.

Urdal, Henrik. 2008. "Population, Resources and Political Violence: A Sub-National Study of India 1956–2002." Submitted to *Development and Change.* http://www.wilsoncenter.org/events/docs/Urdal_2.pdf.

Wolf, Aaron. 2007. "Shared Waters: Conflict and Cooperation." *Annual Review of Environment and Resources* (32): 241–69.

World Bank. 2009. *World Development Indicators.* Washington, DC: World Bank. http://go.worldbank.org/6HAYAHG8H0.

Yardley, Jim. 2007. "Brazil's Leader Speaks Out." *New York Times*, February 7. http://www.nytimes.com/2007/02/07/world/asia/07china.html?pagewanted=all.

# CHAPTER 9

# Simulating the Effects of Climate Change on Poverty and Inequality

## Lykke Andersen and Dorte Verner

Human beings are very versatile and inhabit both extremely cold places (for example, the Inuit people of the Arctic region) and extremely hot ones (for example, the Berber people of North Africa, where air temperatures often exceed 50°C). The Bedouins of the Sahara inhabit a place so extremely dry that it may go for years without a drop of rain. In the extremely wet environment where the Emberá people of Lloró, Colombia, live it usually rains more than 13 meters per year. All other things being equal, however, human development is easier to achieve in places with intermediate climates, and such areas tend to support much higher population densities and more prosperous societies than places with extreme climates.

If temperate climates are more conducive to human development than ones that are either too cold or too hot, then the long-run relationship between temperature and income levels must look something like the one depicted in Figure 9.1, which simply indicates that countries with moderate climates are richer than countries with extreme climates.

If such an inverted-U-shaped relationship between temperature and incomes exists, then poor countries will be more sensitive to climate change than rich countries, simply because the slope of the relationship is steeper at the extremes. A small change in temperatures for countries

**Figure 9.1    Theoretical Relationship between Temperature and Income**

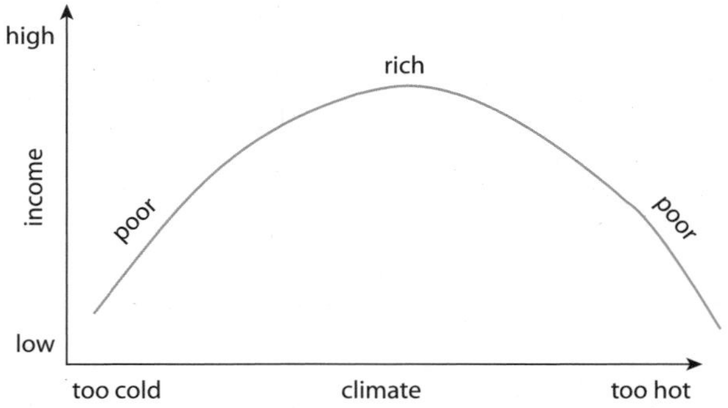

*Source:* Authors.

in the optimal range, where the slope is relatively flat, would cause only a minor change in incomes, whereas a small change at the extremes would cause large changes in incomes.

Apart from this general relationship between average climate and average income, other good arguments exist to suggest that the poor face more serious consequences of climate change and variability than the rich. For example, the rural poor are more likely to depend on agriculture, an activity whose productivity fluctuates strongly in response to variations in rainfall and temperatures. The poor are also more likely to live on marginal lands, which are ecologically fragile and vulnerable to droughts, floods, mudslides, and other natural disasters. The poor are less likely to be served by public infrastructure, such as drainage canals, that can reduce the probability of weather events' turning into natural disasters. In addition, the poor have smaller savings to cope with adverse shocks, which means that they may have to sell productive assets, take children out of school, reduce nutrient intake, or take other measures that have adverse long-run effects on the well-being of the household. A related issue is poor people's inability to insure themselves against adverse shocks, as well as the generally lower level of social insurance in poor countries. If climate change and variability hurt the poor more than the rich, it would imply that poverty and inequality would increase as a consequence of climate change and variability.

There is some evidence to back up this hypothesis. A recent cross-country study by Dell, Jones, and Olken (2008) suggests that income growth in poor countries has been very sensitive to the climate change experienced over the last 50 years, whereas rich countries seem to have been unaffected. However, that study does not control for other factors that may have affected growth, such as investment rates, education levels, war, and AIDS, for example, which implies that differences in all these factors are automatically attributed to climate change. To better control for historical differences that are likely unrelated to future climate change, Horowitz (2006) suggests using data on within-country variation in incomes and climate gathered from large, heterogeneous countries. That approach is used in this chapter. The approach makes it possible to bring relatively detailed, local-level data to bear on the relationship between climate and income in individual countries. Though it is always risky to draw inferences about the future from cross-sectional relationships, these relationships are used to assess the likely direction and magnitude of the effects of anticipated future climate change on poverty and income inequality. In this chapter no attempt is made to estimate the effects of changes in climate variability or changes in the frequency of extreme events, as data to quantify such changes at the local level are not available.

The findings of the analysis show that the impacts of climate change are very situation specific, so that caution should be used before generalizing. For example, in Brazil poverty is more widespread in the hot north and northeast than in the south and southeast. Northern Brazil is expected to warm faster than southern Brazil, and people living in already hot areas are expected to suffer more from additional warming than people in cooler areas. The combination of these three factors implies that the projected warming in Brazil would tend to cause an increase in both poverty and inequality. In neighboring Bolivia, however, the situation is inverted. The poor are concentrated in the cold highlands and would therefore benefit more from additional warming than the wealthier population in the hot lowlands. This means that future warming could tend to reduce both poverty and inequality in Bolivia.

## Contemporary Relationships between Climate and Income in Selected Countries in Latin America and the Caribbean

This section explores the contemporary relationships between climate and income in five countries in Latin America and the Caribbean, all of which have large within-country variation in both climate and incomes: Bolivia, Brazil, Chile, Mexico, and Peru. To capture the heterogeneity within each of these countries, local-level data are used. In Bolivia, Mexico, and Brazil data are available on 311, 2,350, and 5,507 municipalities, respectively; in

Chile on 324 comunas; and in Peru on 1,829 distritos. For simplicity, all these local administrative units are called "municipalities" in the remainder of the chapter.

To estimate a short-run, contemporary relationship between climate and income, rather than a long-run, historical relationship, the study controls for variables that are clearly related to income, but which are probably not affected by climate change in the short run (that is, within a few decades). For example, although climate change is likely to affect agricultural productivity in the short run, it is unlikely to reduce education levels or reverse the urbanization process, except possibly in the very long run. Therefore both education and urbanization rates are included as control variables in the short-run regressions, as outlined in appendix B. The regressions also include rainfall and rainfall squared because rain also may affect development in important ways.[1]

The four explanatory variables included in the regression models— temperature, rainfall, education, and urbanization rates—explain a very large part (73 percent to 92 percent) of the variation in income levels among municipalities in Bolivia, Brazil, Chile, Mexico, and Peru (table 9.1). This suggests that we have included the most important explanatory variables, and that adding additional control variables would likely make little difference.

Education level is the most important explanatory variable in all cases. That is not surprising, given that the economic growth literature has shown education to be a key growth-enhancing variable. The effect of the two climate variables, temperature and rainfall, varies from country to country as shown in figure 9.2.[2]

Panel (a) of figure 9.2 shows that average annual temperatures in the Bolivian municipalities vary between 3°C and 27°C and that average annual consumption per capita ranges between US$245 and US$2,562 (purchasing power parity, or PPP-adjusted US$ of 2001). The relationship between temperature and consumption is nonlinear and statistically significant, but the maximum is reached in the warmest regions of the country, rather than in regions with temperate climates. The coldest regions have consumption levels that are about 40 percent lower than those of the hottest regions, implying that the cold parts of Bolivia might gain significantly if average temperatures were to increase—all other things equal, and not taking into account the issue of water scarcity discussed in chapter 2.

Panel (b) of figure 9.2 shows that the level of income in Brazilian municipalities is significantly related to temperature, with average annual

**Table 9.1    Estimated Short-Term Relationships between Climate and Income**

| Explanatory variables | (1) Bolivia (log per capita consumption) | (2) Brazil (log per capita income) | (3) Chile (log per capita income) | (4) Mexico (log per capita income) | (5) Peru (log per capita income) |
|---|---|---|---|---|---|
| Constant | 5.2046 | 1.4138 | 9.4701 | 5.3729 | 2.5912 |
|  | (65.21) | (8.40) | (248.77) | (25.66) | (14.28) |
| Temperature | 0.0421 | 0.2093 | −0.0592 | −0.0313 | 0.1837 |
|  | (4.53) | (14.25) | (−7.37) | (−1.84) | (11.37) |
| Temperature$^2$ | −0.0007 | −0.0056 | – | 0.0010 | −0.0051 |
|  | (−2.54) | (−17.30) |  | (2.34) | (−11.46) |
| Rainfall | −0.0291 | 0.7086 | −0.3153 | −0.0283 | −0.5562 |
|  | (−0.35) | (9.24) | (−4.30) | (−0.68) | (−17.44) |
| Rainfall$^2$ | −0.0231 | −0.6324 | 0.1195 | 0.0031 | 0.1171 |
|  | (−0.88) | (−9.43) | (3.68) | (0.25) | (10.77) |
| Education level | 0.1940 | 0.3312 | 0.3504 | 0.0326 | 0.0214 |
|  | (19.51) | (119.73) | (30.11) | (24.40) | (15.49) |
| Urbanization rate | 0.4429 | 1.0284 | – | 0.0032 | −0.0095 |
|  | (3.56) | (14.36) |  | (3.31) | (−8.17) |
| Urbanization rate$^2$ | −0.3320 | −0.8880 | −0.6291 | 0.0001 | 0.0001 |
|  | (−2.83) | (−15.57) | (−10.52) | (7.92) | (10.67) |
| Number of observations | 311 | 5,507 | 324 | 2350 | 1,829 |
| $R^2$ | 0.9194 | 0.9275 | 0.8229 | 0.7377 | 0.7449 |

*Source:* Authors' estimations based on assumptions explained in the text.
*Note:* Numbers in parenthesis are *t*-values. When *t*-values are numerically larger than 2, the coefficient is considered to be statistically significant, corresponding to a confidence level of 95 percent.

temperatures of about 18°C being optimal. The warmest municipalities (average annual temperature of 30°C) are estimated to have income levels that are only about half as high as those in municipalities located in the optimal range.

Panel (c) shows that in Chile, people living in cooler regions tend to be more prosperous than people living in hot regions (other things being equal). The differences are both statistically and economically significant, with people in the cooler regions earning at least twice as much as people in the warmest regions.

Panel (d) shows no significant relationship between temperature and income in Mexico. The differences between the poorest and richest municipalities in Mexico are very large, but they are related mainly to differences in urbanization rates, education levels, and other factors unrelated to temperature in the short run. The flat relationship suggests that income levels in Mexico are probably not very sensitive to climate change, at least

**Figure 9.2    Estimated Short-Run Relationships between Temperature and Income**

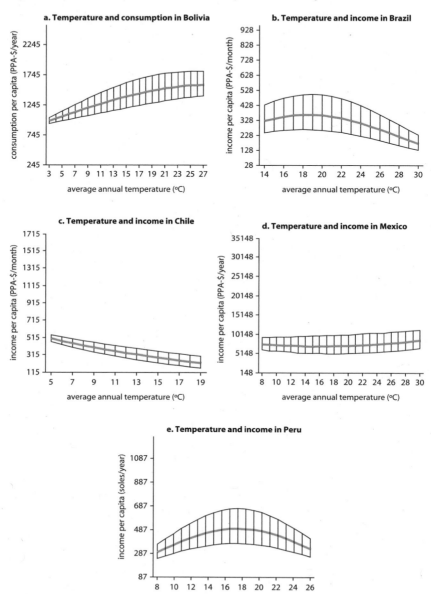

*Sources:* Authors' elaboration based on regression results shown in table 9.1.
*Notes:* The green central line corresponds to the point estimate from table 9.1, while the thin black lines delimit the 95 percent confidence interval, as estimated using Stata's lincom command. Axes are scaled to cover actually observed per capita incomes and temperatures in each country. PPA = purchasing power adjusted.

not to changes in average temperatures such as projected by the Intergovernmental Panel on Climate Change (IPCC). However, Mexico may still be vulnerable to increased climate variability and extreme events, as discussed in chapter 3.

Finally, panel (e) shows that the estimated short-run relationship between temperature and income in Peru is hump-shaped and both statistically and economically significant. When one controls for other differences (education, urbanization, and rainfall), the inhabitants of municipalities that have the most suitable temperatures earn, on average, about 52 percent more than the inhabitants of the provinces with the hottest climates and about 66 percent more than the inhabitants of the coldest regions.

These analyses thus reveal a hump-shaped relationship between temperature and income in Brazil and Peru. In Chile, people are better off in the coldest regions; in Bolivia they are better off in the hottest regions; in Mexico there is little relationship at all between temperature and income.

Rainfall has a significant impact on income in Brazil, Chile, and Peru (table 9.1). In Brazil, incomes show a hump-shaped relationship with rainfall, with moderate amounts of rainfall being more advantageous than extreme amounts. In contrast, in Peru and Chile people seem to be better off with either very little or a lot of rain and worse off with intermediate amounts. In Bolivia and Mexico, rainfall and consumption levels or incomes display no systematic relationship.

Thus the municipality-level regressions indicate that the climate-income relationships differ substantially from country to country. This means that the likely impacts of climate change on incomes, poverty, and inequality also differ substantially from country to country.

In the following section the effects of expected future climate change are simulated for the five case study countries, using the estimated relationships from table 9.1 and the expected climate changes in each region, as projected by the latest generation of IPCC models.

## Simulating the Effects of Projected Future Climate Change at the Municipal Level

To simulate the effects of expected climate change during the next 50 years, the regional projections made by Working Group I of the Fourth Assessment Report from the IPCC (IPCC 2007a) are used. These climate models cannot project specific climate change in each of the thousands of municipalities in the five countries, but some systematic patterns are seen that can be taken into account. In Brazil, for example, warming is expected to be faster in the north than in the south—and that is what has been observed over the last few decades (Timmins 2007). A central projection would be a 2.5°C increase in temperatures in the northern part of Brazil and a 2.0°C increase in the southern part during the next 50 years. Similarly, warming is expected to be faster in the northern part of Chile than in southern Chile. In the following, the effects of 1.5°C of warming in the northernmost part and 0.75°C in the southernmost part, and intermediate values in intermediate regions of Chile, are analyzed. In Peru, warming in the jungle region is expected to be around 2°C, in the mountain region 1.5°C, and in the coastal region 1°C. In Bolivia, warming is projected to be around 2°C in all lowland areas and 1.5°C in highland areas. Finally, in Mexico the coast is projected to warm less than the center, and especially less than the northern part of the central region, but because no statistical relationship was found between climate variables and income in that country, it makes no sense to simulate the effect of changes in climate in Mexico.

With respect to precipitation, there is little agreement as to the direction of change, as the confidence intervals all include zero change (IPCC 2007b). The simulations of the effects of future climate change therefore assume no change in precipitation, except in a region in central Chile for which most models project a decrease in rainfall due to a poleward shift of the South Pacific and South Atlantic subtropical anticyclones (IPCC 2007a). The magnitude of decrease that can be expected in central Chile over the next 50 years appears to be around 10 percent and corresponds to what has been observed over the last 50 years (Andersen and Verner 2008).

The estimated relationships from table 9.1 are used to simulate the effects of the above-described changes in climate by comparing two scenarios: one with climate change (CC) and one with no climate change (NCC). All variables other than climate variables are assumed to be the same in the two scenarios, so as to isolate the effects of expected climate change (see appendix B for the regression equations).

### Simulating the Impact of Future Climate Change in Brazil

In Brazil the expected temperature increases over the next 50 years are estimated to cause a 12 percent decrease in average per capita income, other things being equal. This is based on the northern states losing considerably more than average and the southern states losing less (table 9.2). The north and northeast will lose 22.5 percent and 19.8 percent, respectively, according to this simulation.

Climate change can be expected to cause an increase in inequality among Brazilian municipalities and to increase poverty. The largest simulated loss for any individual municipality is 29 percent for the municipality of Uiramutã, in Roraima in the northern part of the Amazon. The largest gain is 8.6 percent for the municipality of Campos de Jordão, in the State of São Paulo. At the municipal level a strong positive relationship appears ($\rho = 0.58$) between the current level of income and the subsequent gains from climate change, indicating that currently richer municipalities will likely lose less from future climate change than will currently poorer municipalities. The simulations thus suggest that future climate change is likely to contribute to an increase in inequality among Brazilian municipalities. Future climate change would also act to increase poverty, as the regions that are currently the poorest are all to see substantial income reductions due to temperature increases (figure 9.3).

### Simulating the Impact of Future Climate Change in Peru

Temperature increases over the next 50 years are estimated to cause a 2.3 percent decrease in average per capita income in Peru, all other things equal. As in the case of Brazil, the average hides much larger impacts at the state and municipal levels. Incomes in the jungle state of Loreto are estimated to fall by more than 15 percent in 50 years if the region experiences a 2°C increase in temperatures, compared to a situation of no climate change. In contrast, the mountain state of Puno is estimated to experience increases in incomes on the order of 9 percent

Table 9.2    Brazil: Estimated Impact of Climate Change on per Capita Income, 2008–58

| Region | Impact on per capita income (% change) |
|---|---|
| North | −22.5 |
| Northeast | −19.8 |
| Center-west | −15.6 |
| Southeast | −7.2 |
| South | −2.9 |
| Brazil | −11.9 |

*Source:* Authors.

**Figure 9.3    Brazil: Current per Capita Income versus Expected Future Impacts of Climate Change, by Mesoregion**

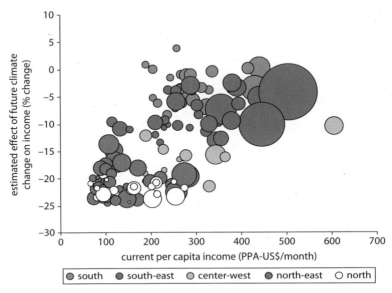

*Source:* Authors' estimation.
*Note:* The size of the bubble indicates the population size in each municipality. PPA = purchasing power adjusted.

if temperatures in the region turn out to be 1.5°C higher than with no climate change (table 9.3).

Climate change is estimated to contribute to increasing poverty in Peru but not to affect the income distribution significantly. In Figure 9.4, the estimated change in incomes for each municipality is plotted against the current level of incomes. It is seen that among the currently poor municipalities, there will both be winners and losers, with losers being in the majority. A very weak negative correlation ($\rho = -0.06$) appears between current income levels and estimated changes due to future climate change, suggesting that climate change will likely have no significant effect on the income distribution in Peru.

*Simulating the Impact of Future Climate Change in Chile*
Expected temperature increases over the next 50 years are estimated to cause a 7 percent decrease in average per capita income in Chile, other things equal. Table 9.4 shows the simulated effects of future climate change on incomes in each of the twelve macroregions in Chile. The negative

**Table 9.3    Peru: Estimated Impact of Climate Change on per Capita Income, 2008–58**

| State | Impact on per capita income (% change) |
|---|---|
| Amazonas | −8.4 |
| Ancash | −0.8 |
| Apurimac | 5.0 |
| Arequipa | 3.0 |
| Ayacucho | 3.3 |
| Cajamarca | −2.5 |
| Callao | −4.8 |
| Cusco | 5.2 |
| Huancavelica | 5.8 |
| Huanuco | −1.9 |
| Ica | −3.3 |
| Junin | 2.3 |
| La Libertad | −3.3 |
| Lambayeque | −6.7 |
| Lima | −4.3 |
| Loreto | −15.5 |
| Madre de Dios | −10.0 |
| Moquegua | 0.5 |
| Pasco | 3.0 |
| Piura | −6.5 |
| Puno | 9.4 |
| San Martin | −12.4 |
| Tacna | −0.5 |
| Tumbes | −7.7 |
| Ucayali | −14.6 |
| Total | −2.3 |

*Source:* Authors' estimations.

effects are stronger in the north, which is expected to experience more warming, than in the south, but all regions are projected to experience a significant negative impact from climate change (holding all else constant).

In Chile, as in Peru, climate change is estimated to contribute to increasing poverty but not to increased inequality between municipalities. Figure 9.5 plots the estimated change in incomes for each municipality against the current level of incomes. A very weak, barely statistically significant, negative relationship ($\rho = -0.12$) indicates that while future climate change is likely to have a negative effect on incomes and increase poverty, it is not going to contribute to increased inequality between municipalities.

**Figure 9.4    Peru: Estimated Change in Incomes Due to Future Climate Change versus Current Incomes, Municipal Level**

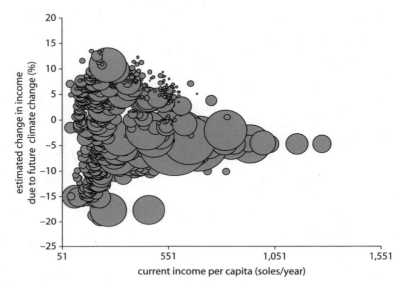

*Source:* Authors' elaborations based on estimations.
*Note:* The size of the bubble indicates the population size in each municipality.

**Table 9.4    Chile: Estimated Impact of Climate Change on per Capita Income, 2008–58**
*(percent)*

| Region | Effect of temperature changes | Effect of changes in precipitation | Total effect of future climate change |
|---|---|---|---|
| Tarapacá | −8.3 | 0.0 | −8.3 |
| Antofagasta | −7.9 | 0.0 | −7.9 |
| Atacama | −7.4 | 0.0 | −7.4 |
| Coquimbo | −7.1 | 0.0 | −7.1 |
| Valparaíso | −6.8 | 0.0 | −6.8 |
| Metropolitan Region | −6.8 | 0.0 | −6.8 |
| O'Higgins | −6.6 | 0.0 | −6.6 |
| Maule | −6.5 | 0.8 | −5.7 |
| Bío Bío | −6.4 | 0.5 | −5.9 |
| Araucanía | −6.2 | 0.7 | −5.5 |
| Los Lagos | −5.9 | −1.6 | −7.5 |
| Aisén | −5.4 | −0.6 | −6.0 |
| Magallanes | −4.6 | 0.0 | −4.6 |
| Total | −6.7 | 0.0 | −6.7 |

*Source:* Authors' estimations.

**Figure 9.5    Chile: Estimated Change in Incomes Due to Future Climate Change versus Current Incomes, Municipal Level**

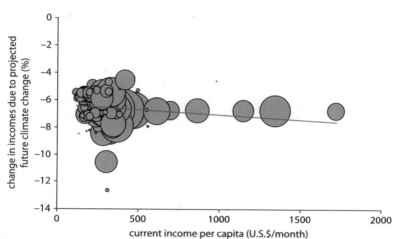

*Source:* Authors' elaborations based on estimations.
*Note:* The size of the bubble indicates the population size in each municipality.

## Simulating the Impacts of Future Climate Change in Bolivia

Expected temperature increases over the next 50 years are estimated to cause a 2.9 percent increase in per capita consumption in Bolivia, other things being equal. Table 9.5 shows the impact of future warming on consumption levels in Bolivia. As the municipality-level regression indicated that people in Bolivia are better off the warmer the climate, the simulations show a positive effect for all regions, but a particularly positive effect in the currently cold highland states of Oruro, La Paz, and Potosí.

In Bolivia, the effect of projected climate change over the next 50 years is an increase in per capita consumption levels, a reduction in poverty, and a reduction in inequality among municipalities. The most positively affected municipality is Bolívar, in Cochabamba, gaining about 6 percent. The least favored municipality is Puerto Guayaramerin, in Beni, which gains only 0.2 percent in per capita consumption. A significant negative relationship ($\rho = -0.33$) appears between the initial level of consumption and the estimated effect of future climate change, indicating that currently poor municipalities are likely to gain more from expected climate change than currently richer ones. This means that future warming would tend to contribute to a reduction in poverty as well as a reduction in inequality between municipalities.

**Table 9.5    Bolivia: Estimated Impact of Climate Change on per Capita Consumption, 2008–58**

| State | Impact of temperature changes on consumption levels (% change) |
|---|---|
| Beni | 0.4 |
| Chuquisaca | 2.6 |
| Cochabamba | 3.8 |
| La Paz | 4.1 |
| Oruro | 4.6 |
| Pando | 0.3 |
| Potosí | 4.0 |
| Santa Cruz | 0.9 |
| Tarija | 2.1 |
| All Bolivia | 2.9 |

*Source:* Authors' estimations.

## Summing Up

In this chapter municipality-level data from five large countries in the LAC region were used to estimate contemporary relationships between average climate and average incomes. The estimated relationships were then used to simulate the impact of the climate changes that the IPCC projects for the next 50 years. The simulation results should not be interpreted as forecasts but are simply indicative of the direction and magnitude of the effects that might be expected from the climate changes suggested by the IPCC models. In the simulations, all other explanatory factors than climate are held constant. That, of course, is not a realistic assumption but simply serves to isolate the effect of climate change from all the other factors that also will affect people's incomes over the next several decades. In reality, many things are going to change over the coming decades, and most of the changes will likely reduce people's vulnerability to climate change. More educated people, for example, are usually less vulnerable to climate change because their occupations are typically less dependent on climate. Carbon dioxide fertilization, technological advances, and expanded irrigation systems may work to make agriculture less sensitive to climate change, too. That such changes are not taken into account implies that the simulations carried out in this chapter may exaggerate the effects of climate change.

The results of the simulations are summarized in Table 9.6. Of the five countries investigated, three are expected to be adversely affected (Brazil, Chile, and Peru), as their average levels of income are estimated to fall,

**Table 9.6    Summarizing the Municipality-Level Analysis**

| Country | Effect of future climate change on average incomes (% change) | Effect on poverty | Effect on inequality |
|---------|---------------------------------------------------------------|-------------------|----------------------|
| Bolivia | 2.9 | Decrease | Decrease |
| Brazil | −11.9 | Increase | Increase |
| Chile | −6.7 | Increase | Neutral |
| Mexico | Neutral | Neutral | Neutral |
| Peru | −2.3 | Increase | Neutral |

*Source:* Authors' estimations.

whereas Bolivia is estimated to benefit slightly, and in Mexico incomes were found to be insensitive to differences in climate. Poverty is estimated to increase in Brazil, Chile, and Peru, decrease in Bolivia, and remain unaffected in Mexico. Inequality is estimated to increase in Brazil, decrease in Bolivia, and be unaffected in the remaining three countries.

In conclusion, although there are good theoretical arguments to the effect that climate change and climate variability would hurt the poor disproportionately and thus increase poverty and inequality, the evidence for that hypothesis is rather mixed. In Bolivia and Peru the poorest parts of the population live in the cold highlands and would likely benefit from future warming, whereas in Brazil and Chile the poorest people live in the warmest regions and would likely suffer from further warming. In Mexico, poor and rich are spread evenly across all climate zones. It should be remembered, however, that this chapter has only investigated differences in average temperatures and average precipitation, not differences in climate variability and extreme events, which are much harder to quantify and model with the currently limited information.

The municipality-level regressions presented in this chapter all show that education is far more important for income than climate. The relatively modest impacts from climate change indicated by the analysis could easily be outweighed by the positive effects of education increases over the next 50 years. In Bolivia, for example, while a 2°C increase in temperatures would cause consumption levels to increase by about 3 percent, a two-year increase in average education levels would increase per capita consumption levels by 47 percent. In Brazil, a 2°C increase in temperatures is estimated to cause a 12 percent reduction in average incomes, but a two-year increase in average education levels is estimated to cause a 94 percent increase in incomes. In Chile, it would take less than 0.2 year of additional education to counteract the estimated negative effect of climate change over the next 50 years. Thus it would appear that a very efficient way of counteracting potential adverse effects of climate change on incomes would be to invest more in education. It is important to keep in mind, however, that in the absence of climate change, incomes would benefit substantially more from investments in the education sector.

## Notes

1. Notice that climate variability is not addressed in this model, only systematic climate change over several decades.
2. All axes span the actual observed values of temperatures and income in each country. More details about the data for each country can be found in the more elaborate, individual country case studies (Andersen and Verner 2008, 2009, 2010; Andersen, Román, and Verner 2008; Andersen, Suxo, and Verner 2009).

## References

Andersen, L. E., S. Román, and D. Verner. 2008. "Social Impacts of Climate Change in Brazil: A Municipal-Level Analysis of the Effects of Recent Climate Change on Life Expectancy, Consumption, Poverty, and Inequality." Draft, World Bank, Washington, DC.

Andersen, L. E., A. Suxo, and D. Verner. 2009. "Social Impacts of Climate Change in Peru: A Municipal-Level Analysis of the Effects of Recent Climate Change on Life Expectancy, Consumption, Poverty, and Inequality." Policy Research Working Paper No. 5091, World Bank, Washington, DC.

Andersen, L. E., and D. Verner. 2008. "Social Impacts of Climate Change in Mexico: A Municipal-Level Analysis of the Effects of Recent Climate Change on Life Expectancy, Consumption, Poverty, and Inequality." Draft, World Bank, Washington, DC.

———. 2009. "Social Impacts of Climate Change in Bolivia: A Municipal-Level Analysis of the Effects of Recent Climate Change on Life Expectancy, Consumption, Poverty, and Inequality." Policy Research Working Paper No. 5092, World Bank, Washington DC.

———. 2010. "Social Impacts of Climate Change in Chile: A Municipal-Level Analysis of the Effects of Recent Climate Change on Life Expectancy, Consumption, Poverty, and Inequality." Policy Research Working Paper No. 5170, World Bank, Washington, DC.

Dell, M., B. F. Jones, and B. A. Olken. 2008. *Climate Change and Economic Growth: Evidence from the Last Half Century.* NBER Working Paper No. 14132. Cambridge, MA: National Bureau of Economic Research.

Horowitz, J. K. 2006. "The Income-Temperature Relationship in a Cross-section of Countries and Its Implications for Global Warming." Department of

Agricultural and Resource Economics, University of Maryland, Submitted manuscript, July. http://faculty.arec.umd.edu/jhorowitz/Income-Temp-i.pdf.

IPCC (Intergovernmental Panel on Climate Change). 2007a. *Climate Change 2007: The Physical Science Basis.* Contribution of Working Group I to the Fourth Assessment Report of the IPCC. Geneva: IPCC.

———. 2007b. *Climate Change 2007: Impacts, Adaptation, and Vulnerability.* Contribution of Working Group II to the Fourth Assessment Report of the IPCC. Geneva: IPCC.

Timmins, C. 2007. "If You Cannot Take the Heat, Get out of the Cerrado . . . Recovering the Equilibrium Amenity Cost of Nonmarginal Climate Change in Brazil." *Journal of Regional Science* 47(1): 1–25.

# CHAPTER 10

# Building Short-Term Coping Capacity and Longer-Term Resilience through Asset-Based Adaptation

## Tine Rossing, Olivier Rubin, and Inger Brisson

Not all natural hazards turn into natural disasters, as highlighted in chapter 3. The link between the two depends on vulnerability, which is a function of the character, magnitude, and rate of climate change to which a system or community is exposed, and the sensitivity of the system and its capacity to cope with the negative impacts that such exposure might bring to individuals or human systems. The most vulnerable to climatic hazards are the poor in high-exposure countries (IPCC 2007; Task Force 2003; WWF 2006).

Following on from chapter 3, which used the sustainable livelihoods framework (SLF) to explore how exogenous shocks might destroy and deplete livelihood assets, the present chapter explores the reverse relationship: how different types of assets (human, social, physical, financial, cultural, and natural) can be strengthened to mitigate the impacts of natural disasters and help people adapt to both rapid and slow-onset climate change. Case studies help identify best-practice ways to strengthen people's ability to adapt their livelihoods to a changing climate. Some cases focus on how building resilience can enable people to withstand both slow-onset hazards such as drought or sea level rise, and rapid onset hazards such as hurricanes or floods. Other cases emphasize adaptation measures that help people cope with their aftereffects.

The chapter focuses particularly on the part that social assets—namely bonding, bridging, and linking social capital, as defined below—can play in vulnerable communities and how they can affect a community's ability to engage with external entities such as nongovernmental organizations (NGOs )and state institutions (local and federal) in times of distress. As Adger (1999) notes, the way in which individuals and groups within a society interact with one another will influence their vulnerability to climate change, including variability and extremes, notably through mechanisms such as risk sharing, mutual assistance, and collective action. Social assets are often under stress just before, during, and after rapid-onset climate-related disasters, but also in the case of slow-onset disasters or changes, such as droughts or sea level rise, that cause continual stress. As a result, social assets can be eroded, particularly when stress is prolonged or recurrent.

Thus far, attention to social assets in disaster management has been rare, and the tendency has been to treat disaster management mostly as an engineering issue calling for technical solutions.[1] But different groups are not exposed to natural hazards in equal measure. The vulnerability, resilience, and livelihood assets of a community depend on its location-specific biophysical, social, and economic circumstances. Vulnerability is defined as the degree to which a system or community is susceptible to, or unable to cope with or adapt to, adverse effects of climate change, including climate variability and extremes (IPCC 2007). The opposite side of the coin is resilience; the less vulnerable a community is, the more resilient it is. Resilience is the ability of a system or community, be it social or ecological, to withstand shocks and surprises and revitalize itself when damaged (Tompkins and Adger 2004). In this chapter, the term "coping capacity" refers to the short-term ability to safeguard oneself and a given asset base, especially during and immediately following an extreme climate-related event such as a flood or hurricane. "Adaptive capacity" refers to the long-term planning and management of assets, before or during recovery from such an event or in response to gradual climate change. Clearly, vulnerability, resilience, and the ability to cope vary from one community to another. This means that interventions to help people prepare for, cope with, and adapt to the effects of climate change need to focus not just on areas and numbers of people at risk, but on who is at risk and the type of risk they face.

The chapter is structured as follows. The first of its three sections outlines the framework used for analysis. The SLF is extended to distinguish among three different types of social capital that, in principle, communities can draw on when faced with natural hazards. Also, recognizing that livelihood assets can be used either to avoid (or mitigate) the impact of climate hazards on livelihoods or to adapt to their consequences, the discussion extends the SLF to accommodate a distinction between these two different coping strategies: (a) enhancing the longer-term resilience of the livelihood to a climate impact through adaptation, and (b) strengthening coping and recovery capacity, as a more immediate-term response to a climate event, also through adaptation. In the second section of the chapter, the augmented SLF is applied to five diverse cases: experiences with droughts in southern Bolivia and in northeast Brazil illustrate the importance of increasing livelihood resilience through adaptation to slow-onset climate-related disasters. Experiences with hurricanes in Nicaragua and floods in Belize provide examples of successful livelihood adaptation measures that improve short-term coping and recovery from sudden extreme events. All five cases highlight the importance of social capital and the role that institutions, especially local ones, can play in generating synergy among the different types of capital and thus increasing the effectiveness of coping strategies. Experiences reported in other chapters of this book support these findings. The chapter concludes by drawing some implications for policy, emphasizing the importance of approaches that focus on the livelihood strategies of poor communities and also on local governance structures, which play a key role in determining the impacts of natural hazards on livelihoods.

## Social Capital and Livelihood Adaptation and Resilience

The following section outlines the framework used for analysis. The SLF is extended to distinguish among three different types of social capital that, in principle, communities can draw on when faced with natural hazards. Also, recognizing that livelihood assets can be used either to avoid (or mitigate) the impact of climate hazards on livelihoods or to adapt to their consequences, the discussion extends the SLF to accommodate a distinction between these two different coping strategies: (a) enhancing the longer-term resilience of the livelihood to a climate impact through adaptation, and (b) strengthening coping and recovery capacity, as a more immediate-term response to a climate event, also through adaptation.

## Bonding, Bridging, and Linking Social Capital

It is generally accepted that "social capital" refers to the trust, social norms, and networks that affect social and economic activities. Nakagawa and Shaw (2004) broadly define "social capital" as "the function of mutual trust, social networks of both individuals and groups, and social norms such as obligation and willingness toward mutually beneficial collective action." That general definition will be used in this chapter, where the "collective action" is either mitigation efforts before a natural disaster or recovery and adaptation efforts afterward.

To assess the resilience and the adaptive capacity of vulnerable communities in response to climatic disasters the analysis relies on the categories of social capital defined by Woolcock (2000): bonding, bridging, and linking. "Bonding social capital" refers to ties among people who tend to be closely connected, such as immediate family members, neighbors, close friends, and business associates sharing similar demographic characteristics. "Bridging social capital" denotes ties among people from different ethnic, geographical, and occupational backgrounds who have similar economic status and political influence. Hence, whereas bonding social capital is based on kinship, friendship, and loyalty, bridging social capital is arguably based mostly on trust and reciprocity. For enforcing obligations, bonding capital tends to rely on rules of enforcement and sanctions through informal collective action, while bridging capital, because its bonds are weaker, tends to rely on legal and formal institutions (Adger 2003). Woolcock (2000) defines linking social capital as the ties between the community and people in positions of influence in formal organizations, such as schools, agricultural extension offices, the police, or local or national government entities.

Figure 10.1 provides an overview of the roles the three types of social capital can play in improving the resilience and adaptation of communities facing natural hazards. It illustrates, for example, how both bridging and linking social capital can provide external support from informal and formal institutions in times of distress.

As Woolcock (2000) observes, poor people tend to have strong bonding social capital and some level of bridging social capital, but generally little linking social capital with formal organizations—often to the detriment of their economic development. This is problematic because linking social capital plays a critical role in actually reducing livelihood vulnerability. It does this by facilitating access to crucial physical and financial assets such as infrastructure, credit, and insurance. People who are marginalized may not be in a position to make effective claims for

**Figure 10.1    Three Types of Social Capital for Communities Faced with Natural Hazards**

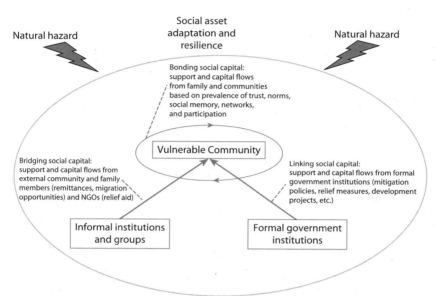

*Source:* Authors' elaboration.

assistance following a natural disaster or other exogenous shock. In such situations intermediary organizations, such as developmental NGOs, may be key in creating links between vulnerable people and public institutions. It should be noted, however, that there might be tensions as well as complementarities between different types of social capital. Linking social capital often requires the presence of a small number of people with the necessary social skills and education to communicate with external agencies. The people with these attributes are often young men, many of whom have spent years outside their communities acquiring those skills. When they exercise the skills in their home communities they may undercut the bonding social capital that is expressed in respect for elders and local and traditional knowledge—as for instance in traditional indigenous communities—and they may also exacerbate rifts between different segments of their communities.[2]

Without linking social capital, poor communities in times of stress (after a natural disaster or a crop failure resulting from sudden climate variability or drought) must to rely mostly on their human and bonding social capital (relatives in the community) as a very fragile safety net.

They must also draw on their other scarce assets to cope. The fabric of bonding capital tends to be strong in a vulnerable community, being related to strong links of reciprocity or empathy, and so bonding capital may be helpful for building specific reciprocity and mobilizing solidarity within the community. But a risk exists that the closed-circuit relationships on which it is based might be broken by a large climate shock. Bridging capital has advantages for building links to external agents and for disseminating information. However, although NGOs can assist in creating links to public institutions and making claims for assistance, both bonding and bridging capital are also important for effectively making claims on public institutions. For instance, community mobilization can be a key factor in obtaining the assets that strengthen resilience through making claims on public officials and political actors. This applies both in emergencies and in achieving access to important services on a routine basis, helping build resilience.

Overinvestment in certain types of social capital could be counterproductive, and diversification among different types of social capital is of great importance for both the individual agent and the larger society. Inclusion and trust among some social groups could lead to (or show themselves in) exclusion and mistrust toward other social groups. In addition, very strong bonding capital can generate islands of microsocieties within the larger society (Adler and Kwon 2002; Putzel 1997). For an individual agent, social capital is not costless to build, so it follows that one can invest too much in maintaining social linkages and trust relations.

In analyzing susceptibility to adverse impacts of climate change, the group of interest need not be an entire community determined geographically, but could be narrower, or could be defined on the basis of ethnicity, religion, age, gender, occupation, and so forth. Hence it will often be necessary to distinguish among the different types of social capital—and in particular to focus on linking capital because vulnerable people are susceptible to discrimination and stigmatization from state institutions at all levels.[3] The following example illustrates the roles played by the different types of social capital in a vulnerable community.

### Social Capital and Adaptive Capacities in Chiapas, Mexico

Experience in Mexico's southern state of Chiapas shows how coping mechanisms and adaptive capacities can be stretched to the limit in poor communities after recurrent or major natural disasters, especially when governmental support is insufficient. Chiapas is particularly vulnerable to climate events because of its geographic exposure, low incomes (more than two-thirds of the population is poor), and great reliance on climate-sensitive sectors such as agriculture (Saldaña-Zorilla 2007).[4]

In the most vulnerable communities in Chiapas—Cacahoatan, Escuintla, and Cintalapa—a qualitative study (Saldaña-Zorilla 2006) asked respondents to rate different possible sources of financing for disaster relief, by scoring them between 1 (*irrelevant*) and 5 (*very important*). The study found that in times of stress community members rely more on their bonding capital (relatives in the community), than on their bridging (neighbor solidarity) or linking (government aid) capital to overcome their social vulnerability. Relatives living within the community were deemed to be the most important source of disaster finance (scoring 332), followed by governmental aid (270) and neighbor solidarity (266). Less important were the sale of property (207); governmental funds (173); sale of land (170); and community loans (161).

The importance of relatives living in the same community and aid from the government underlines the high dependence on ex-post financial instruments and points to a need to implement prevention measures. In communities that had been affected repeatedly by climate shocks but lacked formal financial mechanisms to respond, people tended to seriously consider migration as an adaptive mechanism: 41 percent of the interviewees stated that they had current plans to migrate. As seen in chapter 7, migration is increasingly being used as a coping mechanism, supported by bridging capital; that is, would-be migrants plan to rely on an existing familiar network, but one located elsewhere, whether domestically or abroad.

These responses could be viewed as an expression of people's lack of trust in the government. The respondents also showed a strong perception that the state was not providing effective, long-term, sustainable strategies to strengthen the capacity of subsistence farmers in Chiapas to cope with, and adapt to, external shocks (Conde, Sakdaña, and Magaña 2007). Moreover, the lack of such strategies was perceived as an obstacle to the accumulation of assets.[5]

These concerns are acknowledged by some government officials responsible for social programs. They see the relevance of integrating the goal of disaster reduction into poverty reduction strategies to meet overall goals. The main challenge to doing that seems to consist of combining risk assessments with vulnerability analyses at the municipal level. In Mexico, risk identification and vulnerability analysis at the municipal level are the main legal requirements to apply for funding for mitigation works and insurance. Thus the success of the country's disaster vulnerability strategy depends largely on the initiative shown by the municipal authority.

Compounding the effects of adverse weather, several elements of national policy make life more difficult for poor farmers in Chiapas. The first is national agricultural policy: with domestic agricultural prices pushed down by low-priced farm products from Mexico's main trade partners, incomes have declined for farmers who cannot respond by raising their yields or enlarging their cultivated area (ECLAC 2001 2003; World Bank 1994). Falling incomes reduce these farmers' ability to create a financial reserve for difficult times, creating a vicious circle of low resilience and coping capacity, low savings, lack of climate-adaptive instruments (crop insurance, a reserve fund), and greater vulnerability to disasters. Second, a lack of long-term credit precludes improvements for many farmers. Adaptation measures such as auxiliary irrigation and commercial crop diversification require financing. Although some limited credit windows exist for smallholders, the support is generally not extensive, and most households increasingly depend on alternative income sources to finance their agricultural activities.[6] Third, even though insurance can be a powerful tool for reducing vulnerability to climate change and variability by transferring or sharing risk, available coverage is still insufficient in Mexico for people and enterprises of modest means (Kreimer and others 1999). Weather-indexed insurance can help farmers protect their overall income, reduce their vulnerability to climate variability, and enable them obtain access to credit (OECD 2006). However, governments, households, and businesses in poor countries cannot easily afford commercial insurance to cover their disaster risks.

For an individual household—without bridging or linking social capital support in the form of credit or insurance from banks or the government—disasters aggravate poverty, creating a need to take out high-interest loans (or default on existing loans), sell assets and livestock, or engage in low-risk, low-yield farming to lessen exposure to extreme events (Linnerooth-Bayer, Mechler, and Pflug 2005).

In Chiapas in times of stress, community members rely mostly on their bonding capital (their relatives in the community) as their safety net. This example supports the earlier point that insufficient bridging and linking capital—and the resulting insufficient credit and insurance—can keep a community from reducing its livelihood vulnerability. Their lack of bridging capital is compelling communities to erode their existing asset base to cope.

## Building Livelihood Resilience and Coping Capacity through Adaptation

Livelihood adaptation is a response to climatic change (both gradual and abrupt) that involves changes in livelihood strategies. When adaptation measures focus on enhancing resilience—whether at the community, household, or individual level—they improve the capacity to absorb both climate shocks and more gradual climatic impacts in the longer term. Adaptation can also be directed to helping people cope better in the short term, especially in the face of a shock such as a hurricane, and recover faster afterward.

Just as all types of livelihood assets—natural, human, physical, financial, cultural, and social—can be eroded by gradual climate change and climate-related disasters (as seen in chapter 3), all types can be used to build people's longer-term resilience to both slow-onset climate change effects and rapid-onset hazards. They can also enhance short-term coping capacity in the face of shocks. Hence, livelihood adaption should focus on how to use the target groups' asset base to build effective livelihood resilience and coping capacity.

**Livelihood resilience.** Livelihood resilience is understood as the ability to withstand external climate stresses (both gradual ones and shocks) and as a longer-term revitalization of livelihoods after damage to livelihood assets has occurred. The extent of resilience depends on the physical exposure to natural hazards, as well as ability to apply the various forms of capital to shield livelihoods.[7]

An illustration of lack of resilience is provided by the severe flooding in República Bolivariana de Venezuela in 1999, when unusually heavy rainfall triggered a disastrous chain of events in several states of the country. With an estimated 30,000 people killed, 600,000 affected, and 114,000 left homeless, the flooding was the most severe weather-related natural disaster in contemporary Latin America. In the worst-hit areas, more than twice the average annual rainfall (between 400 and 900 millimeters) fell within 72 hours, generating flash floods, massive landslides, and debris flows in densely populated areas. The disaster was exacerbated by soil erosion on mountainsides, caused by urban development, deforestation, and squatters, particularly illegal immigrants from neighboring Colombia, Ecuador, and Peru.

Poor households living in shantytowns bore the immediate brunt. Their dwellings were located on steep, unstable hillsides or in the dry riverbeds directly below those hillsides outside larger cities. Mudslides buried most of several towns, and others disappeared completely under avalanches of mud. Many poor communities were completely wiped out. These households had few protective livelihood assets with which to offset the effects of their physical exposure: low physical capital (rickety shacks with limited access to infrastructure, such as roads), lack of financial capital (many were poor, possibly illegal migrants), insufficient natural capital (because of the urban setting), limited human capital (making livelihood diversification strategies difficult), low political capital (limited capacity to claim access to basic services on a routine basis or to claim emergency assistance, for lack of state attention and protection), and a stock of social capital too small to protect against such a large covariate shock as this flood.

The policy challenge in improving resilience to natural hazards is that until disaster strikes, vulnerable groups may not appreciate the importance of resilience and even if they are aware of which natural hazards might threaten them, they will not necessarily recognize how to change their asset use to improve their resilience. People living in complete destitution are forced to focus on the assets required for day-to-day survival, but those are typically not the type of assets needed for surviving a crisis. In addition, climate disasters often alter livelihood assets so much that capital accumulated for the former everyday context is rendered useless. Thus the resources invested in strengthening livelihood resilience to natural disasters are likely to be less than optimal.

The same holds true for social investments. As mentioned earlier, bonding social capital plays a vital role for many poor people's livelihoods, as a protection against shocks such as job loss, disease, or a sudden need for credit. But in the wake of large covariate disasters, when entire communities have been shattered, other types of social capital take center stage. Bridging social capital—in this case more distant, cross-cutting connections to people with limited exposure to the disaster—as well as linking social capital—here understood as hierarchical linkages with state agencies capable of implementing disaster mitigation measures or providing emergency relief—become crucial parts of adaptation, resilience, and coping strategies.

Social resilience can be strengthened by the institutional and social memory that is present in societies that face natural hazards frequently (Adger and others 2005). Other things equal, households with more experience of managing disasters are less vulnerable than households with less experience. Social memory draws on reservoirs of practices, knowledge, values, and worldviews from individuals, communities, and institutions.

Such collective memory will often be crucial for preparing a given system or location for change, building resilience, and coping with surprises. For example, flooding is becoming increasingly common in poor urban areas with limited social memory of disasters. As a result, as Moser and Satterthwaite (2008) describe, responses tend to be mostly ad hoc, individualistic, and short term, such as sleeping on furniture, households' digging trenches around their property, and building small-scale water barriers. Vulnerable groups can enhance their resilience if they can use linking social capital routinely to make claims for access to basic services such as sanitation, water supply, and flood defenses. Having rights to basic public services and claiming those rights in ordinary times will strengthen the general social resilience of a community and make it better prepared to cope with adverse weather-based shocks. Citizenship is key in this process because it provides ways to increase linking social capital. By the same token, lack of citizenship (often the case for migrants) may leave people with too little livelihood resilience against adverse events.

*Livelihood adaptation.* Livelihood adaptation is understood as the changes people make to adapt their livelihoods to a new situation, in response to or in preparation for increased climatic variability or climate change (both gradual and extreme events). Successful adaptation depends on the short-term response to disasters (degree of coping), as well as on long-term rebuilding and long-term response to gradual climate change (degree of resilience). In this regard, the adaptation process depends crucially on the assets that people have available and on pro-poor social policies, including safety nets and indexed insurance. However, in some cases people may realize neither their exposure nor their vulnerability to climatic impacts, in which case it tends to take a disaster to make them aware of the need for adaptation.

In the 1999 flooding in República Bolivariana de Venezuela, recovery and adaptation to the changed circumstances was hampered by the erosion of several kinds of livelihood capital:

- Much physical capital was depleted by the disaster, and rebuilding was still going on years later.
- Access to financial capital was limited after the disaster because of lack of money, credit, and insurance.
- Human capital was severely affected or destroyed by both the direct impact of the disaster in injuries and loss of life and the indirect effects in the forms of disease and unemployment.

- The magnitude of the floods tore apart the social fabric of families and whole communities; bridging social capital, in the form of remittances from abroad, was an important source of income for the affected households.

Natural disasters can have very long-term implications for livelihoods and can cause a need for adaptation. Twenty-five years after the 1976 Guatemalan earthquake, for example, people reported that they were still living in homes that the quake had badly damaged (De la Fuente 2007). Likewise, Hurricane Mitch washed away vital topsoil, replacing it with rocks and rubble in numerous places, adversely affecting farming and altering livelihoods for years afterward. The same challenge was also dealt by the floods to the poor communities in Mecapaca, Bolivia (chapter 3, box 3.3).

Adaptive capacity, like livelihood resilience, partly depends on social memory and capital. These can be substantially changed by a natural disaster. Families may be torn apart and communities destroyed. Bonding social capital, for example, could become restricted to immediate family members; individualistic tendencies are not uncommon in a postdisaster environment (Rosenfeld and others 2005). If the rule of law breaks down, people have opportunities as well as a need (extreme scarcities) to engage in individualistic behavior, and if disillusioned by a lack of state help they may decide to take matters into their own hands.

Where social capital is sufficiently resilient to withstand a disaster, it can play an important role in livelihood adaptation, manifested in strong kinship ties, benevolent reciprocity, and supportive relationships. But such benign social behavior is not fostered by the disaster itself; any goodwill displayed in the aftermath of a disaster stems from the existing stock of social capital.[8] In the Nicaraguan community of El Mirador, after Hurricane Mitch, where there was little interaction among the residents, the lack of cohesion had roots going back many years, with many people being poorly integrated (Tomlinson 2006).

Social capital is also the main determinant of whether the postdisaster recovery will unite or divide people. The popular saying that disasters bring communities closer together holds true only if communities are already close, with enough, and the right type of, social capital before the disaster (see the Bolivian case described later in this chapter). Covariate risks, where many members of the community face the same adverse impacts, are likely to promote collective action only when the social infrastructure is already in place; by themselves, those risks might lead to a collapse of social reciprocity (Agrawal and Perrin 2008).

In situations where all community members face similar risks, the state or local institutions can play an important role in (re)building social capital as well as physical infrastructure. These two aspects can be referred to as "soft adaptation" and "hard adaptation," respectively (Agrawal and Perrin 2008). Strengthening linking social capital, where the state is directly involved in the interrelation, is an important first step, as are state efforts to help regenerate bonding social capital at the community level. A logical next step would thus be to enhance the role of local institutions in (a) addressing the potential underinvestment in social capital (in particular bonding and bridging), so as to strengthen livelihood resilience, and (b) regenerating social capital (in particular bonding and linking) as part of an adaptation strategy.

In what follows, the augmented SLF framework is applied to four cases. Two concern the need for enhancing longer-term livelihood resilience through adaptation, and three address the need for enhancing immediate or short-term coping capacity through adaptive measures. All emphasize the importance of strengthening social capital and working with context-specific measures at the local level.

## A Social Approach to Disaster Analysis: Case Studies

The outcomes of natural disasters cannot be captured solely in mortality rates, numbers of affected people, or economic estimates. As noted earlier, vulnerability to climate change–related shocks varies among communities and among individuals. And not only are many impacts of disasters felt over the long term, they also have substantial aggregate social implications—perhaps including deterioration of mental and physical health, an increase in poverty and inequality, migration, and conflict—as well as household implications such as lack of income, homelessness, domestic violence, family breakup, insecurity, destruction of infrastructure, and eroded assets.[9]

To provide a strong foundation for the design of effective, asset-based adaptation that enhances livelihood resilience and coping capacity, a context-specific social analysis of livelihood assets is needed. Such an analysis, inspired by the vulnerability typology developed by Moser and Satterthwaite (2008), could consist of four points:

1. Identifying vulnerable groups living in places prone to climate-induced natural hazards (to identify where to build long-term protection against both slow- and rapid-onset impacts)[10]

2. Examining the extent to which the groups are capable of taking action
   to avoid the effects of climatic impacts (predisaster damage limitation), in anticipation, for example, of future hurricanes, or when faced
   with a gradually worsening situation such as drought
3. Investigating the ability of different groups to cope with the immediate effects of natural hazards, that is, their ability to judge whether an
   immediate response is required, especially when a natural hazard has
   turned into an actual disaster
4. Analyzing how different groups are able to adapt over the long term
   by undertaking planned actions after a slow- or rapid-onset climate
   event (for example, rebuilding in a manner adapted to heightened exposure or threat) in anticipation of similar or worse climatic impacts
   in the future

The first two aspects relate to how different types of assets—physical,
financial, human, social, cultural, and natural capital—can protect against
livelihood shocks caused by climatic impacts before they happen, thereby
possibly preventing the eventual climate event from turning into an
actual disaster. The latter two aspects refer to the ability to cope with the
shocks immediately after they happen, especially once the natural hazard
has developed into a disaster, and to recover and become more resilient
in the longer run and hence better prepared for future stresses.

Social analysis of vulnerability should also consider how needs evolve
over time. People need different constellations of assets during different
stages of a climate change impact: before, during, in the immediate aftermath, and for long-term recovery and adaptation. For instance, investments in early warning systems and training are crucial elements of
enhancing livelihood resilience, whereas financial capital such as credit or
insurance is vital for long-term recovery and adaptation. The focus should
therefore be on enhancing the specific mix of livelihood assets that will
provide the greatest adaptive capacity, resilience, and coping capacity of
communities in the face of climate change affecting a given location.

Good public policies to help vulnerable groups cope and adapt, based
on anticipated impacts, are needed not only ex-post (after a given impact),
but also ex-ante (before an expected disaster strikes or in anticipation of
impacts of slow-onset climate change). Increasingly, it is recognized that
there is scope for creating and expanding social programs covering adaptation as well as ex-ante weather-risk management and ex-post short-term
coping support. Promising approaches include social funds and other support for community-driven adaptation, safety nets for coping with climate

risks and natural disasters, livelihood programs, microfinance, and index insurance. The common feature of such programs is the addition of a climate-risk dimension to social protection programs that have traditionally focused solely on the problems of poverty.[11]

### Applying a Social Approach to Livelihood Resilience: Case Studies in Bolivia and Brazil

Experience with droughts in southern Bolivia and in northeast Brazil illustrates the importance of increasing livelihood resilience when confronted with climate-related disasters. The Bolivian case provides a good-practice example of how social capital can be fostered and used to develop a vulnerable community's livelihood resilience through measures to prepare for climate-related emergencies. The situation in northeast Brazil emphasizes the need to foster adaptation as well as to develop livelihood resilience; despite the adaptation policies that are in place in the state of Ceará, many communities remain vulnerable to climate change and variability. This case supports the argument that the focus on postdrought policies should be supplemented with investments in capacity building for community preparedness.

**Social capital and livelihood resilience in drought-prone Southern Bolivia.** Bolivia's Chaco region is a flat, drought-prone territory of scrubland and thorny trees.[12] The region's existing drought will very likely be aggravated by the changes in climate and climatic variability projected for Bolivia. Until recently, Chaco's rural communities were not prepared for emergencies, resulting in the loss of livelihood assets such as homes, livestock, and crops during climate disasters.

In November 2004, a prolonged drought severely threatened the food security, health, and nutritional status of about 180,000 people, of whom 26,000 were children. The area was declared a national disaster area, as food insecurity worsened until the next harvest six months later. The losses affected the communities' productive base, increased poverty, and in some cases triggered migration. The poorest people were affected the most. To cope with the impact of the drought on their livelihoods, the communities sought assistance from municipal governments and local organizations, especially food and water. Yet, just as in the Chiapas, Mexico, case described above, interfamily solidarity and networks—bonding social capital—were often their first survival mechanisms.

Through a 14-month initiative implemented by CARE International, the communities succeeded in reducing their poverty by minimizing disaster impacts on their economic, social, and natural capital. The initiative was implemented from February 2006 to April 2007 to develop community emergency preparedness with a risk management approach, targeting 5,500 community members and involving 500 people, including municipal and local school officials. The following steps were used:

- Incorporating a risk management approach (RMA) into development planning
- Training and providing equipment to volunteer groups
- Working with teachers to incorporate RMA into classroom teaching
- Undertaking risk analysis of the communities' watershed management practices and helping to improve them
- Carrying out demonstration projects
- Disseminating and exchanging experiences

As a result of the initiative, the communities, government institutions, and other actors in five municipalities of Chaco were enabled to respond to emergencies, mainly caused by drought, and prevent their harmful impacts.

Many of the major challenges that the initiative encountered were directly linked to the lack of social capital in the drought-prone areas. A key lesson learned was the importance of enhancing human capital and different types of social capital to ensure the longevity of the progress achieved. For example, even though their bonding social capital was strong, the communities tended to give priority to concretely visible works (physical capital), rather than to training (investment in human capital) or to establishing committees (important for bridging social capital) to raise awareness about existing regulations and standards. To address the need for training, education—seen as an essential tool for behavioral change—was used to promote a culture of prevention. A guide to the incorporation of risk management as a crosscutting theme was developed with the participation of key actors. Sensitization and training activities were also carried out with teachers and students, who eventually organized educational fairs and theater and puppet festivals and supplied libraries with material on risk management. As to committees, new volunteer groups were formed and trained in first aid, forest and water management practices, and damage and needs assessment.

Bonding social capital was enhanced by encouraging participatory activities among the communities. Demonstration projects were one of the main prevention activities and included improving water reservoirs, building a pedestrian bridge, reforesting riverbanks, protecting an irrigation channel, and channeling a stream—all carried out by community members themselves. Respecting the local processes of participatory planning ensured ownership of the overall initiative, the empowerment of social sectors to make decisions related to local planning, and the legitimacy of the commitments made. These small projects also helped to foster trust and increase participation among the participants. Especially in rural areas, community members became more aware of the need to reduce their vulnerability in ways supported by the demonstration projects.

Efforts to strengthen linking social capital, such as formal collaboration and interaction with government officials, were also promoted. One example was an attempt to properly inform and sensitize municipal governments, which are normally mostly interested in public works, about the importance of funding risk management activities and about the funding process. To enhance governance, alliances were established with key stakeholders such as the vice-minister of civil defense and cooperation of integrated development (VIDECICODI), departmental prefectures, municipal governments, and community-based committees. Moreover, structures were established for emergency preparedness and response through workshops for disseminating legal standards, strengthening existing alliances, and the formation of municipal interinstitutional committees. Establishing agreements with municipal governments and civil defense branch offices was of the utmost importance: it allowed for clear identification and distribution of roles and responsibilities among stakeholders and also encouraged transparent municipal administration in coordination with strategic allies. The linking social capital of these new committees was further enhanced through the participatory development of knowledge, tools, and methods—such as community and municipal contingency plans—for a timely response to emergencies.

***Drought response in northeast Brazil: the need to strengthen livelihood resilience.*** Some 15 million people live in the rural areas of northeast Brazil, and up to 75 percent of the land in this part of the country is at risk of desertification.[13] Climate projections for the area indicate a strong likelihood of higher temperatures and less precipitation, increasing the risk of long-term droughts. Repeated droughts in the area already deplete key assets, particularly for the most vulnerable people. Though the state has initiated livelihood adaptation programs, the programs seem unable to build up the necessary resilience among the rural poor.

Droughts in northeastern Brazil have huge livelihood implications because people rely heavily on natural capital; some 40 percent of the economically active population depends on agriculture (Lemos 2007). In the poor state of Ceará, which has been experiencing a drought every 3½ years, as much as 96 percent of the agriculture is rain fed, and an estimated 90 percent of small-scale farmers (many of whom are tenants) have no sources of income outside their farms (Brant 2007). The strong dependence on natural capital, together with the great risks associated with the flow of natural capital (due to variations in the timing, locality, and amount of precipitation), leaves many families vulnerable to livelihood shocks. More than half the households in the northeast are considered food insecure (IBGE 2004). In this region, dominated by subsistence farming practices, long-term drought has livelihood implications and exacerbates existing tendencies such as lack of long-term planning or investments, neglect or depletion of human capital (as children work on farms instead of attending school), dependence on state transfers, and migration as a means of survival. Adding to these stresses, the opportunity to diversify from natural capital to financial and human capital is under pressure from the mechanization of agriculture on plantations.[14]

Vulnerability to drought is critically defined by the highly unequal distribution of power and resources in the northeast (Lemos 2007), as reflected in extreme inequality of land tenure and in marginalized groups' lack of irrigation and lack of access to government transfers such as pensions and public works schemes.

The Brazilian government is highly involved in livelihood adaptation, implementing multiple projects such as the Zero Hunger and the Sâo Francisco River Transportation projects. Yet the effect of these relief efforts has not always been constructive. The OECD (2006) reports that the situation for the bottom 20 percent of the population has actually deteriorated "because they fell outside the remit of the formal economy and were thus not covered by pensions and other programs." As stated before, weak linking social capital—in this case, lack of access to public entitlements—weakens both the resilience and the adaptive capacity of many vulnerable households in times of stress. Among the eligible households, the extensive government transfers also appear to enhance clientelistic relationships with the state and increase the risk of corrupt misappropriation and misuse of drought relief (Lemos 2007).

It might be advantageous to focus instead on greater efforts more to enhance the resilience of vulnerable groups. While the Brazilian government is attentive to the plight of many small-scale farmers and landless laborers in the region, little has been accomplished thus far in this regard. Some resilience policies have sought to increase water supply, with some success. However, as our use of the SLF forcefully shows, water supply is just one of many factors determining livelihood resilience. In a context still characterized by inequality of access to land and property rights, as well as liquidity constraints, it is not enough to increase the natural resource stock. Duarte and others (2006) concluded that government interventions have consisted of postdrought social security and welfare payments, with little investment in capacity building for community preparedness. Lemos (2007) argues that instead of focusing mostly on drought relief (coping), the government should also address longer-term redistribution policies (adaptation that enhances resilience).

Together these experiences in southern Bolivia and northeast Brazil illustrate the importance of adaptation to increase livelihood resilience in communities confronted with climate-related hazards. From the southern Bolivian case a key lesson is the importance of enhancing human and different types of social capital, to ensure the longevity of project progress. The situation in northeast Brazil emphasizes the importance of focusing on building livelihood resilience through adaptation, as well as enhancing coping capacity. Because many communities remain vulnerable to climate change and variability, the current focus on postdrought policies should be supplemented with investments in capacity building for community preparedness.

### Applying a Social Approach to Adaptation and Coping Capacity: Case Studies in Nicaragua and Belize

Much of the literature on mitigating or preparing for weather-related disasters focuses on hard adaptation strategies, emphasizing investments in infrastructure and technology, rather than on soft adaptation through social and institutional development. The three cases that follow focus on the social aspects of successful coping and livelihood adaptation measures after climate change–related disasters, to draw lessons about the role of local governance, NGOs, and communities in enhancing coping capacity and effecting livelihood adaptation. In Nicaragua, local-level responses to Hurricane Mitch (1998) signaled a change of adaptation focus away from big public infrastructure projects to softer strategies relying on community development approaches with heavy NGO involvement. In a drought-prone region of northwest Nicaragua, a pilot program including nutrition and education interventions has succeeded in reducing the vulnerability of incomes to weather-related risks. In Belize, experience from recent floods illustrates how different types of social capital, combined with strong local governance, can play a vital role in how a community copes with strong climate shocks and subsequently adapts.

***Social capital and nongovernmental organization involvement in coping and recovery in Nicaragua.*** Cyclones have caused significant havoc in Nicaragua for 30 years, but no example is more telling than when Hurricane Mitch brought the country to its knees in 1998. Though Mitch never entered Nicaragua, at least 3,800 people died, and between 870,000 and 2 million, or 20 percent to 50 percent of the country's population, were directly affected (EM-DAT Database 2008; IADB 2000; NCDC 2004). After the disaster the country's poverty rate stood at close to 50 percent.

The hurricane spurred the national government to undertake serious reforms of its disaster mitigation and preparedness policy, which until then had been "highly politicized and homogenized" (Tomlinson 2006), with a big emphasis on public works projects. Particular progress has since been made through collaboration between NGOs and government at the local level.

For several reasons, including their own limited capacity and the broad range of natural hazards that Nicaragua faces, the NGOs active in that country have concentrated most of their disaster preparedness efforts on improving the capacity of local communities. The NGOs have also strengthened institutional collaboration among themselves by establishing the Civic Coalition for Emergency and Reconstruction (CCER), which brings together more than 320 nongovernmental and social organizations and networks (Cupples 2004).

According to NGOs, communication and collaboration with the government after Hurricane Mitch was difficult. Delaney and Schrader (2000), comparing the national government's plan with the findings of the CCER social audit, found markedly different priorities for rehabilitation and adaptation in the wake of the hurricane.[15] Sixty percent of the government's plan was tied to road construction, whereas CCER data suggested that only 5 percent of the population considered roads to be the first priority. Instead, people were more immediately concerned with the rehabilitation of agriculture (natural capital) and housing (physical capital).

An underlying problem was the lack of a legal framework setting out the roles of each institution at the national and the local level. Consequently, confusion arose regarding different agencies' responsibilities and chains of command for assigning different functions during emergencies. In this environment, the leadership assumed by municipal authorities became essential during the emergency and the post-Mitch rehabilitation. Some mayors took decisive steps to restore livelihoods but did so more as an outcome of their natural leadership than based on any formally established arrangements (Rocha and Christoplos 2001).

The importance of social memory in strengthening local capacity can be seen in the disaster response in some of Nicaragua's most conflict-prone regions. For some municipalities, isolation brought on by armed conflict during the 1980s had fostered and consolidated an organizational capacity that the local population could use to tackle other kinds of problems successfully. The evacuation after Mitch could therefore be accomplished rapidly (Rocha and Christoplos 2001). Also, the mobilization of provisions and organization of emergency shelters was very effective in these municipalities, contradicting the common assumption that civil society is weakest in war-torn areas. What seems to have happened is a strengthening of bonding social capital, as a resilience strategy in response to the depletion of other assets—most notably physical capital—already caused by the armed conflict.

Despite difficulties, NGOs were able to re-create both bonding and bridging social capital in certain sectors. Qualitative research by Tomlinson (2006) in the village of El Mirador found deterioration of social capital, but the same study also describes how NGOs have had some success in generating social networks in health and education. The study contrasts El Mirador's experience with that of El Hatillo, another village in the Matagalpa department that was more socially cohesive than El Mirador. El Hatillo's agricultural cooperatives and well-structured women's groups enabled the villagers to organize emergency committees that brought in food and evacuated the sick for treatment in the wake of Mitch. Seeing that they would receive little support from the municipal or national government, El Hatillo villagers not only enhanced their bonding social capital but also had various groups working constructively to combine resources (bridging capital) to improve longer-term community outcomes.

While cooperation between government and NGOs has managed to increase the adaptive capacity of many poorer communities in Nicaragua since 1998, a more sustainable social approach to climate change must involve devolution of power from the central government to municipalities. Closer collaboration between NGOs and the municipalities could prove crucial in filling a current vacuum in the Nicaraguan discourse on disaster mitigation and preparedness. Indeed, NGOs generally can play key roles as intermediaries, creating links between vulnerable communities and public institutions and facilitating the claims of marginalized groups.

*The role of social capital in short-term coping with Tropical Storm Arthur in Belize.* The Stann Creek district in Belize was hit by Tropical Storm Arthur in June 2008.[16] The way in which the community, aided by local government officials and NGOs, responded to this unprecedented flooding disaster illustrates how different types of social capital, combined with strong local governance, can be vital in people's ability to cope with and recover from a strong climate shock.

Arthur was unique in having no significant wind; low air pressure caused the storm to hover over Belize for four days. The torrential rain from the storm compounded the effects of Tropical Storm Alma, which had developed in the eastern Pacific a few days earlier. Together, Arthur and Alma brought more than 10 inches of rain across Belize within 36 hours, which led to the worst flooding the country has ever experienced, affecting 80 percent of its population of some 300,000. As a result of significant land clearing in the hills, insufficient land cover remained to absorb the water. Moreover, trees and vegetation—natural protective buffers—had increasingly been removed from riverbanks to allow better vistas. Rivers in both southern and northern Belize overtopped their banks causing flash floods, particularly in low-lying and coastal areas.

As Tropical Storm Arthur approached, people were expecting the impact to come in the form of wind rather than water. Though people in Belize are psychologically prepared for the storm and wind effects of hurricanes, few had perceived that the country is actually much more vulnerable to flooding, including flooding inland from heavy rainfall not necessarily related to coastal storms, according to the country's National Emergency Management Organization (NEMO).

In Stann Creek, a poor rural district of about 10,000 people, most of the villagers were caught by surprise. In previous floods rivers had risen gradually and only to a few feet, but on this occasion water levels rose several feet within a few seconds, trapping some people in their houses and sweeping others away. The areas hardest hit were around the poorest part of Stann Creek. More than 20 homes were destroyed and 200–300 were severely damaged, displacing about 4,000 people. However, only five people died.

That so few people died, and the prompt start to the postdisaster process, can be directly attributed to high levels of social capital. Despite the poverty of Stann Creek, all types of social capital were displayed during the cleanup process. Bonding and bridging capital served as fragile safety nets immediately following the disaster, and linking capital (ties between the community and formal institutions) also played a critical role. The local mayor was one of the first people on the spot. Possibly because of the considerable trust that community members already placed in him, he played a vital role in establishing shelters and organizing the distribution of emergency supplies, in close collaboration with the army, which sent troops to the area almost immediately.

The bonding capital observed was particularly strong. In the immediate aftermath, community members helped one another to the best of their abilities. The least affected quickly and willingly reached out to people who had lost everything, providing them with temporary shelter or taking care of their children while the cleanup process was initiated. Belongings and necessities such as food, water, and clothes were proactively shared among the affected. The bridging capital with other communities and NGOs was also significant. The fact that fishermen from a nearby village came almost immediately to the rescue in their boats helped to save many lives.

Part of the reason why the recovery process began so promptly was that members of the community, along with municipal officials and local branches of various NGOs, knew how to organize quickly. The Red Cross was already present in the village, undertaking training workshops, when the disaster struck. A pilot program to train the community in emergency preparedness had been started about six months before the disaster. This program was part of a regionwide effort, spearheaded by the Red Cross, to involve local communities in preventing and mitigating natural disasters. The pilot program pioneered the creation of community networks, which were put in charge of village risk and disaster management, coordinated through the Dangriga branch of the Red Cross and the District Emergency Management Organization. Even though the program had not yet installed early warning systems, the necessary networks were already being trained in the geographical mapping of hazards. They had also undertaken an assessment of vulnerabilities differentiated by gender. Hence, when Arthur struck, women participated actively in all relief operations. They carried out rescue missions, rehabilitated local infrastructure (such as schools), and along with men, distributed food.

In summary, these two cases emphasize how different types of social capital, combined with strong local governance, can be vital in longer-term resilience and short-term coping capacity when communities face strong climate-related shocks.

*The insurance role of safety nets—experience from Nicaragua in improving livelihood resilience and adaptation.* Countries operate several types of conditional cost transfer (CCT) programs in response to natural disasters.[17] However, these typically only treat the problem after it has occurred; they are not preventive in their scope. Another approach was tried in Nicaragua. In an attempt to reduce rural income vulnerability due to uninsured weather-related risks, a one-year pilot program (Atención a Crisis) was introduced in 2006 in six municipalities in a drought-prone region in northwest Nicaragua. The main economic activity of most of the residents in the area is subsistence farming of corn and beans. Frequent droughts make this livelihood quite precarious, and harvests are often completely lost. Many households attempt to cope with the resulting deprivation by seasonal migration (Macours and Vakis 2008). Despite the frequent recurrence of adverse weather events, few households seem to rely on ex-ante risk management strategies. One out of every five households reported in the baseline survey that they would do nothing other than pray to God to prevent negative impacts of future climate hazards. An even larger group of households (30 percent) planned to invest more in agriculture, which, given the high risk of drought in the area, arguably would increase their exposure to future shocks. The ex-post coping actions employed following previous droughts included selling off assets, taking children out of school, and reducing dietary intake, particularly affecting developing fetuses, infants, and young children. The Atención a Crisis pilot program was intended not only to reduce the use of inefficient ex-ante risk management and adverse ex-post coping strategies, but also to improve households' upward economic mobility.

The program introduced three different packages so as to evaluate and compare the effectiveness of each in relation to the objectives. Three thousand poor and vulnerable rural households were selected to participate, using a proxy-means approach. They were randomly assigned one of three types of intervention: (a) nutrition and education package (income transfers); (b) nutrition and education package plus a scholarship that allowed one household member to participate in a vocational training course; or (c) nutrition and education package plus a productive investment grant, aimed at encouraging recipients to start a small nonagricultural activity. The basic nutrition and education package was intended to ensure that people did not reduce their food intake or keep their children out of school. The vocational training and the productive investment grant were intended to boost the beneficiaries' income diversification potential.

Baseline data collected before the start of the program showed that early cognitive and social development was severely delayed among children aged 0–7 years, in very poor households. For example 97 percent of those aged 3–7 years were in the lowest quartile with respect to language development, and 85 percent were in the lowest decile, implying that they were at least 21 months delayed in their receptive vocabulary. Large delays were also found in the children's short-term memory and in their social development. Generally, the older children were the most delayed.

In surveys undertaken nine months after the program was initiated, large improvements were found. School attendance rates for children aged 6–15 years were up. Nutrition and preventive health care were greatly improved both in quality and quantity, and significant improvements appeared in cognitive development outcomes of children aged 0–7 years, particularly in language and social-personal skills. While improvements were found among the beneficiaries of all three packages, compared to the control group, which saw no improvements, the most pronounced effects were found in the group that had received the nutrition and education package together with productive investment grants.

In basic terms, all participant groups had their income protected against natural disasters (such as drought, mudslides, and hurricanes), while communities not in the program and hence not receiving any conditional cash transfer would experience loss of income. This effect was even more pronounced for consumption. The households in the group that had received productive investment grants were best able to increase their income and hence increase their savings, enabling them to cope better with natural disasters. In other words, they were able to strengthen their ex-ante risk management strategies. They were also less likely to migrate than other groups, more likely to diversify their income sources away from agriculture, and more likely to invest in improved technology.

A very interesting and positive effect was that people who received the productive grant started planning for the future, whereas previously they had lived a day-to-day existence. In the words of one respondent,

> Some people just dedicate themselves to survival. Others dedicate themselves to moving forward. It's the way of thinking. There are people that don't think about tomorrow. They hope that God will intervene, and that it will fall from the sky . . . But there are people who changed. Before, they didn't think about tomorrow, but now [with the program] they dedicate themselves to moving up.

Another beneficiary, who received the productive investment package, noted,

> Before the program, I just thought about working in order to eat from day to day. Now I think about working in order to move forward through my business. Through experiences, one learns and opens up toward the future. By talking to others, one understands and learns.

The evidence from such interviews suggests that aspirations and perspectives toward the future may be key for improving household welfare and program impact. The interviews also indicate the potential role of social interactions in changing attitudes. Because most of the households in the communities covered—just a select few—received one of the three packages, the community as a whole felt involved. Several information meetings were held over the duration of the program, and a lot of discussion occurred concerning what worked well and less well. Thus the program, in addition to giving monetary assistance, also enhanced bonding capital and fostered the building of bridging social capital. The outcome of the education and nutrition income transfer was better livelihood resilience (improved ex-post coping behavior), and the productive investment grant helped initiate livelihood adaptation by making people who previously had focused on day-to-day survival more forward-looking.

## Policy Perspectives

Although the causes of climate change are global, the adverse impacts take the form of localized natural hazards, often at the subnational level. Although measures to mitigate climate change and variability often involve international negotiations and agreements and call for technocratic, top-down solutions, measures to promote adaptation to the effects of climate change to enhance longer-term resilience and short-term coping capacity are usually conceived at the national or subnational level. Indeed they most often spring from the vulnerable people themselves, either individually or as collective action (Agrawal 2008). The findings in this book repeatedly emphasize the importance of actions with a local focus. Though, as Agrawal (2008) notes, "information is currently lacking on how and under which conditions area-based and decentralized development approaches can help reduce climate change–related vulnerability, enhance adaptive capacity, and promote sustainable livelihoods," the analysis in this chapter has shown that a fruitful way to inform this debate from a social perspective is to use an augmented sustainable livelihoods framework with a focus on social capital. The last section of this chapter presents some implications for policy, drawing also on recurrent themes from earlier chapters.

### Improving Livelihood Resilience and Adaptation

Among the key points highlighted in previous chapters is the need to allocate more resources to adaptation (both ex-ante and ex-post) to increase longer-term resilience, rather than channeling most resources to postdisaster short-term recovery. Slow-onset hazards do not always require humanitarian intervention, particularly when governments and communities plan ahead and work together to reduce the impact on affected people. Even when intervention is needed, it is important to remember that many communities—especially poor or indigenous ones—have been living with periodic or cyclical drought their entire lives. Humanitarian efforts to protect people from the adverse effects of climate change should therefore try to support existing coping strategies, strengthening a community's longer-term resilience, rather than focusing solely on postdisaster recovery measures. A related point from previous chapters is the need to improve integrated resource management (for example, regarding water), tailoring it to the specific local circumstances mainly through formulation of location-specific policies and institutional arrangements.

A second important finding is that local institutions, both governmental and civic, have a large role to play in increasing the resilience of vulnerable groups and in helping with livelihood adaptation. The key to improving resilience is the establishment of small-scale, local programs that embody close cooperation among all in the community and that safeguard existing livelihoods or create new ones. Community-based risk assessment projects are valuable for their ability to facilitate adaptation while helping create social capital. Local institutions are essential facilitators for households and social groups to deploy specific adaptation practices—particularly among the rural poor, for whom this support may significantly enhance various types of capital necessary for adapting their livelihoods. They can also provide local communities with a forum to voice their concerns and make claims.

Across very different sectors and communities, the overarching policy advice for adaptation, livelihood resilience, and coping capacity is thus very similar: complement national-level efforts by scaling down policies and participation to the local level; rely on different types of community social capital to do so; draw on local knowledge when devising adaptation measures; and devise measures that take into account the specific local context. These policy recommendations are further elaborated in chapter 11.

### Supporting Locally Focused Interventions

Since shocks and stresses related to climate change are context specific, and since adaptive capacity depends heavily on local dynamics, it seems appropriate to focus attention not only on the livelihood strategies of poor communities but also on local governance structures, which play a pivotal role in determining the livelihood impacts of natural hazards. Most importantly, policies to mitigate the social consequences of climate change should seek to promote a synergistic relationship between vulnerable groups and both formal and informal state and local institutions because such an emphasis will improve the prospects for adaptation, livelihood resilience, and coping capacity.

National-level government and institutions should support local-level adaptation efforts through enhanced pro-poor social policies at the national level. There is scope for creating and expanding social programs covering both ex-post disaster coping support and ex-ante weather risk management. Promising approaches include social funds and other support for community-driven adaptation, safety nets for coping with climate risks and natural disasters, livelihood programs, microfinance, and indexed insurance. Common to such programs is that they add a climate-risk dimension to social protection programs that have traditionally focused solely on addressing poverty. Climate change community development programs have the potential to be a strong tool for empowering communities, enhancing social capital, and increasing the asset base. Empowering communities will enable them to take development into their own hands, using their knowledge of their own community to inform decisions about how to achieve access and improvements to infrastructure, hence enhancing their assets and improving their livelihoods. An example is the building of drainage channels in St. Lucia to prevent landslides, as mentioned in chapter 3. Under that program, community knowledge was used to map landslide hot spots in order to determine where drainage channels should be built to minimize the risk of landslides. Subsequently community members themselves built the drainage channels with technical assistance from the program engineers. This served two purposes: 90 percent of the funds spent on the program stayed in the community, and being involved in the construction process gave the community ownership of the drainage channels, enabling their continued maintenance without the need for external assistance (Anderson and Holcombe 2008).

As Heltberg, Jorgensen, and Siegel (2008) point out, local actors will increasingly need external support in their adaptation efforts because climate-related risks—large, covariate, and possibly causing irreversible damage—can overwhelm local adaptive capacity and local-level institutions. For example, in Nicaragua after Hurricane Mitch, local NGOs chose to focus their disaster preparedness efforts on improving local community capacity, believing that full-blown disaster management, with its numerous technical challenges, would be too difficult for the community to achieve.

An issue for national-level government institutions is therefore to identify the most effective means to support local adaptation. For example, while the decentralized, soft adaptation measures promoted in this chapter are crucial, they will need to be complemented by more costly, hard adaptation efforts—often spearheaded by national government and perhaps requiring international donor assistance—typically involving building roads and other communications infrastructure and investing in technological development (Agrawal and Perrin 2008). Also needed are social policy approaches that are responsive to climate risks, such as livelihood programs, social safety nets, and indexed insurance.

There is still much room and much need for including local institutions in resilience and adaptation strategies. Agrawal (2008) documents the importance of local institutions in livelihood adaptation, resilience, and coping strategies, extracting from 118 cases of good-practice adaptation from around the globe. In 77 of the cases, the primary structuring influence for adaptation comes from local institutions, and in all cases of external support, the support is channeled through local formal and informal institutions. But despite this clear reliance on local structures in day-to-day adaptation strategies, only 20 percent of the NAPA documents—national adaptation programs of action—prepared by environmental ministries incorporate local institutions as the focus of adaptation projects. And only 20 out of 173 projects described in the national adaptation programs of action reports identify local-level institutions as partners or agents in facilitating adaptation projects. NAPAs provide a process for least-developed countries (LDCs) to identify priority actions that respond to their urgent and immediate needs to adapt to climate change—those for which further delay would increase vulnerability or costs at a later stage. As part of his paper, Agrawal reviewed that process by assessing the various country program efforts.[18]

In designing local-level approaches, the policy challenge is not merely to enhance or increase assets across the board, but first to identify which assets people need the most in the circumstances at hand—that is, which assets will provide the greatest resilience and coping and adaptive capacity—and as explained above, how those needs are likely to evolve over time. Assistance should focus on enhancing the specific mix of livelihood assets that will provide the greatest resilience, along with developing the longer-term adaptive and short-term coping capacities of communities in the face of climate change affecting a given location. To achieve this requires knowledge of how to promote and enhance collective action, not only among local communities themselves but also between communities and local formal and informal institutions.

Agrawal (2008) notes how equitable access to government institutions and their resources, coupled with transparent communication, is likely to reduce the ill effects of natural hazards, in contrast to a situation characterized by centralized decision making and stratified access to the relevant institutions. While his argument echoes the more traditional calls for devolution (not just decentralization) of power in emergency responses, Agrawal also highlights the important role that local institutions play in creating "the incentive framework within which outcomes of individual and collective action unfold." Thus, well-functioning local institutions can reinforce social capital, just as strong social capital can underpin local institutions. Conversely, a community can be caught in a vicious cycle of inadequate social capital and ineffective local institutions.

Like the other types of capital, social capital functions most productively when combined with other resources (Tomlinson 2006), and it can increase the efficiency of other types of capital. Adger (2003) emphasizes the importance of generating synergy between community social capital and local state institutions. He notes that "state-society linkages are important both for wider sustainable development and for the co-management of resources. States can facilitate sustainable and resilient resource management and enhance adaptive capacity."

In attempting to strengthen linking social capital, a key element is the existence of community nodes: strong informal institutions or agents, capable of increasing linking capital for the whole community or social group. For instance, faith-based organizations can work with communities to increase ties to local governments. Similarly, an individual with strong human capital can increase the linking capital for whole communities by informing villagers of their rights and helping them to make relevant claims. To enable nodes to function in this way requires the existence of public schemes and certain rights. The national government plays an important role in generating formal rights and setting out a clear set of rules that will create opportunities at the micro level. Notably, Moser and Norton (2001) identify the state as the "primary duty-bearer" in enabling livelihood-related human rights.

A prerequisite for strong local governance institutions is, of course, an effective national government that can devolve the necessary power down through the system while retaining governing efficiency. Strong local governance institutions also need to interact with the many informal groups and organizations that are so important in social capital formation. Agrawal (2008) found that informal civic institutions were highly involved in arrangements for adaptation, but that public institutions, when involved in adaptation practices, tended to form relationships only with other formal institutions. Both adaptation and the resulting resilience could be enhanced through the identification of ways to encourage informal processes through formal interventions at the local level, to facilitate cooperation and cohesion.

Against this background, the World Bank is exploring how its operations can strengthen local adaptive capacity and resilience to climate change–related risks through context-specific analysis. Examples are a project to support area-based development initiatives to enhance adaptive capacity and maintain resilience of local actors and institutions to climate change (led by Nicolas Perrin, of the Bank's Social Development Department, and Eija Pehu, of the Agriculture and Rural Development Department)[19] and a research initiative to shed light on the role of social capital in addressing climate hazards, put forward by Arun Agrawal and Nicolas Perrin.

## Notes

1. However, as noted by Nakagawa and Shaw (2004), the Kobe Earthquake in 1995 in Japan encouraged a new movement toward multidisciplinary emergency responses, with clear links between technological and social solutions. Consequently, in recent years, disaster management has become increasingly connected to other sectors, particularly environment, city planning, and community participation. The change in approach stems from a growing realization of the importance of incorporating people and communities in predisaster mitigation and/or postdisaster recovery initiatives.

2. The classic study in this area is Conklin and Graham 1995.

3. Thanks to A. Norton for bringing our attention to this point.

4. The majority of social indicators for Chiapas are much lower than elsewhere in the country. More than two-thirds of its total population is poor, compared with 47 percent of the national population. Consequently, while the national mean of population without school attendance in 2005 reached 8 percent, in this state it exceeded 20 percent. The child mortality rate is 30 per 1,000 children, compared to the national average of 22 per 1,000 children, and is around 1.5 times the rate in Belize. Also, the state reported the highest agricultural losses in Mexico due to climatic events over the past three decades (Conde Saldaña, and Magaña 2007). The 2005 hurricane season was particularly destructive, causing more than US$400 million in losses in the southeast of Mexico, including Chiapas (Saldaña-Zorilla 2007).

5. See also the Nicaragua case study later in this chapter, which found that social income transfers in the form of productive investment grants provided an opportunity to start accumulating assets and hence improve the coping and adaptive capacity of subsistence farmers.

6. Off-farm jobs and emigration are often the coping and adaptation strategies (Conde, Saldaña, and Magaña 2007). In addition, nearly all farmers receive a direct payment per hectare planted through government programs (PROCAMPO); however, this program was not designed as a rural finance program (Wehbe and others 2006).

7. As argued in chapter 3 and supported by Stern (2007); Mendelsohn, Dinar, and Williams (2006); and Heltberg, Jorgensen, and Siegel (2008), the impacts of climate change fall disproportionately on poor and vulnerable people who have contributed the least to cause the problem and who have the fewest resources to cope with the impacts. While this issue is increasingly considered one of social justice, it is beyond the scope of this chapter. However, an interesting discussion of the social justice and human rights aspects of climate change can be found in Mary Robinson's *Barbara Ward Lecture* (Robinson 2006).

8. Chapter 3 elaborated how social assets are often under stress just before, during, and after climate-related disasters and how these assets can be eroded as a result.

9. The poor tend to recover much more slowly than the nonpoor following an exogenous shock, perpetuating their poverty and increasing inequality (Baez and Mason 2008).

10. A step-by-step method for carrying out this identification is illustrated in chapter 3.

11. Heltberg, Jorgensen, and Siegel (2008) provide a more detailed discussion of social policy measures, such as "no-regrets" social policy and social protection for adaptation, social funds for community-based adaptation, and social safety nets for coping with natural disasters and climate shocks.

12. This case study is based on information from OCHA 2004; ISDR 2008; and project information from CARE Bolivia's Web site, http://www.carebolivia.org/site_ingles/_pproyectos1.asp?id_pro=22.

13. Also see chapter 2.

14. This causes a substantial outflow of young people from the area.

15. CCER commissioned CIET (Centro de Investigación de Enfermedades Tropicales) to conduct a three-phase social audit of the process, in an effort to build the community's voice into reconstruction efforts.

16. This case study is based on data obtained during a field visit to Belize in June 2008. Interviews were carried out with people affected by the disaster and people assisting them in their recovery. Additional sources include Red Cross/Red Crescent 2007; ReliefWeb 2008; and Woolcock 2000.

17. This case study draws on Baez and Mason 2008; Macours, Schady and Vakis 2008; Macours and Vakis 2008; and the World Bank Web site "Learning from the 'Atención a Crisis' Pilot Program in Nicaragua's Drought Region." http://web.worldbank.org/wbsite/external/countries/lacext/extlacregtoppovana/0,,print:y~iscurl:y~contentmdk:21762443~pagepk:34004173~pipk:34003707~thesitepk:841175,00.html.

18. For more info on process see http://unfccc.int/national_reports/napa/items/2719.php.

19. This project was funded by the Trust Fund for Environmentally and Socially Sustainable Development (TFESSD) under the auspices of the World Bank.

## References

Adger, W. Neil. 1999. "Social Vulnerability to Climate Change and Extremes in Coastal Vietnam." *World Development* 27 (2): 249–69.

———. 2003. "Social Capital, Collective Action, and Adaptation to Climate Change." *Economic Geography* 79 (4): 387–404.

Adger, W. Neil, T. P. Hughes, C. Folke, S. R. Carpenter, and M. J. Rockstro. 2005. "Social-Ecological Resilience to Coastal Disasters." *Science* 309: 1036.

Adler, Paul S., and Seok-Woo Kwon. 2002. "Social Capital: Prospects for a New Concept." *Academy of Management Review* 27 (1): 7–40.

Agrawal, Arun. 2008. "The Role of Local Institutions in Adaptation to Climate Change." Paper presented at Workshop on Social Dimensions of Climate Change, Social Development Department, World Bank, Washington, DC.

Agrawal, A., and N. Perrin. 2008. *Climate Adaptation, Local Institutions, and Rural Livelihoods.* IFPRI Working Paper W08I-6. Washington, DC: International Food Policy Research Institute.

Anderson, M. G., and L. Holcombe. 2008. "Community-Based Landslide Risk Reduction: Proof of Concept." June 16. Presentation at the World Bank, Washington, DC.

Baez, J. E., and A. D. Mason. 2008. "Dealing with Climate Change: Household Risk Management and Adaptation in Latin America." Background chapter for 2008 World Bank Flagship Report *Climate Change in Latin America and the Caribbean.* Washington, DC: World Bank.

Brant, Simone. 2007. "Assessing Vulnerability to Drought in Ceará, Northeast Brazil." Master's thesis. University of Michigan.

CARE Bolivia. Undated. Project Note BOL 072 DIPECHO. http://www .carebolivia.org/site_ingles/_pproyectos1.asp?id_pro=22.

Conde, C., S. Saldaña, and V. Magaña. 2007. *Thematic Regional Paper: Latin America.* Occasional paper for UNDP 2007, *Fighting Climate Change: Human Solidarity in a Divided World. Human Development Report 2007/08.* United Nations Development Program. Hampshire and New York: Palgrave Macmillan.

Conklin, B. A., and L. Graham. 1995. "The Shifting Middle Ground: Amazonian Indians and Eco-Politics." *American Anthropologist* 97 (4): 695–710.

Cupples, J. 2004. "Rural Development in El Hatillo, Nicaragua: Gender, Neoliberalism, and Environmental Risk." *Singapore Journal of Tropical Geography* 25 (3): 343–57.

De la Fuente, A. 2007. *Climate Shocks and Their Impacts on Assets.* Occasional paper for UNDP 2007, *Fighting Climate Change: Human Solidarity in a Divided World. Human Development Report 2007/08.* United Nations Development Program. Hampshire and New York: Palgrave Macmillan.

Delaney, P., and E. Schrader. 2000. "Gender and Post-Disaster Reconstruction: The Case of Hurricane Mitch in Honduras and Nicaragua." Decision review draft, World Bank, Washington, DC.

Duarte, Mafalda, Rachel Nadelman, Andrew P. Norton, Donald Nelson, and Johanna Wolf. 2006. "Adapting to Climate Change—Understanding the Social Dimensions of Vulnerability and Resilience." In *Environment Matters, Annual Review July 2005–June 2006 (FY2006).* Washington, DC: World Bank. http:// siteresources.worldbank.org/INTENVMAT/64199955-1203372965627/ 21652254/Adapting_to_Climate_Change.pdf.

ECLAC (Economic Commission for Latin America and the Caribbean). 2001. *Instituciones y pobreza rurales en México y Centroamérica.* LC/MEX/L.482. p. 30. Mexico City: ECLAC.

———. 2003. *Panorama Social de América Latina 2002-2003. Pobreza y Distribución del Ingreso.* Santiago de Chile: ECLAC.

EM-DAT (Emergency Events Database). 2008. http://www.emdat.be.

Heltberg, R., S. L. Jorgensen, and P. B. Siegel. 2008. "Climate Change, Human Vulnerability, and Social Risk Management." Paper presented at Workshop on

Social Dimensions of Climate Change, Social Development Department, World Bank, Washington, DC.

IADB (Inter-American Development Bank). 2000. *Development beyond Economics: Economic and Social Progress in Latin America, 2000 Report.* Washington, DC: Inter-American Development Bank.

IBGE (Instituto Brasileiro de Geografia e Estatistica). 2004. "IBGE Releases Previously Unseen Profile of Food Security in Brazil." Press Release, May 17, 2006. http://www.ibge.gov.br/english/presidencia/noticias/noticia_visualiza .php?id_noticia=600&id_pagina=1.

IPCC (Intergovernmental Panel on Climate Change). 2007. *Climate Change 2007: Impacts, Adaptation, and Vulnerability.* Contribution of Working Group II to the Fourth Assessment Report of the IPCC. Geneva: IPCC.

ISDR (International Strategy for Disaster Risk Reduction). 2008. "Environmental Sustainability and Disaster Risk Reduction." For consideration at the 10th Special Session of the Governing Council/Global Ministerial Environment Forum. Principality of Monaco, February 20–22, 2008.

Kreimer, A., M. Arnold, P. Freeman, R. Gilbert, F. Krimgodl, R. Lester, J. D. Pollner, and T. Voigt. 1999. *Managing Disaster Risk in Mexico—Market Incentives for Mitigation Investment.* Disaster Risk Management Series. Washington, DC: World Bank.

Lemos, M. C. 2007. *Drought, Governance, and Adaptive Capacity in Northeast Brazil: A Case Study of Ceará.* Occasional paper for UNDP 2007, *Fighting Climate Change: Human Solidarity in a Divided World. Human Development Report 2007/08.* United Nations Development Program. New York: United Nations Development Program.

Linnerooth-Bayer, J., R. Mechler, and G. Pflug. 2005. "Refocusing Disaster Aid." *Science* 309 (5737): 1044–46.

Macours, K., N. Schady, and R. Vakis. 2008. "Cash Transfers, Behavioral Changes, and the Cognitive Development of Young Children: Evidence from a Randomized Experiment." Manuscript. Johns Hopkins University and World Bank.

Macours, K., and R. Vakis. 2008. "Changing Households' Investments and Aspirations through Social Interactions: Evidence from a Randomized Transfer Program in a Low-Income Country." Working Paper Report No. 45211. Johns Hopkins University and World Bank.

Mendelsohn, R., A. Dinar, and L. Williams. 2006. "The Distributional Impact of Climate Change on Rich and Poor Countries," *Environment and Development Economics* 11: 159–178.

Moser, C., and A. Norton. 2001. *To Claim Our Rights: Livelihood Security, Human Rights, and Sustainable Development.* London: Overseas Development Institute. http://www.odi.org.uk/resources/details.asp?id=1192&title=claim-our-rights-livelihood-security-human-rights-sustainable-development.

Moser, C., and D. Satterthwaite. 2008. "Pro-poor Climate Change Adaptation in the Urban Centers of Low- and Middle-Income Countries. Paper for Social Development Workshop on Climate Change, March 5–6, World Bank, Washington, DC.

Nakagawa, Y., and R. Shaw. 2004. "Social Capital: A Missing Link to Disaster Recovery," *International Journal of Mass Emergencies and Disasters* (22) 1: 5–34.

NCDC (National Climatic Data Center). 2004. *Mitch: The Deadliest Atlantic Hurricane since 1780. Storm Review.* http://www.ncdc.noaa.gov/oa/reports/mitch/mitch.html.

OCHA (United Nations Office for the Coordination of Humanitarian Affairs). 2004. "Bolivia—Drought, Situation Report No. 1." Issued November 16, GLIDE Number: DR-2004-000123-BOL, ReliefWeb, www.reliefweb.int.

OECD (Organization for Economic Cooperation and Development). 2006. *Domestic Policy Frameworks for Adaptation to Climate Change in the Water Sector. Part II: Non-Annex 1 Countries, Lessons Learned from Mexico, India, Argentina, and Zimbabwe.* International Energy Agency and OECD. http://www.oecd.org/dataoecd/46/15/37671630.pdf.

Putzel, James. 1997. "Accounting for the Dark Side of Social Capital: Reading Robert Putnam on Democracy." *Journal of International Development* 9 (7): 939–49.

Red Cross/Red Crescent. 2008. "Climate Guide." Red Cross/Red Crescent Climate Center. The Netherlands: International Federation of Red Cross and Red Crescent Societies. http://www.climatecentre.org/downloads/File/reports/RCRC_climateguide.pdf.

ReliefWeb. 2008. "Tropical Storm Arthur Causes Flooding in Southern Belize." June 9, 2008. www.reliefweb.int.

Robinson, Mary. 2006. "Climate Change and Justice." *Barbara Ward Lecture*, December 11, Chatham House, London. http://www.realizingrights.org/pdf/Barbara_Ward_Lecture_12-11-06_FINAL.pdf.

Rocha, J., and I. Christoplos. 2001. "Disaster Mitigation and Preparedness in Nicaragua after Hurricane Mitch." Report for NGO Natural Disaster Mitigation and Preparedness Projects: An Assessment and Way Forward. ESCOR Award No. R7231. http://www.redcross.org.uk/uploads/documents/Nicareng.pdf.

Rosenfeld, L., J. Caye, O. Ayalon, and M. Lahad. 2005. *When Their World Falls Apart: Helping Families and Children Manage the Effects of Disasters.* Washington, DC: NASW Press.

Saldaña-Zorrilla, Sergio O. 2007. *Socioeconomic Vulnerability to Natural Disasters in Mexico: Rural Poor, Trade, and Public Response.* Disaster Evaluation Unit. Mexico City: CEPAL and United Nations.

————. 2006. "Stakeholders' Views in Reducing Rural Vulnerability to Natural Disasters in Southern Mexico: Hazard Exposure, Coping and Adaptive Capacity." Working paper of the Advanced Institute of Vulnerability to Global Environmental Change. Washington, DC: STARTIIASA. http://www.start .org/Program/advanced_institute3_web/Final%20Papers/SALDANA%20FIN AL%20PAPER%20AIVGEC.pdf.

Stern, Nicholas. 2007. *Stern Review on the Economics of Climate Change.* Cambridge, U.K.: Cambridge University Press.

Task Force on Climate Change, Vulnerable Communities, and Adaptation (IUCN, SEI, International Institute for Sustainable Development, and InterCooperation). 2003. *Livelihoods and Climate Change. Combining Disaster Risk Reduction, Natural Resource Management and Climate Change Adaptation in a New Approach to the Reduction of Vulnerability and Poverty.* Winnipeg, Canada: International Institute for Sustainable Development.

Tomlinson, R. H. 2006. "Community Development in El Mirador, Nicaragua, Post–Hurricane Mitch: NGO Involvement and Community Cohesion." Master's thesis. Department of Geography, University of Canterbury.

Tompkins, E. L., and W. N. Adger. 2004. "Responding to Climate Change: Implications for Development. *Insights* 53 (December): 4–4.

Wehbe, M., H. Eakin, R. Seiler, M. Vinocur, C. Ávila, and C. Marutto. 2006. "Local Perspectives on Adaptation to Climate Change: Lessons from Mexico and Argentina." Working Paper No. 39. Assessments of Impacts and Adaptation to Climate Change.

Woolcock, M. 2000. "Social Capital in Theory and Practice: Where Do We Stand?" Working paper of the Advanced Institute of Vulnerability to Global Environmental Change, STARTIIASA, Washington, DC.

World Bank. 1994. "Staff Appraisal Report of Rainfed Areas Development Project June 1994." Agriculture Operations Division, Country Department II, Latin America and the Caribbean Regional Office, World Bank, Washington, DC.

WWF (World Wildlife Fund). 2006. *Up in Smoke? Latin America and the Caribbean—The Threat from Climate Change to the Environment and Human Development.* Third Report from the Working Group on Climate Change and Development. London: New Economics Foundation.

# CHAPTER 11

# Conclusion

## Sanne Tikjøb and Dorte Verner

Climate change is the defining challenge of our time. More than an environmental issue, climate change and variability threaten to reverse recent progress in poverty reduction and economic growth. This book links carbon dioxide ($CO_2$) emissions and climate change to environmental degradation and to implications for social and economic opportunities for development. Environmental degradation affects water availability, land, fisheries, and wildlife, with social implications affecting food security, livelihoods, health, and habitat. Excessive stress from these factors may cause additional impacts such as conflict, migration, and increased poverty and inequality. Hence, climate change is a threat to poverty reduction and if not addressed will further exacerbate the vulnerability of the poor.

Social implications of climate change are already being felt in Latin America and the Caribbean, and it is the poorest who are most affected. Even if national and global efforts to address climate change improve, current and future climate trends have considerable momentum, and they will dramatically affect economic, human, and social development for years to come. Poverty, inequality, water access, health, and migration are and will be measurably affected by changes in the climate. The study reported in this book found that many already poor regions are becoming poorer; traditional livelihoods are being challenged; water scarcity is increasing, particularly in poor arid areas; human health is deteriorating; and climate-induced migration is already taking place and may increase.

In a vicious cycle, poverty makes people vulnerable to the effects of climate change, and in turn, climate change makes people vulnerable to poverty. The impact of climate change depends as much on socioeconomic vulnerability as on biophysical exposure. Biophysical impacts are superimposed on existing vulnerabilities determined by socioeconomic factors such as an individual's age, gender, and ethnicity; a household's asset base and degree of integration with the market economy; and a community's capacity to tap into social capital among local residents and to link to national support systems that will help build local resilience. In local communities with little resilience, climate change may compound existing vulnerabilities by eroding the asset base of the poor.

Good adaptation policy is good development policy. The policy recommendations offered in this volume focus on reducing social vulnerability to climate change and variability by (a) enhancing good governance and technical capacity on key issues at the local and national levels; (b) developing social capital in local communities to provide for the exercise of voice, representation, and accountability; and (c) strengthening the asset base of the poor by building strong and lasting physical infrastructure, safeguarding the natural resource base, expanding access to financial services, investing in human capital, and supporting the development of social and cultural assets in the community.

This chapter first presents an overview of the social implications of climate change in the key areas addressed in each chapter. Second, it outlines the main recommendations for incorporating climate change adaptation measures into development planning at the government, community, and household levels, as summarized in table 11.1. It concludes with perspectives for future research on the social dimensions of climate change.

## The Social Implications of Climate Change

The findings in this book suggest that climate change may push the poor in Latin America and the Caribbean (LAC) beyond their ability to cope. Poverty, inequality, water stress, disease incidence, and migration patterns are and will be measurably affected by climate change, which will affect people's livelihoods in unprecedented ways.

The LAC region is one of the most ecologically diverse regions in the world, and the close proximity of different eco-zones means that climate

Table 11.1 Summary of Policy Recommendations

*Building the asset base of the poor: Enhancing local livelihoods and access to public services*

| Objective | Policy action | Level of implementation |
| --- | --- | --- |
| *Actions to strengthen physical capital* | | |
| Improve access to public services. | Invest in public works to increase access to water, electricity, and sewerage connection. | Central and local government |
| Prevent erosion and landslides and manage standing water. | Build and promote the use of safe structures such as gabion baskets and drainage channels. Re-vegetate unstable slopes, redirect settlers to safe "invasion zones," employ custodians to prevent settling and denuding of unstable slopes. | Central and local government |
| Enhance governance and management of water. | Implement integrated water resources management at the river-basin level. Increase collaboration among water, climate, and development specialists. Decentralize water management to local level and include all stakeholders in dialogue. | Central and local government and NGOs |
| Improve water access and use. | Invest in canals, dams, and water-saving technologies and improve institutional arrangements, including water rights, water user associations, water pricing. Improve access to mechanisms for purifying water and improve water-storage practices. | Central and local government, civil society |
| Improve the protection of infrastructure vital to the tourist sector, such as coastal zones. | Incorporate climate considerations into tourism planning and development. Enhance building codes and implement policies that restrict development in coastal zones. Undertake environmental impact assessments prior to tourism development. Respect existing natural structures (for example, do not replace existing, natural rocky beaches with white sand). | Central and local government |

*(continued)*

**Table 11.1** Summary of Policy Recommendations *(continued)*

| | *Building the asset base of the poor: Enhancing local livelihoods and access to public services* | |
|---|---|---|
| Objective | Policy action | Level of implementation |
| Support agricultural livelihoods and increase food security. | Implement policies that ensure long-term-sustainable solutions to improve agrochemicals, genetic diversity, energy use, and infrastructure. | Central government |
| | Improve infrastructure, including food safety storage, crop storage, and transportation. | |
| | Expand agricultural extension service. | |
| Build lasting infrastructure: roads, housing, and buildings. | Improve and enforce building codes and provide assistance to improve building structures. | Central and local government and NGOs |
| | Build or upgrade public buildings, such as schools, to storm shelters and hurricane-proof essential public buildings such as hospitals, police stations, and government buildings. | |
| | Build infrastructure such as roads and other communications infrastructure and invest in technology development. | |
| Protect assets against natural disasters. | Implement vulnerability analysis to identify and map vulnerable populations and areas where climate hazards occur. | Central and local government |
| | Enhance early warning systems to reach remote populations. | |
| | Establish and strengthen early warning systems with the participation of indigenous peoples and poor farmers; for example, set up a network of radio transmitters in different communities to prepare for climate-related events in coordination with community radio broadcasting. | |
| | Develop awareness campaigns and provide disaster risk reduction training to communities. | |
| | Set up information services to facilitate access to resources and assets for adaptation to hurricanes. | |
| | Enhance government policies that place legal requirements on national institutions (including local governments) to prepare and update emergency and disaster risk reduction plans. | |
| | Increase awareness of disaster risk through broadcasting programs. | |

## Actions to strengthen human capital

| | | |
|---|---|---|
| Reduce health risks. | Improve access to health services especially for the poor. Reduce malnutrition by implementing or expanding nutritional programs and education. Increase knowledge of household hygiene and environmental factors. Develop models for projecting disease patterns, and improve data collection. Invest in or improve health surveillance systems. Assess the relative burden of future climate change–related health impacts and vulnerabilities at the national and subnational levels. | Central and local government and NGOs |
| Increase the living standards of the poor to help them adapt. | Increase access to safety nets, health care, and education, including environmental education. Improve the quality of education and health care systems. | Central and local government |
| Raise local awareness of climate change and adaptation programs. | Educate communities about climate change to increase local capacity to read and interpret local climate change patterns and trends. Provide local access to relevant information and analysis. Improve information systems and policies related to livelihood programs, social safety nets, and indexed insurance. | Central government, local government, universities, and NGOs |

## Actions to strengthen social capital

| | | |
|---|---|---|
| Increase community participation in local projects and policy development. | Enhance different types of community social capital: Involve community in preparations for climate change and extreme events. Decentralize decision making. Include local communities in decision and implementation processes. Increase community participation, voice coalition, and local governance. Increase support for household and community initiatives. Strengthen existing bonds such as community nodes and other informal institutions. Create awareness campaigns on disaster risk reduction that will further cooperation within communities. | Local government, NGOs, and communities |

*(continued)*

**Table 11.1  Summary of Policy Recommendations** *(continued)*

*Building the asset base of the poor: Enhancing local livelihoods and access to public services*

| Objective | Policy action | Level of implementation |
|---|---|---|
| Create local, national, and regional vulnerability maps, including hazard-prone areas, and increase knowledge about vulnerable communities. | Carry out vulnerability analysis at the local, national, and regional levels, taking into account asset availability and distinctions between long-term trends and natural hazards. The vulnerability analysis should apply a context-specific framework to ensure a strong match between the needs of target groups and planned interventions. | Local government, NGOs, and universities |
| Increase social knowledge. | Undertake poverty and social impact analysis to understand the social implications of climate change, including violent and nonviolent conflict, migration, inequality, and poverty across social groups (age, gender, ethnicity, location). | Central and local government, NGOs, and communities |
| Improve dialogue. | Implement policies that promote synergistic relationships between vulnerable groups and institutions. | Local government |
| Improve relationship between local institutions and indigenous peoples. | Develop programs and capacity building to increase and improve the dialogue between indigenous communities and local governments: | Local government, indigenous peoples |
| Enhance state capacity to deal with conflict. | Implement crime-, violence-, and conflict-prevention programs. Develop plans for deploying military and police forces. | Central and local government, NGOs, and communities |
| *Actions to strengthen cultural capital* | | |
| Implement policies that ensure long-term-sustainable solutions to improve and strengthen cultural capital. | Catalogue traditional knowledge and broaden access to the information. | Local governments and NGOs |
| Improve livelihood resilience and adaptation. | Draw on local knowledge when devising adaptation measures. | Central and local government |

310

| | | |
|---|---|---|
| Enhance the use of traditional knowledge in indigenous communities. | Mobilize the use of traditional indigenous peoples' institutions (as distinct from indigenous formal, political organizations). Develop strategies and concrete support oriented toward helping indigenous communities adapt to climate change and variability that capture the diversity of livelihood strategies and the role and efficiency of cultural institutions. Development agencies and national meteorological services must begin with respect for traditional knowledge and use it as a starting point for their projects. Extend technology and agricultural advice that take local cultures into consideration. | Local government, NGOs, and community |

### Actions to strengthen natural capital

| | | |
|---|---|---|
| Improve livelihood resilience and adaptation. | Improve integrated natural resource management and tailor it to local circumstances through formulation of location-specific policies and institutional arrangements. | Central and local government |
| Increase water sources and access to them. | Increase conservation of water supplies or implement water pricing in urban areas. Shift to less-water-intensive agriculture in rural areas. Create new highland reservoirs. Facilitate workshops to create local cooperation on water management. | Central and local government, communities |
| Preserve the natural resource base. | Strengthen agricultural practices to account for timing and location of crop activities; improve crop protection practices to include crop rotation and diversification of farm activities. Alter inputs such as crop varieties, fertilizer rates, or irrigation. Use technologies that harvest and conserve water and soil moisture. Manage water to prevent flooding, waterlogging, erosion, and nutrient leaching. Use weather forecasting to reduce production risk. Conserve entire hillsides and watersheds rather than individual plots. Prevent deforestation in indigenous territories. | Central and local government, NGOs, communities, individual farmers |
| Improve land and soil quality. | Create awareness and build sense of stewardship among farmers. Raise risk awareness regarding the quality of land. Strengthen policies that protect natural resources. | Central and local government and NGOs |
| Build environmental resilience. | Expand protected areas, such as marine protected areas or parks, using integrated coastal zone management principles. | Central and local government and NGOs |

(continued)

**Table 11.1 Summary of Policy Recommendations** (continued)

| | Building the asset base of the poor: Enhancing local livelihoods and access to public services | |
|---|---|---|
| Objective | Policy action | Level of implementation |
| Prevent climate impacts on crop productivity in indigenous communities. | Enhance research on the potential effects of increased migration, such as diseases affecting crops and decreased productivity. | Universities and community |
| Support agricultural livelihoods. | Improve agricultural research capacity taking into account climate change. Implement long-term-sustainable solutions to improve quality and access to land, soil, water, and nutrients and improve information systems. | Central government |
| Improve land and soil to enhance agricultural productivity in indigenous communities. | Promote alternative productive systems that can better cope with climate change and variability. | Local government, NGOs, and community |
| *Actions to strengthen financial capital* | | |
| Create alternative livelihood opportunities. | Establish marine protected areas and train and employ local populations as rangers, researchers, guides, and park managers. | Local government, private sector, local communities, and NGOs |
| | Create programs that offer training in alternative livelihoods to populations at risk, such as fishermen. | |
| | Develop marketing plans that include activities such as cultural and community tourism, as alternatives to natural resources–based tourism. | |
| | Increase access to credit and land titles for the local population. | |
| | Increase local agriculturists' market access by applying small-scale fair trade principles. | |
| | Ensure sustainable aquaculture by employing stringent environmental legislation. Build safeguards against climate change into aquaculture, for example, by building deeper ponds and identifying climate-resilient species. | |
| Strengthen indigenous peoples' management of natural resources. | Enhance government policies that place legal requirements on national institutions (including local governments) to increase access to land titles for indigenous peoples' territories, so that the territories are left in the hands of indigenous peoples to control and manage. | Central and local government |

change will affect local communities in different ways. Social implications will differ significantly between—and even within—communities, as the vulnerability context may vary greatly from neighborhood to neighborhood. Some households will be able to rely on remittances to cope and rebuild following a disaster; others may rely on educational attainment in taking preventive, adaptive measures against climate change; and still others may use their mobility to their advantage and seek new economic opportunities elsewhere.

Climate change presents both threats and opportunities for existing and new livelihoods. The poor, who lack the necessary resilience to withstand climate shocks, may see their asset base depleted and their livelihood models threatened with extinction. However, households and communities with disposable assets that they can reallocate and invest in adaptation are better placed to improve their livelihoods in the face of climate change. By turning challenges into opportunities, they are more likely to experience the potentially positive effects of climate change.

Policy makers have an obligation to address underlying social issues that exacerbate the effects of climate change on the poor. Adaptation initiatives alone will not be enough to protect the livelihoods of poor communities against climate change in the future. Equally important will be to fully consider climate change and strategically incorporate it in social development programs, as well as to ensure that social dimensions are included in planning and implementing climate adaptation and mitigation initiatives.

### A Dynamic Overview: From Climate Change to Social Implications

Communities across Latin America and the Caribbean are already experiencing adverse consequences from climate change and variability. Precipitation has increased in the southeastern part of South America and now often comes in the form of sudden deluges, leading to flooding and soil erosion that endanger people's lives and livelihoods. Southwestern parts of South America and western Central America are seeing a decrease in precipitation and an increase in droughts. Increasing heat and drought in northeast Brazil threaten the livelihoods of already-marginal smallholders and in the eastern Amazon threaten to turn parts of the rainforest into savannah. The Andean intertropical glaciers are shrinking and expected to disappear altogether within the next 20–40 years, with large consequences for water availability.

Rising atmospheric temperatures have major social impacts in the LAC region. They are already causing the melting of tropical glaciers, with implications for the amounts of water available for farming and live-stock husbandry, domestic use, agriculture, power generation, and indus-trial use. Warmer air temperatures affect the geographical range of disease vectors such as malarial mosquitoes, with severe implications for human health. They also affect the range and yields of crops, with implications for the viability of traditionally grown crop varieties and for agricultural practices, food production and trade, and food security. In adults, tempo-rary malnutrition reduces body mass, immunity, and productivity, but the results are rarely permanent. In children it can stunt growth, impede brain development, or cause death. Hence, risk is increased that climate change may cause an intergenerational downward spiral in human potential. Higher air temperatures also cause human health problems directly, including raising mortality rates among infants, the elderly, and other vulnerable groups. Further, higher temperatures com-bined with decreases in soil moisture lead to deforestation, adversely affecting people's livelihoods in multiple ways.

Rising sea-surface temperatures and higher levels of carbon in seawa-ter affect the size of fish populations, the viability and migration patterns of fish stocks, and the health of coral reefs and mangroves. Damaged ecosystems affect the livelihoods of people who depend on their sustain-able exploitation.

Increases in the intensity of natural hazards such as hurricanes, and changes in their geographical distribution, lead to higher death tolls and more damage to livelihoods, property, and production systems. Weather-related disasters affect human health not only through encouraging the transmission of diseases but also through higher inci-dences of food insecurity, caused by the erosion of crucial environ-mental and physical assets. Changes in the predictability of seasonal weather patterns have big implications for both commercial and sub-sistence agriculture, sometimes rendering traditional routines obsolete and wiping out crops, with implications for food prices, nutrition, and food security.

Changes in precipitation amounts and patterns lead to more droughts—affecting rural livelihoods and food security—and more floods—affecting livelihoods, property, production systems, and food security for both rural and urban populations. Already the widening incidence of drought has been a key reason why one-third of the region's rural young people have migrated to towns and cities over the past 20 years. Both droughts and floods also augment the risk of water- and vector-borne diseases.

Rising sea levels lead to more floods and storm surges, affecting livelihoods, property, and production systems; more beach erosion, affecting the habitability of coastal settlements; salinization of soil, limiting or eliminating its use for agriculture; and the intrusion of saltwater into aquifers relied on for drinking water.

Clearly, the social impacts of climate change depend not just on biophysical exposure but also on the vulnerability of people and institutions to shocks. Vulnerability is shaped by the wider political economy of resource use, and as would be expected, is particularly pronounced among the region's poor.

Below are reported the main dimensions of the social impact of climate change in Latin America and the Caribbean. For each dimension, the starting point is the climatic changes outlined above. Each section reviews the trends and dynamics and summarizes the key policy recommendations that also appear in table 11.1.

## Water Scarcity

Although abundant in the region as a whole, water is scarce at the local level in the same areas. Increased glacier melt, reductions in rainfall, rising sea levels, and more frequent extreme weather events reduce both the availability and quality of water for human use. Water scarcity in turn has three major social implications: first, it affects domestic water use and agricultural output, thereby increasing the risk of food insecurity; second, it affects the range and transmission of vector- and waterborne diseases, which adversely affects health; and third, it disrupts entire livelihood models, which may lead to migration and conflict. The social impacts of water scarcity are growing and are being felt at a local level.

By 2050, the number of people facing water scarcity in Latin America and the Caribbean could rise from more than 20 million today to more than 75 million. Some areas are particularly vulnerable to water scarcity. Arid or semiarid rural subregions in northeast Brazil, Bolivia, Argentina, and Chile will experience a 20 percent reduction in water runoff due to reduced rainfall by 2050. In the Andean region, the melting of intertropical glaciers will severely restrict water availability, putting close to 40 million people, or 70 percent of the Andean population, at risk of losing water supply for drinking, farming, and energy generation by 2020. In urban centers with high concentrations of poverty, where freshwater availability is already low, population growth and dense urbanization patterns, along with growing pressure for economic development, push up the demand for existing water. In many downstream rural communities, it is not uncommon for contaminated waters to be tapped for household uses and irrigation, and as water becomes scarcer, more water will be withdrawn from low-quality sources.

Concerted action is needed to devise integrated water resource management methods through decentralization and community participation. Because both water scarcity and its social impacts are local, integrated solutions should be tailored to local circumstances and should give voice and representation to marginalized groups, whose livelihoods are affected the hardest by climate change. Greater technical and institutional cooperation is needed between water managers and climate and development specialists to devise better solutions to local water scarcity. Specific policy recommendations to address water scarcity include the following:

- Invest in canals, dams, and water-saving technologies and improve institutional arrangements, including water rights, user associations, and pricing.
- Improve mechanisms for purifying and storing water in local communities.
- Increase conservation of water supplies, and shift to less-water-intensive agriculture.
- Support interventions to create new highland reservoirs.
- Implement integrated water resources management at the river basin level.

## Natural Disasters

The incidence of natural disasters is rising, and poor areas are hit the hardest. The incidence of major floods, droughts, and storms in Latin America and the Caribbean has been rising, from roughly 100 in 1970–79 to more than 400 in 2000–08. Patterns of actual and predicted natural hazards show that disasters are more likely to develop in poor areas, even if the hazard frequency there is lower. While weather-related hazards are more frequent in South America, they cause disproportionately more fatalities in Central America and particularly in the Caribbean.

Poverty makes people vulnerable to natural disasters, and natural disasters make people vulnerable to poverty. More than 8.4 million people in the LAC region live in the path of hurricanes, and roughly 29 million live in low-elevation coastal zones, making them highly vulnerable to sea level rise and saline intrusion into groundwater supplies, storm surges, and coastal flooding. On one hand, the impacts of natural disasters are socially differentiated, and the poor are most affected by fatalities and injuries because they tend to live in areas with high risk of floods, landslides, or droughts; because the quality of their housing is too poor to withstand severe weather events; and because they lack resources to help them quickly recover lost assets. The high density of urban slums makes the urban poor more susceptible to disease outbreaks following natural disasters, while the rural poor are vulnerable because of their high dependence on natural resources. On the other hand, natural disasters erode the asset base of poor households by destroying natural and physical assets, diverting human capital, depleting financial resources, and straining social assets, pushing the poor deeper into poverty. Thus, the impact of disasters is superimposed on existing vulnerabilities and may compound the difficulties faced by the poor.

Disaster adaptation policies should combine a mix of hard and soft adaptation measures to strengthen public infrastructure and protect the asset base of the poor. Supporting the infrastructure needs in essential public service areas, such as schools, hospitals, and police buildings, as well as safeguarding access to water, electricity, and sewerage connections, will help build resilience in local communities and thus prevent natural hazards from turning into disasters. In addition, applying a strategic focus to the infrastructure needs of the poorest, for example, by building food and feed storage and safe livestock facilities, will help protect their asset base during extreme weather events. In particular, hazard risk management frameworks should focus on developing social capital in the community by incorporating participation and voice coalition in the design of natural disaster adaptation initiatives. Key policy recommendations include the following:

- Decentralize decision making, draw on local knowledge, and involve communities in planning local responses to natural disasters.
- Apply asset-based vulnerability analysis to both long-term climate trends and natural hazards at the local, regional, and national levels.
- Enhance policies that place legal requirements on national institutions and local governments to keep updated emergency and disaster risk reduction plans.
- Upgrade public buildings, such as schools, so that they can serve as storm shelters when needed; hurricane-proof government buildings such as hospitals and police stations.
- Increase awareness of disaster risk and risk reduction through television and radio broadcasting.

## Rural Livelihoods

The rural poor are particularly vulnerable to climate change because of their high dependence on natural resources. Environmental deterioration of the natural resource base—for example through global warming, which affects crop yields and viability as well as fish migration patterns—will directly affect families and communities that depend on these food sources for nutrition as well as income generation. For the rural poor, who often lack the human and financial capital to diversify their livelihoods, the depletion of existing assets increases the risk of poverty and internal migration. Particularly vulnerable are agrarian communities and artisanal fishermen, as well as communities dependent on ecotourism.

**Agrarian communities.** The expected impacts of climate change and climatic variability will directly affect food supplies, and for millions they will endanger food security. By eroding natural resources and physical assets, climate change will make farming more difficult and unpredictable. Poor small-scale cultivators, pastoralists, and day laborers tend to live in arid and semiarid regions, mountain slopes or plateaus, and tropical rainforests, all fragile environments that often suffer from environmental degradation. Poor dryland farming is particularly at risk. Some of the major effects will come from increased frequency and severity of droughts and floods, which degrade farmland through erosion and desertification and damage farm property as well as public infrastructure such as roads and irrigation channels, with consequences for production capacity and market access. Reduced water availability will particularly affect grain crops and livestock production in Central America (Costa Rica, Mexico, and Panama), the Andes, and parts of Argentina, Brazil, and Chile. Floods may also damage facilities for food and feed storage, with immediate effects on food security and safety.

Atmospheric warming will affect crop yields and viability. For cereal crops, an analysis of global production shows that yields in most of the region will be reduced with increasing temperatures. Changes in the range of crops can also be expected, so that some crops and crop varieties will no longer be viable in areas where they have traditionally thrived. These expected changes will directly affect food supplies and endanger food security. Within a larger vulnerability context, climate change may also affect food security indirectly by increasing demand for bioenergy to replace fossil fuels, affecting world food prices. Key policy recommendations to support climate change adaptation in agrarian livelihoods include the following:

- Raise risk awareness among farmers and build a sense of stewardship to strengthen social capital in local communities.
- Implement water conservation systems, and promote alternative productive systems with higher resilience to climate change.
- Support transport infrastructure to improve market access for local farmers.
- Enhance agricultural research capacity, taking into account climate change.

***Artisanal fishing.*** Climate change and variability, in the form of storms, increasing sea surface temperatures, and rising sea levels, will significantly worsen current environmental problems that threaten the livelihood and sometimes the food supply of artisanal fishing communities. Changes in the migration patterns of fish stocks due to changing sea-surface temperatures, and the destruction of fishermen's physical capital during natural disasters threaten the livelihoods of artisanal fishermen. The particular vulnerability context of small-scale fishermen relates to their lack of access to insurance and property rights, as the globalization of trade and the privatization of access rights undermine their reliance on traditional areas for fishing. In addition, early warning systems often do not reach the remote location of small-scale fishing communities, leaving fishing villages and their assets vulnerable to sudden extreme weather events.

Better planning and improved management of natural resources can help build the resilience of small-scale fishing communities. Aquaculture—if designed to be pro-poor, sustainable, and environmentally friendly—can potentially provide an important source of livelihood for fishermen losing their jobs. Compared with what is typical at present, however, such aquaculture projects require the establishment of much clearer land rights that support the interest of indigenous and impoverished communities, and they should ensure that the local community holds the right to manage production. Marine protected areas offer another alternative livelihood for communities previously dependent on fishing, while at the same time safeguarding coastal marine habitats. Key policy recommendations to support climate change adaptation in artisanal fishing communities include the following:

- Manage fish stocks better by means of tradable quotas, and so forth.
- Protect aquaculture against climate change by building deeper ponds and selecting species resilient to saltwater intrusion, temperature changes, and sea level rise.
- Establish marine protected areas and support local livelihoods by retraining and employing former local fishermen in safeguarding coastal marine habitats.

*Ecotourism.* The fastest-growing segment of the tourism industry, with 6 percent annual growth, ecotourism is especially vulnerable to climate change because of its close reliance on the integrity of ecosystems. Particularly popular among poor communities, ecotourism offers an alternative to farmers and fishermen whose livelihoods may already be threatened by climate change. While promoted for its pro-poor focus and local ownership, development of ecotourism projects is too often implemented without consideration for safeguarding the investment against climate change and variability. In the case of an extreme weather event, that could put entire communities at risk of losing their jobs and their investment.

Development of sustainable ecotourism will require the timely implementation of specific environmental measures to protect natural assets, physical infrastructure, and local jobs against climate change. Building the resilience of poor communities economically dependent on ecotourism will call for better zoning for new developments, the consistent application of environmental impact assessments, and the involvement of local communities in protecting the natural resource base. Key policy recommendations to support climate change adaptation in the tourism sector include the following:

- Develop and implement new building codes and policies to restrict development in near-shore zones and areas at high risk of damage from climate change.
- Incorporate detailed environmental impact assessments to identify climate risks and assess viability of planned resorts.
- Strengthen the protection of environmental resources with the support and assistance of local communities.

## *Urban Livelihoods*

Their housing conditions make the urban poor vulnerable to extreme weather events, especially hurricanes. Incoming migrants typically settle in flimsy housing in densely packed shantytowns, on marginal lands such as floodplains, unstable slopes, and low-lying coastal land, which other people consider too risky for settlement. The location of the 20 largest cities in the LAC region near coasts, slopes, swamps, and other areas prone to flooding exposes them to natural hazards. Hurricanes, torrential rainfall, floods, landslides, and other climate-related events may cause severe damage to shantytowns or destroy them altogether. The same disasters may also destroy vital roads and utilities, cutting off poor neighborhoods from help. As discussed in the section on health, after such disasters, lack of clean water and sanitation may provoke disease outbreaks. Warmer mean temperatures also raise the risk of disease.

The vulnerability of the urban poor is exacerbated by their lack of assets and public services needed to cope with climate change. In the megacities of Latin America and the Caribbean, few of the poor have education beyond the primary level. Often lacking regular jobs, most make their living in the informal sector. They typically lack adequate access to public services such as piped water, sanitation, and electricity— some because connection is not available, others because connection is too expensive.

Although the urban poor do not directly depend on natural resources for a living, their poor housing conditions, job insecurity, and lack of human and other assets put them at risk. For example, large shares of the urban poor use their dwellings for productive activities and income generation. Today more than one in four people in the region (27 percent), or 65 percent of all the region's poor, live in urban slums, and the number is expected to rise as the threat to rural livelihoods forces more and more people to migrate to urban areas in search of new ones.

Thus far, urban authorities in Latin America and the Caribbean have been slow to adopt measures for adaptation to climate change. City authorities need to identify and map where climate hazards are greatest, target the most vulnerable groups, and design adaptation measures accordingly. In the short term, policies that focus on disaster preparedness will help build local resilience to natural hazards. In the long term, policies that address building the asset base of the poor through improved housing, education, health, and infrastructure will develop the adaptive capacity needed to protect livelihoods affected by climate change. Key policy recommendations include the following:

- Enhance human capital of the urban poor, notably the provision and quality of education, health care, and other social protection services.
- Improve property rights of the poor.
- Upgrade high-risk housing and utilities infrastructure such as safe water supply, sanitation, and electricity, and enforce proper building codes.
- Update disaster preparedness plans to establish roles and responsibilities, and develop plans of action for immediately before and after an extreme weather event.
- Implement relocation programs to safe, alternative housing, with clean water, sanitation, and electricity at affordable prices. Programs should inform residents of the importance of vacating high-risk areas, while being sensitive to the costs borne by the people affected, especially if they have to move away from current income opportunities.

## Human Health

Changes in temperature, precipitation patterns, and extreme weather events have growing direct and indirect impacts on health. Climate change is likely to expand the geographic range of vector-borne diseases and lengthen their transmission season. Dengue fever could become one of the major health risks resulting from climate change and variability, and new, more virulent strains are emerging. This disease is already extending its reach in Mexico and central South America, and by the 2050s its transmission rates are likely to have grown two to five times in most parts of South America, putting many cities at risk, including São Paulo, Caracas, and Mexico City. The risks from malaria are particularly serious when it spreads into new areas where immunity levels are low. Local changes in temperature, rainfall, and humidity are expected to cause malaria to advance into areas not previously affected, while the disease is projected to become less common in some areas where it is currently endemic.

The expected increase in the incidence of floods will lead to more outbreaks of waterborne diseases, and increasing temperatures will create conditions allowing pathogens to multiply faster. Waterborne diarrheal diseases, including cholera, thrive where the lack of safe drinking water and sanitation makes good hygiene difficult. This situation is much exacerbated in the chaotic conditions following natural disasters. These are by far the deadliest group of diseases associated with the environment, killing almost 20 times as many people in the LAC region in 2002 as malaria and dengue fever combined.

Projections of the biophysical effects of climate change and variability suggest that an additional 1 million people in Latin America and the Caribbean could be short of food by 2020. Malnutrition in the region will be affected by droughts as well as by floods, which not only destroy croplands but also cause vector- and waterborne diseases that increase the risk of malnutrition. Malnutrition is the main source of mortality from both malaria and diarrhea; a body weakened by malnutrition is more susceptible to disease. Conversely, malaria and diarrhea can inhibit the body's ability to absorb nutrients, causing malnutrition. In Guatemala natural disasters were found to have permanent effects on child development. Poor health and malnutrition can decrease human capital by affecting learning ability and labor productivity, increasing the risk that asset-poor households will fall into poverty, while making it more difficult for others to escape poverty.

Looking ahead 50 years, the simulated effects of climate change on life expectancy in five countries are mixed. In Peru, they suggest a reduction in average life expectancy of about 0.2 years. This average, however, hides a dramatic difference: Peru's currently poor and cold highland regions would benefit, while the currently hotter and richer coastal regions would lose, narrowing overall health inequality among Peruvian districts. In Brazil, the simulated average loss of life expectancy is 1 year, and it is expected to widen the inequality of health outcomes across municipalities, while in Chile climate change is estimated to reduce overall life expectancy by six months and to reduce health inequality. Bolivia appears to be the least affected by expected climate change, and Mexico also appears quite insensitive.

Health risks from climate change and variability differ depending on the locality. Small changes in temperature and precipitation can have a large impact on public health, and factors such as inadequate health infrastructure, migration, lack of safe water and sanitation, and poor water storage practices exacerbate the risks. Populations at increased risk tend to live (a) in border regions with endemic diseases that are sensitive to climate, (b) in regions with an observed correlation between epidemic diseases and weather extremes such as El Niño episodes, (c) in areas where multiple climate-change impacts are projected to affect health simultaneously (for example, stress on food and water supplies and risk of coastal flooding), and (d) in areas with high socioeconomic stress and low adaptive capacity (for example, areas with poor land use or underdeveloped health infrastructure).

Education seems to be a very effective, "no-regrets" policy to counter adverse health effects from climate change. Regression analyses show that the relationships between climate and health are outweighed by strong, unequivocally positive relationships between education and health. In Brazil, for example, the estimated adverse health effects of climate change over the next 50 years could be countered by an increase in average education level of just 1.1 year. Building adaptive capacity and reducing health risks, especially among asset-deprived groups, are together an important development issue facing the region today. Policies to reduce climate-induced health risks should focus on improving capacity to monitor and predict disease patterns, building the asset base of the poor, and reducing the vulnerability of groups facing the greatest health risks:

- Develop models for projecting disease patterns, and research the relationships among health outcomes, climate variables, and socioeconomic factors.
- Implement functioning health surveillance systems to monitor the incidence and spread of diseases and guide the development and evaluation of adaptation strategies.
- Combat water scarcity by preventing groundwater from being contaminated by the use of sewage or wastewater for irrigation or by runoff during floods and heavy rain.
- Implement simple, community-based mechanisms for purifying and storing water that prevent parasites from thriving.
- Implement low-cost programs to promote household hygiene, and expand nutritional programs for poor households at risk of malnutrition.
- Improve access to health care for migrant populations, and improve access to safe water and sanitation in urban shantytowns.

## Migration

Internal migration in response to climate change is already occurring in the LAC region and seems likely to increase. Across the region during 1988–2003, one-third of the rural population aged 15–29 migrated to urban areas. The widening incidence of drought has been a key factor. Currently most of the LAC region's migrants are young men, but the region now also has one of the world's highest migration rates for women. Climate change projections indicate that many communities across Latin America and the Caribbean—from poor rural areas to city slums—are at risk of being forced to move because their environment no longer supports their livelihoods. However, physical and natural assets constitute only two aspect of a person's vulnerability to environmental hazards; many other economic, political, and social factors at the individual, community, and national levels are considered in the decision to migrate.

An econometric analysis of current migration patterns between Bolivian municipalities suggests that about 2,000 internal migrants per year can be attributed to climate change over the last 50 years. People abandon both the coldest and the hottest rural areas, but rather than move to areas with more pleasant climates, they move to urban areas, where economic opportunities are less climate sensitive. Migration is lowest from the richest and poorest municipalities (people in the latter may be unable to afford the expense of migrating), highlighting the risk of some poor people being trapped in areas with increasingly adverse climates and diminishing economic opportunities.

An econometric analysis of international migration between Mexico and the United States did not indicate that the expected gradual climate change over the next 50 years would increase migration flows. However, given the complexity of the decisions involved in planned or distress migration, it was not possible to estimate the flow or intensity of future international migration patterns that climate changes may induce in the LAC region as a whole. Specific policy recommendations to address future climate-induced migration include these:

- Discourage settlement in areas prone to persistent climate hazards by replanting unstable slopes, redirect settlers to safe zones, and employ custodians to prevent settling and denuding of unstable slopes.
- Support agrarian livelihoods by implementing policies to improve agrochemicals, genetic diversity, water conservation, energy use, and infrastructure in ways that ensure long-term, sustainable solutions.
- Educate local populations about climate change and create programs that offer training in alternative livelihoods to populations at risk of migration.
- Develop strategies to help indigenous communities adapt to climate change and variability; the strategies must capture the diversity of their livelihood strategies and respect the role and efficiency of their cultural institutions.
- Enhance research on the potential effects of increased migration, including diseases affecting crops and decreased agricultural productivity.

## Conflict

Climate change may exacerbate existing risks of conflict rooted in socioeconomic factors. That may happen for several reasons, for example, if climate change diminishes the supplies of food, water, forests, energy, and land or encourages migration. Other dangers will occur if climate change undermines the capacity of the state, for example, by raising the costs of infrastructure in remote areas, thus limiting the reach of the state; by eroding fiscal resources; or by weakening the legitimacy of a state if it cannot provide basic needs for affected populations. If natural disasters are more frequent and severe, chaotic conditions may provide opportunities for rebel groups to challenge the government's authority.

Whether violent conflict is induced by climate change will likely depend on how scarcities are perceived and communicated by key political actors. Though most LAC countries are relatively strong, middle-income societies with accountable governments, the adverse impacts of climate change and variability are likely to put further stress on those already suffering from high income inequality and crime rates. That could lead some factions within weak political systems to exploit the situation rhetorically and mobilize groups for conflict over scarce resources.

The political economy should be considered to the same degree as environmental degradation as an underlying reason for violent conflict due to climate change. Adapting effectively to climate change and avoiding violent conflict will require both local and national governments to be inclusive of the most vulnerable groups in society, ensuring that they are heard and represented in dialogues with other stakeholder groups. Addressing underlying political tensions and socioeconomic inequalities constitutes a "no-regrets" policy that will allow local communities and individual households to build resilience against climate change and conflicts that it might induce. Key policy recommendations to prevent violent conflict from climate change include the following:

- Decentralize decision making in preparations for climate change and extreme events, to increase community participation, voice coalition, and local governance.
- Improve integrated natural resource management and tailor it to local circumstances through formulation of context-specific policies and institutional arrangements.
- Develop capacity-building programs to increase and improve the dialogue between indigenous communities and local governments.

- Implement crime, violence, and conflict prevention programs, and develop plans for deploying military and police forces to maintain security during natural disasters.

### Poverty and Inequality

Climate change will undoubtedly affect the livelihoods of poor people and may increase poverty rates in Latin America and the Caribbean. Using municipality-level regressions to analyze the climate-income nexus over the next 50 years in five large LAC countries, researchers found that climate change would tend to cause a reduction in average income levels in Brazil, Chile, and Peru and cause an increase in the poverty levels (all else equal). However, the climate-income relationships differ substantially from country to country, implying that the likely impacts of climate change on incomes, poverty, and inequality also differ substantially from country to country.

In Bolivia, where consumption levels in the coldest regions are as much as 40 percent lower than the national average, consumption is estimated to increase by an average of 3 percent because of climate change, while inequality declines. In Brazil, people in the warmest regions, i.e., the North and Northeast, have the lowest income levels, at 50 percent lower than the average. Simulating the effects of projected climate change indicates that the North and Northeast will likely bear the brunt of negative climate change effects with a decrease in income of 20 percent to 23 percent, compared to a national average decrease of 12 percent. For Peru, where people currently living in a moderate climate earn 30 percent to 90 percent more than those living in the hot or cold extremes, no effect on income distribution is detected, only a relatively small decline in income levels (about 2 percent) by 2058. In Chile, the most prosperous people live in the cold regions. Over the next 50 years, income distribution is not expected to change because of climate change, but simulations suggest that climate change will cause average incomes to fall by about 7 percent (all else equal). For Mexico, no systematic relationship was found between climate and income levels, implying that Mexico will be quite insensitive to the expected moderate changes in average temperatures.

Addressing the long-term effects of gradual climate change on poverty and inequality is most effectively done by building a resilient human capital asset base through the established development practices of increasing access to, and quality of, education, health care, and social protection. The impacts from climate change indicated by this analysis could easily be overcome by the positive effects of increases in education over the next 50 years. In Brazil, a 2-year increase in average education level is estimated to cause a 94 percent increase in incomes, whereas in Chile, it would take less than 0.2 years of additional education to counteract the estimated negative effect of climate change over the next 50 years. "No-regrets" policy recommendations to reduce poverty and inequality include these:

- Improve access for the poor to quality education, health care, and social safety nets.
- Undertake poverty and social impact analysis to understand the social implications of climate change, including violent and nonviolent conflict, migration, inequality, and poverty across social groups (age, gender, ethnicity, location).
- Implement policies that promote dialogues between vulnerable groups and local and national government institutions.
- Expand the agricultural research agenda to account for the impact of climate change on agricultural output and productivity.

## Adaptation: Good Governance, Social Capital, and Local Assets

Right now, families in southern Belize are concerned with how to rebuild their homes after an earthquake in May 2009. Rural populations in the high Andes are considering how to adapt their livelihoods to a future with disappearing glaciers, and small fishing communities in the Amazon are deciding whether to stay another season or try their chances in the big city.

This book provides evidence of the social implications of climate change. Even though the debate about the scale and impact of climate change may not be settled, the case is very strong for immediate action to prevent both immediate and longer-term social costs from escalating. As this book has documented, climate change threatens especially to overwhelm the local adaptive capacity in poor communities. Societies will bear the burden of individuals, families, and communities who suffer the negative impacts of climate change.

To increase resilience by reducing vulnerability to climate change is typically seen as the responsibility of households and communities, through livelihood adaptation and asset collection and allocation. This book shows that local decision making, social capital, and partnerships are essential to building resilient communities. But increasingly, climate change adaptation requires a greater role for national and global structures to support local processes—thereby making climate change adaptation an integral part of future development work.

### Good Adaptation Policy Is Good Development Policy

Policies to address the social consequences of climate change are not limited to the social sector. Many issues must be addressed as "hard adaptation issues," with solutions found in infrastructure, technology, and finance. In the following, however, emphasis is placed on incorporating social aspects of climate change adaptation into existing development policies and practices.

Social development tools can prove indispensable in the formulation of adaptive measures because they tend to be context specific, multidimensional, and disaggregated across groups and are often conducted upstream. This is an advantage when dealing with the social impacts of climate change, as the design and implementation of adaptation measures rely strongly on the integration of inclusive and participatory approaches. In particular, this volume recommends the use of community-specific social analysis focusing on improving livelihood outcomes, careful attention to building social assets within and between stakeholder groups, and strengthening resilience through asset-based adaptation at the local level.

Good adaptation policy is also good development policy. Across different sectors and communities, the overarching advice for building livelihood resilience through adaptation is the same: complement national-level efforts by scaling down policies and participation to the local level; rely on different types of community social capital to do so; draw on local knowledge when devising adaptation measures; and devise measures that take into account the specific local context.

### A Three-Pronged Approach to Reducing Social Vulnerability to Climate Change

Successfully addressing social vulnerabilities to climate change requires action and commitment at multiple levels. In a three-pronged approach, this book offers key operational recommendations for strengthening the commitment and broadening the scope of climate change adaptation at the government, community, and household levels. Emphasis is on enhancing good governance and technical capacity in the public sector, building social capital in local communities, and protecting the asset base of poor households.

***Prong 1. Enhance good governance and technical capacity in the public sector.*** At the government level, specific attention should be paid to building both the institutional and the human capacity to better identify and address the needs of populations highly vulnerable to the effects of climate change. By strengthening the institutions and infrastructure designed to respond to rapid-onset and slow-onset climate-related disasters, along with enhancing the human capital involved, the government is better equipped to protect citizens and their assets, as well as the country's recent development gains and future prospects.

Better targeting of populations at risk is critical to addressing the needs of those who are the most vulnerable. While different groups may be equally exposed to natural hazards, they are not equally affected. Vulnerability is as much a function of socioeconomic indicators as of biophysical exposure, and for that reason resilience to climate change varies between and within communities. The unequal distribution of vulnerability to disasters within and across communities is visible even at the household level. Therefore interventions to help people adapt to the effects of climate change and rebuild following a disaster need to focus not just on areas and numbers of people at risk, but on who is at risk and the types of risk they face. One promising approach is the development of climate indicators to be applied to conditional cash transfer programs.

Strengthening governance and responsiveness in the public sector is required especially to address the needs of the most vulnerable groups. These groups often suffer disproportionately from weak and unresponsive public institutions. One key aspect concerns governments' disaster preparedness. Although many community-driven adaptation projects are successful in building local resilience, the best practices urgently need to be scaled up. The necessary institutional framework for developing and implementing such initiatives is often lacking. It is therefore important to promote a synergistic relationship between vulnerable groups and both formal and informal state and local institutions, thereby developing the linking social capital required for more equitable access to local, national, and international institutions and their resources. Another key aspect of effective responsiveness relates to innovation in financial products that give social protection for climate change–affected households and communities. Elaborated in more detail in the section on financial capital below, these products enhance the policy options available to local and national governments to help families rebuild their livelihoods after a natural disaster. Finally, an equally critical aspect of good governance is related to slow-onset disasters, such as increasing water scarcity. What is often needed is a comprehensive strategy to build technical capacity across key public actors, especially to enhance the ability to integrate climate change aspects into sectoral approaches such as integrated water resource management. By analyzing the institutional integration of, for example, water issues and the impact of climate change on water resources in different areas of government, it is possible to identify what gaps need to be filled in human resource skills, technical capacity, and organizational processes. International donors can support this learning process to improve governance and facilitate improved decentralization and development of partnerships.

Infrastructure must be developed that is designed to withstand climate-induced tension, such as increased soil erosion or mudslides, to secure emergency access, and to protect health and assets. Building lasting physical infrastructure is a fundamental element in any national and local adaptation plan and is essential to fulfilling the objectives of reaching the target groups and delivering an effective response. Given the scarcity of resources, the urgency of the matter, and the longevity of large-scale infrastructure projects, the World Bank and other donors can effectively incorporate participatory climate adaptation into projects in water and sanitation, roads and bridges, and electricity, whether these are already in progress or in the pipeline.

### Prong 2. Develop social capital in local communities: voice, representation, and accountability.

The ways in which individuals and groups within a community interact with each other constitute the community's social capital and influence how vulnerable a community is to climate shocks and variability. Although often overlooked or underestimated, social assets provide the foundation that allows other assets to be generated and allocated appropriately. Just like any other asset, social assets are often too weak to withstand the pressure of a natural disaster, which may unravel the bonds that hold a community together and allow for risk sharing, mutual assistance, and collective action. However, social capital can also be used to build resilience that may enable a community to strengthen its response to climate change. This is effectively done by empowering people and the community itself, drawing on local knowledge, and encouraging local participation.

People and communities should be empowered to give voice not only to their challenges, but also their solutions. The augmented sustainable livelihoods framework, with a particular focus on social capital, is helpful as the basis for planning because when applied to a given setting, the findings provide an understanding of which groups are vulnerable and why; to what extent they can rely on their relatives, neighbors, and government agencies as a coping mechanism; which assets will provide the greatest resilience and adaptive capacity; and how needs are likely to evolve over time. Community-based risk assessment projects are valuable for their ability to involve local participation in adaptation while helping to create social capital. Local institutions can provide communities with a forum to voice their concern and seek representation and accountability; they function as facilitators for households and social groups to inform, design, and implement adaptation practices. However, for local institutions to function effectively, strong institutional ties with the national government are required to ensure a continued exchange of information. For example, incorporating the data and knowledge collected at the local level into regional and national adaptation strategies is contingent on integrating power down through the system, while at the same time keeping governance efficient.

Drawing on local knowledge and institutions in designing adaptation measures is essential to achieving sustainable adaptation. The findings in this book repeatedly emphasize the importance of actions that are conceived and executed locally, using area-based, decentralized approaches to enhancing resilience where livelihoods are irrevocably changed. Social capital is essential to facilitate this kind of representation of local interests and knowledge, yet it is also an outcome of the process. Although tensions between different types of social capital can develop, often in rural and traditional settings, the goal is to address underinvestment in social assets by regenerating bonding social capital among stakeholders at the local level.

Involving local stakeholders in adaptation initiatives is a practical means to increase institutional and project accountability while building the local asset base. At the institutional level, local civil society organizations can play a key role in tracking the allocation of funding for adaptation projects at the regional and local level. At the project level, retraining local agents to fill jobs in project management and field monitoring and evaluation can provide a safety net by promoting climate-resilient jobs while working toward reducing community risk from climate change. This holds potential to strengthen the physical and financial capital of those whose livelihoods are threatened by climate change. It will also build the social capital needed for communities to voice and represent their own interests, so that national institutions can be made more responsive to, and accountable for, the needs of local communities.

***Prong 3. Build household resilience through asset-based adaptation: a "no-regrets" approach.*** People living in poverty are particularly vulnerable to the erosion of their asset base. During a sudden decline in assets, such as during a climate-related disaster, poor households often cannot achieve even low consumption levels without having to deplete productive assets even further, whether livestock, family health, or children's education. Hence, building the asset base of the poor is a "no-regrets" approach to good adaptation while working toward local development goals.

As part of an asset-based vulnerability analysis, it is helpful to distinguish between the asset protection needed during a natural disaster and the asset building needed to withstand projected long-term, gradual climate changes. Different constellations of assets are needed during different stages of an impact of climate change: before, during, in the immediate aftermath, and for long-term recovery and adaptation. For instance, early warning systems and training are crucial elements of enhancing livelihood resilience, whereas financial capital, such as credit or insurance, is vital for recovery and long-term adaptation. Interventions should focus on enhancing the specific mix of livelihood assets that will provide the greatest resilience and adaptation in the local climate-related vulnerability context.

Table 11.1 provides a summary of policy recommendations regarding how to build the asset base of the poor and enhance local livelihoods and access to public services. Highlights of the recommendations are presented below in each of the following asset categories: physical, human, social, cultural, natural, and financial. All these categories are interconnected, and actions aimed at the same adaptation objective will be seen to overlap.

### Physical Capital

Recommended actions to strengthen physical capital focus on improving public works and infrastructure, with the wider goal of creating access to services for the most vulnerable people. They include preventing erosions and landslides, particularly for urban settlers; decentralizing water management to the local level; protecting productive infrastructure such as tourist areas and storage facilities for local harvest; separating sewer and storm drain systems and increasing their capacity; and devising early warning systems to save lives. Such efforts will have multiple effects and will help protect against the spread of diseases and the loss of jobs and other productive means, as well as increase food security and integration with agricultural markets. An often overlooked dimension of adaptation is the transfer of technology. Climate-smart communication technologies offer opportunities to implement innovative adaptation measures for resource-strained or geographically isolated communities.

## Human Capital

Improving the health and education of a population is the safest "no-regrets" approach to enable long-term adaptation to climate change. As seen above, studies from Brazil show that an extra two years of schooling can mitigate the negative effects of climate change on income. As a short-term coping strategy, the use of climate indicators in conditional cash transfer programs can help ensure adequate nutrition levels in chronically poor populations affected by climate change. Other programs to strengthen human capital include raising awareness of climate change and its associated impacts, as well as of the adaptation programs that are available to families.

## Social Capital

Building and preserving social capital are ongoing processes. As an intangible asset, social capital can be difficult to measure and document. Yet investments in social capital should be a priority because they can be a foundation for the allocation of other assets and can create important positive spillover effects. Building social capital is a "no-regrets" approach to ensure that an adaptation measure chosen for implementation is targeted, timely, and effective. Through participatory approaches that facilitate the dialogue among local communities and the sharing of local concerns with local and national institutions, social capital can develop within and among stakeholder groups (bonding and linking). And it can help to ensure that local concerns are considered in devising the appropriate resources and strategies. Existing institutional frameworks can assist in promoting this asset, for example, through poverty and social impact analysis, which helps achieve a deeper understanding of how local climate change affects inequality, poverty, and migration among different socioeconomic groups. Establishing partnerships with local stakeholders will help promote the sustainable implementation of an adaptation project.

## Cultural Capital

Drawing on cultural capital is critical to designing adaptation strategies with sustainable outcomes. Cataloguing traditional knowledge, broadening access to that information, and mobilizing its integration with local adaptation plans is a win-win activity for both traditional communities and national climate change planners: it serves to strengthen livelihood resilience as well as to promote sustainable outcomes through local participation.[1]

## Natural Capital

Preserving the natural resource base must be part of both urban and rural adaptation strategies. While this objective covers a broad and important field, certain priorities are worth highlighting for their direct effect on livelihood resilience: water, agriculture, and environmental resilience. First, to conserve water as an asset requires action to implement water pricing, switching to less-water-intensive agriculture, and creating new highland reserves. Second, to sustain livelihoods that depend on agricultural production, adaptation measures should seek to improve crop rotation and diversification of farm activities. Farming techniques that promote adaptation to new environmental circumstances should complement traditional farming methods. Finally, promotion of sound environmental policies, particularly for watersheds, marine protected areas, and coastal zones, is needed to support the development of the environmental resilience on which natural capital depends.

## Financial Capital

Expanding access to financial services must be a priority in adaptation programs. In Nicaragua, for example, a conditional cash transfer program was introduced with a productive investment grant as part of a package with basic nutrition and education. This enabled recipients to begin adapting their livelihoods to the growing threat of drought. The innovative part of the program consisted of adding a climate-risk dimension to social protection programs that traditionally focus on the chronically poor. The same principle can be applied to other adaptive measures, such as making transfers contingent on the recipients' dwelling in less-exposed or less-vulnerable areas. Other noteworthy approaches to strengthening financial capital include social funds and support for community-driven adaptation, safety nets for coping with climate risks and natural disasters, improving access to credit and land titles, and microinsurance and indexed insurance. Finally, offering local people training geared toward employment in climate-resilient jobs will help support the development of much-needed alternative livelihoods. A successful program in Belize trained local fishermen who could no longer make a living from fishing as rangers, researchers, tourist guides, and park managers, with the added benefit of building resilience for their entire communities.

## Perspectives on Future Research

Conventional indexes of water well-being often fail to fully measure the complexity of water scarcity, which encompasses not only water availability and use, but also water quality and environmental demand. Data problems and constraints tend to be the most important limitation on the development of more comprehensive and useful indexes. Particularly concerning climate change, scaling remains a severe challenge for converting data from global climate models and scenario results into information for operational use at the local level. New research initiatives should aim at developing more sophisticated water indicators to address the limitations of relying on single-factor indicators and to capture water stress and vulnerabilities at the subregional and local levels. To develop a set of best-practice indicators of specific relevance to Latin America and the Caribbean, it would be worthwhile to focus future case studies on how integrated water resource management (IWRM) can be tailored to local circumstances and how scientific knowledge about climate change and IWRM principles is being applied on the ground.

Current research on rural livelihoods tends to focus on how extreme events affect poor rural households. An increasing need exists to better understand how climate change and variability affect the long-term sustainability of agricultural systems in marginal environments. Research should focus on how complex agrarian and livestock systems can adapt to climate changes and variability and assess the coping capacity of rural communities in different regions. Such research will not only improve knowledge of social impacts but, most importantly, aid in building adaptive capacity at all levels within the farming community.

A need exists to strengthen interaction between research teams and local communities. Many farming communities in the LAC region, including indigenous groups, rely on knowledge of the local environment for their livelihoods, and their capacity to cope and adapt to climate change and variability depends on how quickly their knowledge can be adjusted to reflect the changing climate. That stresses the need for better education of rural populations and the need to target research and knowledge development and dissemination more toward them. It may also be advantageous to involve rural communities in developing new ways of learning that better incorporate both traditional knowledge and science-based results and that compensate for the poor conventional learning skills associated with high illiteracy rates.

The impact of climate change on coastal livelihoods remains severely understudied. Research is also needed for different subregions and sectors to better understand the impact of climate change on the urban poor. For coastal industries such as fisheries and tourism, up-to-date time-series statistics on employment and value added are vital to developing adaptation measures to help safeguard these industries against climate change.

In regard to fisheries, more detailed analysis is needed on the potential for small-scale aquaculture as well as the vulnerability context for small fishing villages. Particular attention should be paid to understanding the needs and sensitivities of artisanal fishing communities, some of which are more accepting of ideas from outside than others. Given the speed at which the tourist industry is growing in Latin America and the Caribbean, research is needed on how to safeguard the industry and how to establish best-practice examples for the region on protecting natural resources and preventing job losses. Ecotourism has received surprisingly little attention in the context of climate change, and more research is needed to understand the specific adaptation needs of the often impoverished communities that depend on it.

The analyses reported in this book of the effects of climate change and variability on health, poverty, and inequality only considered changes in mean temperatures and precipitation. Research the effects of increased climate variability, extreme weather events, and disasters on population well-being and income distribution is also urgently needed. While those are much harder to quantify with the limited information currently obtainable, future research should focus on building better data sets and frameworks to improve modeling of disease and poverty.

The analysis of climate-induced migration has shown that it is possible to quantify the amount of internal and international migration attributable to climate change. Yet to get an overall sense of future climate-induced migration, similar types of analyses should be repeated for other countries in the LAC region and beyond. Great variations in climatic conditions, levels of vulnerability, and projected climate change among different countries and subregions will require that future research focus on the specific dynamics of the local context. The study reported in this book did not consider migration due to changes in extreme events because the impacts of extreme events at the municipal level are hard to quantify. Yet research inclusive of this level of detail would be valuable in designing better policies.

Climate change and variability may undermine human security for certain groups, but whether they will lead to violent conflict will depend on other key socioeconomic and political conditions—and most importantly on how they are perceived and communicated by key political actors. Most studies of climate-related factors have focused on physical consequences such as soil degradation, deforestation, and water scarcity. Inasmuch as these physical consequences seem inevitable, it would be beneficial to expand the focus to encompass the political economy conditions that might perpetuate conflict. Also, instead of stressing the environmental scarcity aspect, a fruitful social research perspective would place more emphasis on opportunities posed by environmental change as well as on the sociopolitical channels that translate environmental consequences into different outcomes across various groups.

Perhaps the greatest challenge in capturing the social implications of climate change is to analyze the comprehensive effects of the interacting social dimensions affected by it. For example, research initiatives on rural livelihoods are hindered by the lack of appropriate frameworks to account for the complex dynamics of changes in precipitation and the effect on food security, and the resulting implications for future migration patterns. Whereas this book has looked in depth at various social dimensions of climate change, the challenge remains to address deeper systemic and structural dynamics of human well-being in a changing climate context for future generations.

## Note

1. For a deeper look at the impact of climate change on indigenous people and on the use of traditional knowledge systems and institutions in adaptation planning, see J. Kronik and D. Verner, *Indigenous Peoples and Climate Change in Latin America and the Caribbean* (Washington, DC: World Bank, 2010).

# Climate Change and Climatic Variability in Latin America and the Caribbean

## Jens Hesselbjerg Christensen

Warming of the climate system is unequivocal, as is now evident from observations of increases in global average air and ocean temperatures, widespread melting of snow and ice, and rising global average sea level. Continued greenhouse gas emissions at or above current rates would cause further warming and induce many changes in the global climate system during the 21st century that would very likely be larger than those observed during the 20th century (IPCC 2007).

Information about long-term variations as well as more recent changes in climate in Latin America and the Caribbean (LAC) is essentially absent from the international literature. Such data series as exist are traditionally safeguarded by national weather services and other regional authorities, which do not necessarily have routine procedures to make data easily accessible to external users. Only observations from a sparsely distributed network of stations are available for analysis in depth. The paucity of information has so far limited not only studies of observed climate change and variability, but also projections of climate change and variability in the region, and the quality of climate models pertaining to the LAC region is difficult to assess.

This appendix summarizes knowledge on recent climate change and variability and most likely future climate change and variability in Latin America and the Caribbean, building primarily on the 2007 Intergovernmental Panel on Climate Change (IPCC) Fourth Assessment reports and recent literature (see also box A.1).

**Box A.1**

## Climate Definitions

Often there is a considerable gap between common perceptions of what is meant by "climate" and a more scientific, meteorologically based definition. That mismatch gives rise to misleading statements and assessments of the role of humankind in observed changes, even in high-level documents. To better guide the reader of the present volume, the following provides a basic meteorological understanding of what climate is.

**Annual Global Mean Temperatures**

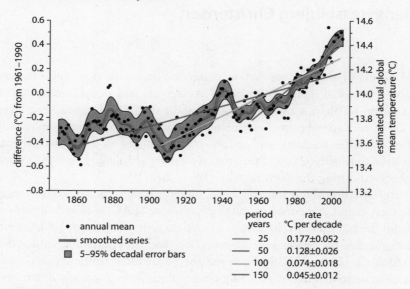

| | period years | rate °C per decade |
|---|---|---|
| • annual mean | | |
| ▬ smoothed series | —— 25 | 0.177±0.052 |
| ▨ 5–95% decadal error bars | —— 50 | 0.128±0.026 |
| | —— 100 | 0.074±0.018 |
| | —— 150 | 0.045±0.012 |

*Source:* Adapted from IPCC 2007.
*Note:* Annual global mean temperature (black dots) with linear fits to the data. The left-hand axis shows temperature anomalies relative to the 1961–90 normal period, and the right-hand axis shows estimated actual temperatures, both in °C. Linear trends are shown for the last 25 (green), 50 (blue), 100 (orange), and 150 years (red). The smooth blue curve shows decadal variations, with the decadal 90 percent error range shown as a pale blue band about that line. The total temperature increase from the period 1850–99 to the period 2001–05 is 0.76°C ± 0.19 °C.

*(continued)*

**Box A.1** *(continued)*

Climate is basically the statistical properties of weather over a long period. Its simplest and best-understood elements are mean temperature and mean precipitation amounts, whether on a monthly, seasonal, or annual basis. No law of physics dictates exactly what time frame to use. That weather changes from one year to another and possibly shows significant trends over some years does not fall within this definition of climate. A long time period must be defined in such a way that comparisons between two climate periods (two equally long periods) are close to invariant. That means that year-to-year, or even decadal variations should show up only marginally. However, even at longer time scales the statistics of weather do not seem to be entirely invariant. In practice, climatologists in the first part of the 20th century decided to use a period of 30 years as a compromise to balance the need for invariance in the conditions from one period to another. This led to the definition of 30-year climate norms, which started with the period 1901–30. The latest norm is for the period 1961–91. For practical reasons, these periods are still used with rigor. Many climate variables, such as annual temperature and precipitation, are compared with respect to the latest reference period. As climate turns out to vary significantly even between these so-called norm periods, it also has become customary to look at long-term trends.

The figure in this box depicts a time-series of annual global mean temperatures from 1850 to 2005. The overall increasing trend is portrayed in a number of ways (see figure note). It is important to observe that the underlying data points show a clear scatter around the overall trend. Trying to interpret the time evolution over too short a period is made difficult and misleading by the considerable interannual "noise." For example, trying to make a linear trend line through the last 8–10 years (from 1998 to the present) would result in a trend close to zero, or even a negative trend, suggesting that the overall warming has stagnated. Such interim periods have occurred previously. It is only with a longer time span that the real picture of change clearly shows itself. Change need not be a constantly evolving phenomenon; even a reverse signal at times need not contradict the long-term evolution.

The definition of "climate change and variability" is not just a question of comparing climate periods; in general it must also involve regions large enough to show a coherent picture of change. Only then is it possible to relate the climate trend of the site to any large-scale change taking place in the region of concern. The geographical scale on which such coherence is found appears to be not much finer than continental. Therefore, clear-cut and robust statements about observed climate change and variability on a detailed country basis would be dubious if they are not balanced with a view to the larger-scale trends both geographically and temporally. Opposite trends at nearby stations may still accord with changes on a larger scale.

*Source:* Author.

Latin America and the Caribbean are warming largely in line with the global trend and are likely to continue doing so. Local exceptions to the general tendency will appear, but available models do not give robust results for every part of the region. In particular, many issues related to changes in the Amazonas are still unresolved because important aspects of the interaction between vegetation and climate are still little understood. Different models also tend to behave differently in simulating the present climate within the region, and that limits the ability to take the simulated responses to anthropogenic forcings at their face value for the region. For this reason, the best estimate of climate change and variability in the region comes from assessing the results from many models, as was done by Christensen and others (IPCC 2007). The present analysis tries to go behind some of the statements provided there, by undertaking a comparative assessment of the results of general circulation models (GCMs).

## Images of Present Change

### Temperature

As noted above, LAC is projected to continue to warm at a rate little different from that of the world as a whole. Figure A.1 shows the geographical distribution of linear temperature trends for the periods 1901–2005 and 1979–2005, respectively. The long-term trend shows larger geographical variations and noticeable hot spots in southeast Brazil, Uruguay, northeast Argentina, and northwestern Mexico, where the warming has been more than double the global increase. For the more recent period, the warming trend shows less geographical variance.

### Precipitation

Data are less widely available for precipitation than for temperature, but the information that is available shows that precipitation varies widely within the LAC region. Figure A.2 depicts the geographical distribution of change. The most noticeable large-scale, coherent pattern is a long-term tendency toward drying in the tropics and subtropics (the Caribbean and northern South America), while temperate (southern South America) climates experience more precipitation.

For large parts of the region, data are not readily available for analysis. In some countries efforts have been made to rescue and collect observational data, though not necessarily to make them publicly available. Analysis of these data confirms the broad tendencies, but trends may vary quite widely even at the community level (scale of 1–200 km), with quite large positive

## Figure A.1    Linear Trend of Annual Temperatures

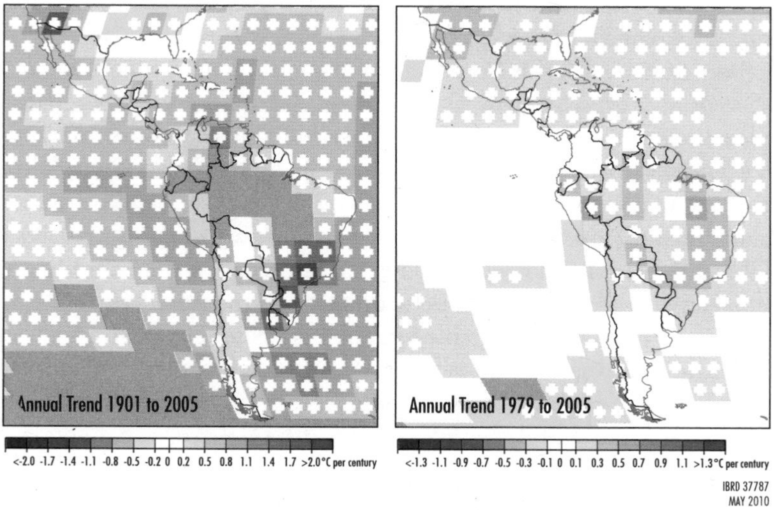

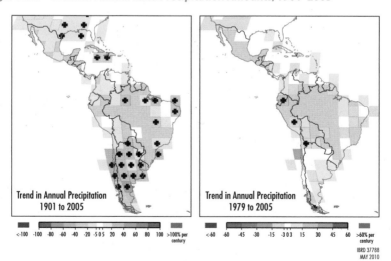

*Source:* Adapted from IPCC 2007.
*Note:* Left panel shows °C per century, 1901–2005. Right panel shows °C per decade, 1979–2005. Areas in grey have insufficient data to produce reliable trends. Trends significant at the 5 percent level are indicated by white + marks.

## Figure A.2    Trend in Annual Land Precipitation Amounts, 1901–2005

*Source:* Adapted from IPCC 2007.
*Note:* Areas in grey have insufficient data to produce reliable trends. Note the different color bars and units in each plot. Trends significant at the 5 percent level are indicated by black + marks.

trends occurring next to negative ones. It is only when data are aggregated over large regions that changes can be described in the context of global warming (Zhang and others 2007).

## Sea Level

Sea level in many regions of the world varies considerably on many time scales. The shortest time variation and in some regions by far the largest sea level changes are induced by tides. As these occur on a regular basis, however, natural and managed environments are adapted to this variability.

Globally, sea level has been rising for centuries, and during the 20th century an average increase of 0.17 meters was measured (IPCC 2007). Regionally, however, considerable variation appears (Figure A.3). Oceanic circulation is complex; there is little reason to believe that the observed geographical distribution of these changes can be directly scaled to climate change and variability scenarios because changing ocean conditions (currents, salinity, and temperature) may cause less-obvious results along

**Figure A.3    Geographic Distribution of Long-Term Linear Trends in Mean Sea Level**

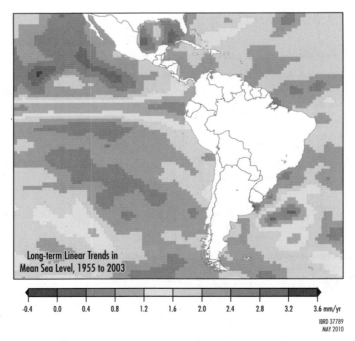

Long-term Linear Trends in
Mean Sea Level, 1955 to 2003

-0.4    0.0    0.4    0.8    1.2    1.6    2.0    2.4    2.8    3.2    3.6 mm/yr

IBRD 37789
MAY 2010

*Source:* Adapted from IPCC 2007.
*Note:* For 1955–2003, based on past sea level reconstruction with tide gauges and altimetry data.

the way. However, past experience often serves as a reference point against which to compare projections of the future.

## Recent Events

Most of Latin America and the Caribbean has experienced several instances of severe weather in recent decades, damaging property, infrastructure, and natural resources. Fatalities in connection with torrential rain and hurricanes have been counted in thousands in this century alone. The overall perception of these events is that they signal climate change and variability. However, no formal scientific detection of these changes at the regional level has been made. That makes it difficult to assess whether or not those extreme events—whether individually or collectively—are the results of general climate change and variability, as indicated by the overall warming of the continent and adjacent areas.

People who have experienced severe and damaging weather events generally wish to be able to assign them to some special factor. Yet changes in climate or the recurrence of extreme events need not be related to global warming. To formally attribute change or the occurrence of particular events to a cause, a statistically sound number of events must normally be considered. By nature, extreme events are rare at any given location. Even events that occur over a large geographical region such as Latin America and the Caribbean cannot simply be aggregated and studied as a whole because the chain of physical events leading to any one of them is likely to differ from event to event and from location to location. That precludes a simple statistical treatment of the data. Therefore, recent reports of the IPCC have had very little to say about recent changes at a regional level, and even less about the national or provincial level, and their possible links with global climate change and variability.

Despite these limitations, people relate the impact of new events to their memory of past occurrences. And as adverse changes create a need for adaptation and mitigation measures, it is important to use past experience in preparing for potential hazards.

## Images of the Future

Global climate models have long been the main tool used to project climate change and variability under certain assumptions about future emissions of greenhouse gases and other anthropogenically induced drivers of the climate system. The climate variable most commonly used to illustrate anthropogenically induced climate change is the global annual mean temperature. Observations show that the average global temperature rose by 0.74°C over the last century (IPCC 2007). Model experiments show that even if all human-induced forcing agents were held constant at their year 2000 levels, further warming would occur in the next two decades at a rate of about 0.1°C per decade, mainly reflecting the slow response of the oceans. About twice as much warming (0.2°C per decade) would be expected if emissions are within the range of the IPCC's *Special Report on Emissions Scenarios* (SRES) (IPCC 2001). Best-estimate projections from models indicate that the average warming over each inhabited continent by 2030 is insensitive to the choice among scenarios, and according to the IPCC (2007) it is very likely to be at least twice as large as the corresponding model-estimated natural variability that took place during the 20th century.

Further to this, best estimates and likely ranges for global average surface air warming for six SRES emissions marker scenarios were provided in the IPCC's *Climate Change 2007* (2007). For example, the best estimate for the low scenario (B1) is 1.8°C (likely range 1.1°C to 2.9°C), and the best estimate for the high scenario (A1F1) is 4.0°C (likely range 2.4°C to 6.4°C).[1]

These estimates, referring to the entire planet, hide large regional variations. The spread among different model results, as indicated by the ranges of the two marker scenarios, B1 and A1F1, adds to this uncertainty. Although different models tend to respond to an anthropogenic forcing rather similarly on the global scale, the spread is a clear signal of large discrepancies across regions.

Climate varies from region to region. This variation is caused by the uneven distribution of solar heating; available atmospheric moisture; the individual responses of the atmosphere, oceans, and land surface; the interactions among these; and the physical characteristics of the regions. The perturbations of the atmospheric constituents that lead to global changes affect certain aspects of these complex interactions. Some anthropogenic forcings are global in nature, while others are regional. For example, carbon dioxide, which causes warming, is distributed evenly around the globe regardless of where the emissions originate, whereas sulfate aerosols (small particles), which offset some of the warming, tend to be regional in their distribution. Furthermore, the response to forcings is partly governed by feedback processes that may operate in regions other than those in which the forcing is greatest. Thus, the projections of changes in climate will also vary from region to region.

Against this backdrop, the array of models that yield robust projections of the global rise in mean temperature yields much less robust estimates of regional climate change and variability (IPCC 2007). To the extent that assessments of future climate change and variability can be made, they are the following:

- All of Central and South America is very likely to warm during this century. The annual mean warming is likely to be similar to the global mean warming in southern South America but larger than the global mean warming in the rest of the area.

- Annual precipitation is likely to decrease in most of Central America, with the relatively dry boreal spring becoming drier. Annual precipitation is likely to decrease in the southern Andes, with relative precipitation changes being largest in summer. A caveat at the local scale is that changes in atmospheric circulation may induce large local variability in precipitation changes in mountainous areas. Precipitation is likely to increase in Tierra del Fuego during winter and in southeastern South America during summer.

- It is uncertain how annual and seasonal mean rainfall will change over northern South America, including the Amazon forest. In some regions, qualitative consistency is seen among the simulations (rainfall increasing in Ecuador and northern Peru, and decreasing at the northern tip of the continent and in southern Northeast Brazil).

- The systematic errors in simulating current mean tropical climate and its variability, and the large differences among models in future changes in El Niño amplitude, preclude a conclusive assessment of the regional changes over large areas of Central and South America. Most models are poor at reproducing the regional precipitation patterns in their control experiments and have a small signal-to-noise ratio, in particular over most of Amazonia. The high and sharp Andes Mountains are unresolved in low-resolution models, affecting the assessment over much of the continent. As with all landmasses, the feedbacks from land use and land cover change are not well accommodated and lend some degree of uncertainty. The potential for abrupt changes in biogeochemical systems in Amazonia remains as a source of uncertainty. Large differences in the projected climate sensitivities in the climate models incorporating these processes and a lack of understanding of processes have been identified. Over Central America, tropical cyclones may become an additional source of uncertainty for regional scenarios of climate change and variability, since the summer precipitation over this region may be affected by systematic changes in hurricane tracks and intensity.

The following section interprets these findings on a regional scale.

## Aspects of Observed Climate

Though, ideally, general circulation models (GCMs) should be able to provide information at the regional scale on which they resolve, efforts to improve such models have concentrated on the ability to describe specific geophysical phenomena such as El Niño, monsoon systems, and sea ice. That focus has precluded paying specific attention to certain aspects of model performance on a regional level for many parts of the world, including Latin America and the Caribbean.

Therefore, alternative methods have been developed to derive detailed regional information in response to geophysical processes at finer scales than resolved by GCMs. Nested regional climate models (RCMs) and empirical downscaling have yielded new ways to assess important regional processes that are central to climate change and variability. These assessments allow the development and validation of models to simulate the key dynamic and physical processes of the climate system. Until recently, few scientific programs were aiming at providing high-resolution information on climate change in developing nations, including those in the LAC region. One exception is the Japanese initiative, the Earth Simulator. There information from a dedicated simulation, with a high-resolution GCM, is providing fine-scale information globally and hence also for the LAC region. However, results from this simulation should be carefully compared with the collective information from other GCMs (IPCC 2007) as information from a single simulation will only provide indicative information about possible future change.

Within the community studying climate change impacts and adaptation a move is growing toward integrated assessment, yielding projections of regional climate change and variability that provide key inputs into decision support systems aimed at reducing vulnerability (Bales, Liverman, and Morehouse 2004). At present, the regional projections are perhaps the weakest link in this integrated assessment, and the bulk of information readily available for policy and resource managers (such as via the IPCC

Data Distribution Center)[2] largely derives from GCMs, which have limited ability to accurately simulate local-scale climates, especially with regard to the key parameter of precipitation. GCM data are commonly mapped as continuous fields, which do not convey the low skill of the models for many regions, or are area aggregated, which renders the results of little value for local application.

To help understand the potential accuracy of climate change and variability projections derived from climate models, it is imperative to compare the performance of these models against the observed climate. Below are summarized some of the main LAC climate characteristics, which a model should be able to describe with some realism if its projections for the future are to be credible.

## Mexico and Central America

Most of the American isthmus (for example, central-southern Mexico and Central America) and the Caribbean has a relatively dry winter and a well-defined rainy season from May through October (Magaña, Amador, and Medina 1999; Taylor and Alfaro 2005). The progression of the rainy season largely results from air-sea interactions over the Americas' warm pools (such as the Gulf of Mexico) and the effects of topography over a dominant easterly flow, as well as the temporal evolution of the Inter-Tropical Convergence Zone (ITCZ).[3] The mountain range running the length of the American isthmus limits moist air moving across it, resulting in a precipitation pattern characteristic of mountain ranges. As moist air that has formed over the sea moves in over land, it is forced up the mountainside, and as it cools down, releases rainfall on the ascent. The high mountains and the lee side of the mountains remain largely dry. This effect is present whether the air flows from east to west or vice versa. During the boreal winter, the atmospheric circulation over the Gulf of Mexico and the Caribbean Sea (the Intra Americas Seas, or IAS), is dominated by the seasonal fluctuation of the Subtropical North Atlantic Anticyclone, with invasions of extratropical systems that affect mainly Mexico and the western portion of the Great Antilles. The so-called Nortes, or Tehuantepecers, produce some precipitation and changes in temperature over the coastal regions of the IAS (Romero-Centeno and others 2003).

The rainy season in the American isthmus and the Caribbean is mainly the result of easterly waves and tropical cyclones that contribute to a large percentage of the precipitation. When low vertical wind shear coincides with warm sea surface temperatures, two of the necessary preconditions exist that may result in easterly waves' maturing into storms and hurricanes, generally within the 10°N to 20°N latitudinal band.

Over most of the IAS, warm pool precipitation is weak because subsidence (descending air masses) inhibits convective activity, making this region a climate paradox. Over the Caribbean Sea, winds are strong because of a low-level jet (LLJ), which seems to play a key role in the distribution of precipitation over Central America. A unique characteristic of boreal summer precipitation over the Pacific side of the American isthmus and the adjacent warm pool is its bimodal structure, with maxima in June and September and a relative minimum during the middle of the boreal summer, late July and early August. This relative minimum, known as Canicula or Midsummer Drought (MSD), has been attributed to air-sea interactions and teleconnections between the IAS and the eastern Pacific warm pool (Magaña, Amador, and Medina 1999). This characteristic in convective activity on these time scales appears to influence tropical cyclone formation in the eastern Pacific.

Most climate variability in the region is related to the El Niño–Southern Oscillation (ENSO), a phenomenon that influences the distribution, frequency, and intensity of many of the regional atmospheric phenomena. The signal of El Niño over the American isthmus and the Caribbean is contrasting, not only among seasons but also in relation to coast (Pacific or Caribbean). For instance, during El Niño boreal winters, precipitation over northwestern Mexico, the Greater Antilles, and part of the Pacific Coast of Central America increases, and less rainfall than usual is observed in parts of Colombia and República Bolivariana de Venezuela and the Lesser Antilles. During El Niño boreal summers, most of the American isthmus experiences negative precipitation anomalies, except along the Caribbean coast of Central America where positive precipitation anomalies are observed. Tropical cyclone activity over the IAS diminishes during El Niño summers (Gray 1984; Tang and Neelin 2004).

The mean state of ENSO and its global pattern of influence, amplitude, and interannual variability and frequency of extreme events have varied considerably in the past. Many of these changes appear to be related to, though definitely not entirely due to, changes in global climate and the history of external forcing agents, including recent anthropogenic forcing (Mann, Bradley, and Hughes 2000). Interannual and decadal ENSO-like climate variations in the Pacific Ocean basin are important contributors to the year-to-year (and longer) variations of the climate in South (and North) America. Despite potentially different source mechanisms, both interannual and decadal ENSO-like climate variations yield wetter subtropics (when the ENSO-like indexes are in positive, El Niño–like phases) and drier midlatitudes and tropics (overall) over the Americas, in response to equatorward shifts in westerly winds and storm tracks in both hemispheres (Dettinger and others 2001).

### South America

A complex variety of regional and remote factors contributes to the climate of South America (Nogués-Paegle and others 2002). The tropospheric upper levels are characterized by high pressure centered over Bolivia and low pressure centered over northeast Brazil. At low levels the Andes effectively block air exchanges with the Pacific Ocean, but a continental-scale gyre transports moisture from the tropical Atlantic Ocean to the Amazon region and then southward toward extratropical South America. The South American low-level jet starts a regional intensification of this flow, channeling it along the eastern foothills of the Andes into the so-called Chaco low.[4] The LLJ carries significant moisture from the Amazonas toward southern South America, and it is present throughout the year, but strongest during the austral winter season (Berbery and Collini 2000; Vernekar, Kirtman, and Fennessy 2003).

A clear warm season precipitation maximum, associated with the South American Monsoon System (SAMS), dominates the mean seasonal cycle of precipitation in tropical and subtropical latitudes. The rainfall over northern South America is directly influenced by east-west circulation patterns, and consequently tropical sea surface temperature anomalies affect regions such as the Ecuador coast and north-northeast Brazil.[5] The SAMS is also modulated by incursions of drier and cooler air from the midlatitudes over the interior of subtropical South America (Garreaud 2000; Vera and Vigliarolo 2000). Rainfall anomalies over subtropical South America are associated with regional feedback processes and interactions among the topography, the SAMS, and the midlatitude systems.

Another important feature, a regional part of the ITCZ, is the South Atlantic Convergence Zone (SACZ)—a southeastward extension of cloudiness and precipitation from southern Amazonas toward southeast Brazil and the neighboring Atlantic Ocean. The SACZ reaches its easternmost position during December, in association with high precipitation over much of Brazil, a southeasterly flow over eastern Bolivia, and low precipitation in the Altiplano. The variability of precipitation during the austral summer on a variety of time scales (intraseasonal, interannual) has been related to changes in the position and intensity of the SACZ (Liebmann and others 1999). Variability of the SACZ also influences the atmospheric circulation and rainfall anomalies over eastern South America between about 20°S and 40°S, through a dipole pattern in the vertical motion field that reflects changes in the intensity of the SACZ and "compensating" descent over southern Brazil, Uruguay, and northeastern Argentina (Doyle and Barros 2002; Robertson and Mechoso 2000).

The leading mode of the interannual variability in the Southern Hemisphere is the Southern Annual Mode (SAM) (Kidson 1988; Thompson and Wallace 2000). For southeastern South America, results indicate that the SAM activity modulates the regional variability of the precipitation, especially during the late austral spring, when the SAM index correlates well with ENSO (Silvestri and Vera 2003).

## Global Climate Change Issues

The scientific attribution of climate change and variability to causes calls for a comprehensive understanding of natural variability and large-scale external drivers such as greenhouse gas concentrations and aerosol loads. By combining information on observed changes with information from specially designed experiments using climate models with and without the external drivers, it becomes possible to distinguish a possible effect originating from the driver, and hence to attribute the cause of an observed change to the driver. However, just as models cannot perfectly represent real climate, observations cannot precisely portray climate evolution. For that reason, it is very difficult to precisely define what models should depict, and what should be understood as merely an artifact of natural variability, reflecting processes that are basically unaffected by an external driver.

Climate variability exists on many time scales and on many geographical scales. The occurrence of El Niño provides a good example of the difficulty of detecting and attributing climate change and variability. El Niño occurs at approximately 4-year intervals. The physical chain of reactions leading to its occurrence is well understood (IPCC 2007), and state-of-the-art climate models are generally able to simulate both El Niño and La Niña episodes with fidelity. But to simulate the observed sequence of events, models need to know the precise ocean state at certain intervals. Given the chaotic behavior of the climate system, the ocean component of a climate model will drift away from the exact development within a few years of simulations, even if it is initialized with the best observed state of the ocean. The same behavior is seen in the atmospheric component because categorical weather forecasts beyond 2–3 weeks are impossible.

Since it is impossible to capture the precise evolution of important regional climate features such as El Niño and La Niña, models will not generally be able to describe the transient trends and variability associated with such large-scale natural drivers. This should not be seen as a limitation, but acknowledging such general facts provides some insight on how to assess the results from climate change models.

For models, the depiction of regional features of change becomes increasingly demanding, the smaller the scale one wants to study. It is a combination of the limited resolution presently used in most models and the nature of natural climate variability, which need not be as well characterized as El Niño or La Niña either in temporal regularity or in extent. It is therefore interesting that, collectively, models can simulate the temperature trends of the 20th century not only on the global but also on a continental scale. Figure A.4 makes that clear and also shows that the rising trends can only be interpreted as the result of anthropogenic drivers of the climate system.

Figure A.4 also shows that the models do not capture the ocean surface temperatures during a short period in the early part of the 20th century very well, and as a consequence, even though the models are better at simulating land surface temperatures for that period, the observed increase in global surface temperatures for the period are less well captured. This discrepancy should therefore be looked for in models' ability to simulate the accurate ocean surface temperatures. Because detailed ocean temperatures are the result of a complex dependency on the centuries-old oceanic circulation and the immediate climate forcing, it is hardly a surprise that there is a mismatch at a time when the external climate forcing was small.

To what extent can or should models be expected to simulate the climate evolution in Latin America and the Caribbean? Given the arguments above, there is no reason to expect models to be able to reproduce natural variations in detail, with the exception of the large-scale temperature increase. It is therefore also important to keep in mind that only if the externally induced climate change and variability are sufficiently large will the models depict a signal above "climate noise" or natural variability. For this reason alone, most efforts to model regional climate change and variability focus on scenarios wherein a large climate response should be expected. Moreover, they focus on the century-long time scale and not on

**Figure A.4    Comparison of Observed Continental- and Global-Scale Changes in Surface Temperature**

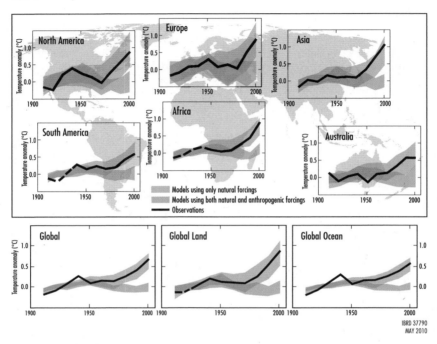

*Source:* Adapted from IPCC 2007.
*Note:* Results simulated by climate models using natural and anthropogenic forcings. Decadal averages of observations are shown for the period 1906–2005 (black line) plotted against the center of the decade and relative to the corresponding average for 1901–50. Lines are dashed where spatial coverage is less than 50 percent. Blue-shaded bands show the 5 percent to 95 percent range for 19 simulations from five climate models using only the natural forcings due to solar activity and volcanoes. Red-shaded bands show the 5 percent to 95 percent range for 58 simulations from 14 climate models using both natural and anthropogenic forcings.

the near term, even though the near-term horizon is where the need to know the evolution may seem most urgent.

The implication of these findings for interpreting locally observed recent changes in temperature and other climate variables is that they are subject to doubt. Observations may even suggest that local change is opposite to the global or continental-scale behavior. This does not disprove global or regional climate change and variability, but it does remind us that even if global temperatures should increase by 3–4°C within this century, many sites may not experience any of the change, whereas others will see an even greater change.

## Projections of Regional Climate Change

Region-by-region projections of climate change and variability need to be interpreted based on a proper understanding of key regional processes and of the skill of models in simulating current regional climate (Box A.2).

It is imperative for the use of the information provided at the regional level, and even more so at a country level, to see the statements offered in this appendix against the background of uncertainties related to regional climate change and variability. Largely because observational records do not have a long and uninterrupted time-series (going back to the beginning of the 20th century, for example), there is no evidence that what researchers have seen will be less or more severe in comparison to what may be experienced within this century. Observed climate trends and the occurrence of extreme events in the LAC region cannot generally be used

---

Box A.2

### Sources of Uncertainty in Regional Climate Change Projections

Climate models are constructed to match observed climatic conditions, which they manage to do with varying degrees of accuracy. When applied in century-long climate change and variability projections, those differences result in deviating responses to the prescribed climate forcing resulting from specified levels of greenhouse gases. This introduces uncertainty in the climate projections.

We do not know with any certainty the future emissions of greenhouse gases; that also introduces uncertainty in projections of climate change and variability on time scales beyond a few decades.

Natural variability in present climate conditions introduces a certain background "noise," which any projection of climate change and variability has to be compared with. Depending on the amplitude of the variability, the actual change may be difficult to filter out, as the "signal" remains below the noise. This is particularly relevant on the regional and local scales because the amplitude of variability generally increases as the scale gets smaller.

Weather phenomena that are climate specific but only appear irregularly or very rarely, for example, extreme events happen so rarely that a formal analysis of change is impossible because a proper statistical basis for analysis is lacking. This is particularly relevant on the regional and finer scales.

*(continued)*

**Box A.2** *(continued)*

Whereas climate change and variability projections for the distant future are independent of the precise state of current climate, that is not the case for projections covering the next 1–2 decades. Currently, imperfect knowledge, primarily about the state of the oceans, precludes prediction of seasonal weather patterns much beyond 6 months, even on the very large scales.

Climate change and variability projections for the near future—despite being independent of the emission scenario—generally become very uncertain as a result of the signal-to-noise issue described above and the limited knowledge that researchers possess about the exact current state of the climate.

Robust statements about change on the regional scale are therefore only possible if the model projections also are physically sound, meaning that a certain effect is expected because of large-scale changes in atmospheric circulation, moisture content, or temperature change. These measures of change should be captured by most, if not all, climate models. In most cases, it is therefore not possible to provide formal, quantitative estimates of error for the projected values of change.

*Source:* Author.

to deduce information about things to come. On the other hand, models clearly suggest that changes will take place over this century that will generally be in the direction of more of the extremes—more intensive precipitation, longer dry spells, warm spells, and heat waves with higher temperatures than generally experienced up to now.

It has often been stated that observed changes in adverse climate events already show characteristics that models predict for the future. However, this cannot be statistically confirmed because the intrinsic rarity of extreme events prevents them from being properly sampled. The conclusion drawn must be that we may not have seen the worst yet, and that preparations to deal with existing climatic hazards are needed as a first step toward coping with things to come. That statement is generic and particularly applicable to the LAC region, given the so-far-limited scientifically scrutinized analysis.

The following discussion is organized according to the regions used in IPCC (2007), largely covering Central and South America and some of the islands in the Caribbean. These regions are continental-scale (or based on large oceanic regions with a high density of inhabited islands) and may have a broad range of climates and be affected by a large range of climate processes. As they are generally too large to be used as a basis for conveying quantitative information on regional climate change and variability, the LAC region is further divided into Southern South America (SSA), Amazon (AMZ), and Central America and Mexico (CAM).

This regionalization is very close to that initially devised by Giorgi and Francesco (2000), but it includes additional oceanic regions and some other minor modifications similar to those of Ruosteenoja and others (2003). The objectives behind the original Giorgi and Francesco regions were that they should have simple shape, be no smaller than the horizontal wave length typically resolved by GCMs (judged to be a few thousand kilometers), and where possible, should recognize distinct climatic regimes. Although these objectives may be met with alternative regional configurations, as yet no such options are well developed in the regional climate change and variability literature.

Several common processes underlie climate change and variability in a number of regions, and before discussing LAC countries individually, it is relevant to summarize some of them. The first is a fundamental consequence of warmer temperatures and the increase in water vapor in the atmosphere. Water is continually transported horizontally by the atmosphere, from regions of moisture divergence (particularly in the subtropics) to regions of convergence. Even if the circulation does not change, these transports will increase because of the increase in vapor, and regions of convergence will get wetter and regions of divergence drier. We see the consequences of this increased moisture transport in plots of the global response of precipitation—where, on average, precipitation increases in the intertropical convergence zones, decreases in the subtropics, and increases in subpolar and polar regions (Meehl and others 2007). Regions of large uncertainty often lie near the boundaries between robust moistening and drying regions, with different models placing those boundaries differently.

Two other important themes in the extratropics are the poleward expansion of the subtropical high-pressure systems responsible for the overall long-term stable and dry weather conditions, and the poleward displacement of the midlatitude westerly wind bands with the associated storm tracks. This atmospheric circulation response is often referred to as the "excitation of the positive phase" of the Southern Annular Mode. Superposition of the tendency toward subtropical drying and poleward expansion of the subtropical highs create especially robust drying responses on the equatorward boundaries of the subtropical oceanic high centers.

The generally poor ability of climate models to simulate present climate for the region, together with incomplete analyses of climate change and variability in global climate model results for Latin America and the Caribbean and a serious lack of detailed studies of downscaled climate information, enables relatively few robust statements to be made (IPPC 2007).

Nonetheless, based on the analyses of many climate models (IPPC 2007; Meehl and others 2007) some clear indications about regional changes appear to be rather robust. Figure A.5 summarizes the collective information about temperature and precipitation changes toward the end of the current century, compared with the current climate, based on a large number (21) of GCMs used by the IPCC (2007).

**Figure A.5    Temperature and Precipitation Changes over Central and South America**

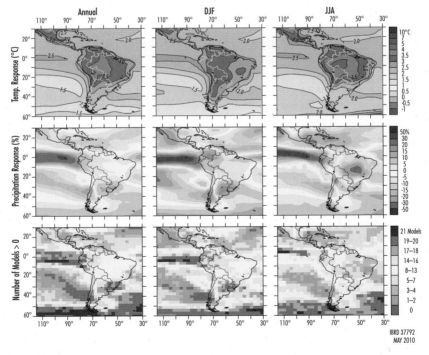

IBRD 37792
MAY 2010

*Source:* Adapted from IPCC 2007.
*Note:* Top row: Annual mean, December-January-February and June-July-August temperature change 1980–99 and 2080–99, averaged over 21 models. Middle row: same as top, but for percentage change in precipitation. Bottom row: number of models out of 21 that project increases in precipitation.

## Temperature

The balance of evidence assessed by the IPCC leads to the following statement:

All of Central and South America is very likely to warm during this century. The annual mean warming is likely to be similar to the global mean warming in southern South America but larger than the global mean warming in the rest of the area (IPCC 2007).

This implies that temperatures in all seasons will continue to rise during the 21st century. Some aspects of temperature-related events are also expected to change. Unless the temperature increase is a result of entirely new circulation patterns, changes in heat wave frequency and intensity are expected. Likewise, there is an enhanced risk of changes in the seasonality of some severe events because a higher temperature level in general will tend to favor a longer warm season with possible related extreme events. A clear example is the hurricane season (Box A.3).

---

**Box A.3**

### Hurricanes

An important driver for tropical cyclones and hence hurricanes is the sea surface temperature. A threshold of about 26°C determines whether a hurricane can be formed. In a warmer world with everything else being equal, the hurricane season is likely to be prolonged and the area prone to hurricane development may expand. However, a zone free of such systems will always remain near the equator, where atmospheric motions cannot support the development of intensive storms. March 2004 saw the first hurricane ever identified in the southern Atlantic. This hurricane severely affected the coastal zone of northeast Brazil. Such systems could become more frequent because of higher sea surface temperatures in the region.

In the tropics, particularly when the sun is close to its zenith, outgoing longwave radiative cooling from the surface to space is not effective in the optically thick environment caused by the high water vapors over the oceans. Links to higher latitudes are weakest at this time, and transport of energy by the atmosphere, such as occurs when the sun is less strong, is not an effective cooling

*(continued)*

**Box A.3** *(continued)*

mechanism, while monsoonal circulations between land and ocean redistribute energy in areas where they are active. However, tropical storms cool the ocean surface through mixing with , deeper ocean layers and through evaporation. When the latent heat is realized in precipitation in the storms, the energy is transported high into the troposphere where it can radiate to space.

As the climate changes and sea surface temperatures continue to rise, the environment in which tropical storms form is changed. Higher sea surface temperatures are generally accompanied by increased water vapor in the lower troposphere, and thus the moist static energy that fuels convection and thunderstorms is also increased. As noted above, hurricanes generally form from preexisting disturbances only where the sea surface temperatures exceed about 26°C; a rise in those temperatures therefore potentially expands the areas over which such storms can form. However, many other environmental factors, including wind shear in the atmosphere, also influence the generation and tracks of disturbances. El Niño and variations in monsoons, as well as other factors, also affect where storms form and track. The potential intensity, defined as the maximum wind speed achievable in a given thermodynamic environment, similarly depends critically on sea surface temperatures and atmospheric structure (Emanuel 2003). Many factors in addition determine whether convective complexes become organized as rotating storms and develop into full-blown hurricanes.

Although attention has often been focused simply on the frequency or number of storms, the intensity and duration likely matter more. The energy in a storm is proportional to velocity squared, and the power dissipation of a storm is proportional to the wind speed cubed. Consequently, the effects of these storms are highly nonlinear, and one big storm may have much greater impacts on the environment and climate system than several smaller storms.

From an observational perspective, then, key issues are the tropical storm formation regions; the frequency, intensity, duration, and tracks of tropical storms; and the associated precipitation. For land-falling storms, the damage from winds and flooding, as well as storm surges, is especially of concern but often depends more on human factors, including whether people place themselves in harm's way, their vulnerability, and their resilience achieved through such measures as building codes.

In summary, some models project hurricane intensities to increase, but there is little to document how their frequency and number of landfalls may develop.

*Source:* Authors.

## Precipitation

The IPCC's *Climate Change 2007: The Physical Science Basis* (2007) considered only a few features of future precipitation change in Latin America and the Caribbean to be robust. It stated:

- Annual precipitation is likely to decrease in most of Central America, with the relatively dry boreal spring becoming drier. Annual precipitation is likely to decrease in the southern Andes, with relative precipitation changes being largest in summer.

- Area mean precipitation in Central America decreases in most models in all seasons. It is only in some parts of northeastern Mexico and over the eastern Pacific during June, July, and August that increases in summer precipitation are projected. However, since tropical storms can contribute a significant fraction of the rainfall during the hurricane season in this region, these conclusions might be modified by the possibility of increased rainfall in storms not well captured by these global models.

- The annual mean precipitation is projected to decrease over northern South America near the Caribbean coasts, as well as over large parts of northern Brazil, Chile, and Patagonia, while it is projected to increase in Colombia, Ecuador, and Peru, around the equator, and in southeastern South America. The seasonal cycle modulates this mean change, especially over the Amazon basin where monsoon precipitation increases in December, January, and February and decreases in June, July, and August. In other regions (for example, the Pacific coasts of northern South America and a region centered over Uruguay and Patagonia) the sign of the response is preserved throughout the seasonal cycle.

As shown in the bottom panels in figure A.5, models foresee a wetter climate near the Rio de la Plata and drier conditions along much of the southern Andes, especially in December, January, and February. However, when estimating the likelihood of this response, the qualitative consensus within this set of models must be weighed against the fact that most models cannot reproduce the regional precipitation patterns in their control experiments with much accuracy.

The poleward shift of the South Pacific and South Atlantic subtropical anticyclones is a very marked response across climate models. Parts of

Chile and Patagonia are influenced by the polar boundary of the subtropical anticyclone in the South Pacific, and they experience particularly strong drying because of the combination of the poleward shift of circulation and increase in moisture divergence. The strength and position of the subtropical anticyclone in the South Atlantic is known to influence the climate of southeastern South America. The projected increase in rainfall in southeastern South America is related to a corresponding poleward shift of the Atlantic storm track.

Some projected changes in precipitation (such as the drying over east-central Amazonia and northeast Brazil and the wetter conditions over southeastern South America) could be a partial consequence of the El Niño–like response that is projected by many models (Meehl and others 2007).[6] That would directly affect tropical South America and affect southern South America through extratropical teleconnections (Mo and Nogués-Paegle 2001).

### Extreme Events

Little research is available on extremes of temperature and precipitation for the LAC region. Christensen and others (2007) estimated how frequently the seasonal temperature and precipitation extremes as simulated in 1980–99 are exceeded, using the A1B scenario from a large ensemble of GCM simulations.[7] They found that essentially all seasons and regions will be extremely warm by the end of the century. In Central America, the projected decrease in precipitation is accompanied by more frequent dry extremes in all seasons. In Amazonia, models project extremely wet seasons in about 27 percent of all summers (December, January, February) and 18 percent of all autumns (March, April, May) in the period 2080–99, with no significant change for the rest of the year. For southern South America significant changes are not projected in the frequency of extremely wet or dry seasons. However, more careful analysis is required to determine how often these wet and dry extremes are projected by individual models in the large ensemble of GCM simulations producing these results before making definitive conclusions about the likelihood of these changes in extremes.

On the daily time scale, an ensemble of simulations from two atmosphere-ocean-coupled general circulation models was analyzed; both models simulate a temperature increase on the warmest night of the year that is larger than the mean response over the Amazon basin but smaller than the mean response over parts of southern South America (Hegerl and others 2004). Concerning extreme precipitation, both models foresee a stronger wettest day per year over large parts of southeastern South America and central Amazonia and weaker precipitation extremes over the coasts of northeastern Brazil.

Changes in extremes were analyzed based on multimodel simulations from nine global coupled climate models (Tebaldi and others 2006). Figure A.6 depicts projected changes in extreme precipitation and the number of consecutive dry days (Meehl and others 2007). The general pattern of change suggests that nearly everywhere precipitation intensity is increasing, enhancing the risks of flash floods. At the same time, a clear tendency appears toward an increase in the number of consecutive dry days almost everywhere. This suggests an increased risk of droughts. These seemingly opposite tendencies are physically consistent: in a warmer atmosphere, more moisture is available for precipitation in moist air than under present-day conditions. This will lead to an enhanced possibility for more intensive precipitation events. On the other hand, higher temperatures will also make the less-moist air masses even drier, leading to fewer rainy days overall, and will thus increase the chance of longer dry periods.

In Central America and in the Caribbean, a substantial contribution to the intensive precipitation events is connected with tropical cyclone activity. It is therefore relevant to keep in mind how the incidence of such systems may change. Recent studies with improved global models, ranging in resolution from about 100 to 20 km, suggest future changes in the number and intensity of tropical cyclones (hurricanes). A synthesis of the model results to date indicates, for a warmer future climate, increased peak wind intensities and increased mean and peak precipitation intensities in future tropical cyclones, with the possibility of fewer relatively weak hurricanes and more numerous intense hurricanes. However, the total number of tropical cyclones globally is projected to decrease. The observed increase in the proportion of very intense hurricanes since 1970 in some regions is in the same direction but is much larger than predicted by theoretical models.

## Sea Level
Projected rises in the global average sea level at the end of the 21st century (2090–99) relative to 1980–99 were estimated by the IPCC for the six SRES marker scenarios. The results, given as 5 percent to

**Figure A.6    Changes in Extremes Based on Multimodel Simulations
from Nine Global Coupled Climate Models**

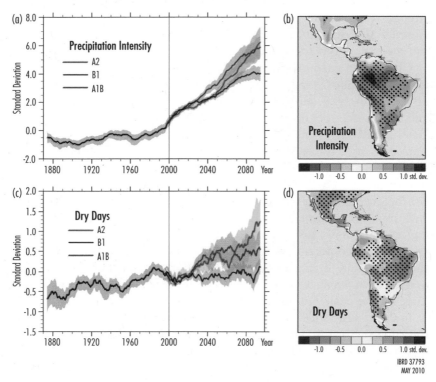

*Source:* Adapted from IPCC 2007.

*Note:* Panel a: Globally averaged changes in precipitation intensity (defined as the annual total precipitation divided by the number of wet days) for a low (SRES B1), a middle (SRES A1B), and a high (SRES A2) scenario. Panel b: Changes in spatial patterns of simulated precipitation intensity between two 20-year means (2080–99 minus 1980–99) for the A1B scenario.

Panel c: Globally averaged changes in dry days (defined as the annual maximum number of consecutive dry days). Panel d: Changes in spatial patterns of simulated dry days between two 20-year means (2080–99 minus 1980–99) for the A1B scenario.

Solid lines in panels a and c are the 10-year smoothed multimodel ensemble means; the envelope indicates the ensemble mean standard deviation. Stippling in panels b and d denotes areas where at least five of the nine models concur in determining that the change is statistically significant. Extreme indexes are calculated only over land following Frich and others (2002). Each model's time-series was centered on its 1980–99 average and normalized (rescaled) by its standard deviation computed (after de-trending) over the period 1960–2099. The models were then aggregated into an ensemble average, both at the global and at the grid-box level. Thus, changes are given in units of standard deviations.

95 percent ranges based on the spread of model results, are shown in table A.1. Thermal expansion contributes 70 percent to 75 percent to the best estimate for each scenario. The use of atmosphere-ocean-coupled general circulation models to evaluate ocean heat uptake and thermal expansion makes these estimates more credible than previous ones. It has

**Table A.1    Projected Global Average Surface Warming and Sea Level Rise at the End of the 21st Century**

| Case | Temperature change (°C, at 2090–99 relative to 1980–99)[a] | | Sea level rise (meters, at 2090–99 relative to 1980–99) Model-based range (excluding future rapid dynamical changes in ice flow) |
|---|---|---|---|
| | Best estimate | Likely range | |
| Constant year 2000 concentrations[b] | 0.6 | 0.3 – 0.9 | NA |
| B1 scenario | 1.8 | 1.1 – 0.9 | 0.18 – 0.38 |
| A1T scenario | 2.4 | 1.4 – 3.8 | 0.20 – 0.45 |
| B2 scenario | 2.4 | 1.4 – 3.8 | 0.20 – 0.43 |
| A1B scenario | 2.8 | 1.7 – 4.4 | 0.21 – 0.48 |
| A2 scenario | 3.4 | 2.0 – 5.4 | 0.23 – 0.51 |
| A1F1 scenario | 4.0 | 2.4 – 6.4 | 0.26 – 0.59 |

*Source:* IPCC 2007.

a. These estimates are assessed from a hierarchy of models that include a simple climate model, several Earth Models of Intermediate Complexity (EMICs), and a large number of Atmosphere-Ocean Global Circulation Models (AOGCMs).

b. Year 2000 constant composition is derived from AOGCMs only.

also reduced the projections compared to previous estimates. In all the SRES marker scenarios except B1, the average rate of sea level rise during the 21st century very likely exceeds the 1961–2003 average rate of 1.8 ± 0.5 mm per year. For an average model, the scenario spread in sea level rise is only 0.02 meter by the middle of the century, but by the end of the century it is 0.15 meter.

In the IPCC best estimate assessment, recent claims of changes in accelerating melting from Greenland and ice mass loss from Antarctica were not considered, largely because of incomplete knowledge. However, values considerably higher than the best estimates—up to a factor 2 higher by the end of the century—cannot be ruled out. Continuing documentation of the observed changes supports this possibility. Therefore, the IPCC estimates may be somewhat conservative, although there is still great uncertainty concerning the future changes in both the Greenland and Antarctica ice masses as a consequence of continued rise in temperatures.

### Other Time Horizons

To document the proof of concept, most of the climate change and variability projections in the IPCC report (2007) focused on the late

21st century. Moreover, most attention was given to the A1B SRES marker scenario.[8] To give some qualitative information about likely changes in time periods closer to the present, that is, the 2020s or the 2050s, it is necessary to introduce scaling arguments, as a full utilization of the available general circulation model information entering the IPCC assessment work is not practically possible for the present book. As demonstrated in Christensen and others (2007), using a scaling approach is reasonable, provided the climate parameters being addressed are robust and vary slowly over time. This further implies that one should address decades rather than a particular year. Thus, investigating 2020 should inform one about the period 2010–30, and estimates for 2050 would refer to the period 2040–60; but even longer periods would be preferable (see box A.1).

Taking a scaling approach would imply that the local amplification factor, compared to the global mean temperature change, is known. Figure A.5, in combination with the global temperature increase for the A1B scenario, provides the foundation for such a scaling argument.[9] The mean warming between the periods 1990–99 and 2090–99 can be estimated to be 3°C. For the period representing 2020, this number equals 0.5°C, and for 2050, the number is about 1.5°C. These figures correspond to scaling numbers of 0.17 and 0.5 respectively. This can be immediately used to interpret the top row of temperature maps in figure A.5. Precipitation is somewhat more subtle to deal with; as a gross measure, one may still use the same scaling numbers, keeping in mind that the boundaries between positive and negative change could well change because of nonlinearity in the climate system and natural variability, which tend to dominate as long as the temperature signal is weak.

Considering the projected change in extreme events, as characterized by the analysis of consecutive dry days and precipitation intensity discussed in connection with figure A.6, some indications about the scaling can be found from inspecting the time evolution of the globally averaged change. Note that little or no change is found for the 2020s, whereas by the 2050s the first clear signs of the climate change and variability signal toward the end of the century could be expected, although perhaps only showing through with about 50 percent of the power.

## Further Research

To advance the scientific knowledge and awareness of past, contemporary, and future climate for Latin America and the Caribbean, it is vital that efforts are made to secure or rescue existing databases of climate records in the region. All such data archives, no matter how small, are potentially useful for this purpose. Databases should be made available to the international scientific community—at best free of charge, but if not, then through collaborative international projects to analyze and exploit them.

It is also essential to encourage greater involvement by LAC region scientists in climate analysis at the international level. At present, too many efforts in climate analysis go only as far as the archives of national meteorological services, although a huge effort has been undertaken by the World Meteorological Organization, and in particular the U.S. National Oceanic and Atmospheric Administration/National Climatic Data Center, to push LAC climatologists forward.[10] Only by entering the international science arena will regional findings become useful in assisting humankind to mitigate and adapt to anthropogenic climate change and variability.

Regional climate change and variability projections for Latin America and the Caribbean based on comprehensive analysis of climate only exist in the form of output from cause-resolution GCMs, with a few exceptions. A strong demand exists for coordinated research exploring a range of possibilities to provide information on regional or local climate change and variability. That could be achieved by enabling concerted actions, with participation from the LAC region as well as developed nations, using the approach applied in such research programs as the North American Regional Climate Change Assessment Program (NARCCAP) (Mearns and others 2005) and the Prediction of Regional Scenarios and Uncertainties for Defining European Climate Change Risks and Effects (PRUDENCE), for example (Christensen and others 2007). These initiatives focus on skillful projections of regional climate change and variability, providing not only estimates of changes in mean properties and their variation, but also the scientific knowledge to permit a quantifiable assessment of the uncertainties associated with the projections.

## Notes

1. A1F1 describes a high-emission scenario with high economic growth, where global population peaks around 9 billion in 2050 and declines to about 7 billion by 2100, with continued high greenhouse gas (GHG) emissions resulting in cumulative emissions from 1990 until 2100 of 2,182 GtC. B1, in contrast, describes a low-emission scenario, with a similar development in population figures, also with high economic growth, but where the gains of the economic growth to a large extent are invested in improved efficiency of

resource use ("dematerialization"), equity, social institutions, and environ-mental protection. Even in the absence of explicit interventions to mitigate climate change, the proactive environmental policies lead to relatively low GHG emissions, reaching cumulative levels by 2100 of 976 GtC since 1990. For a more detailed description of the SRES marker scenarios see Nakicenovic and others (2000).

2. The IPCC Data Distribution Center can be found at http://www.ipcc-data.org.

3. The ITCZ is an air mass located directly under the sun at the equator and bounded by the two tropics, in which air that ascends because of the heat of the sun is replaced by air from below originating to the north as well as the south of the equator.

4. A jet is a well-confined strong wind that efficiently transfers air masses from one region to another. Other examples of jet streams are found over the Atlantic at high altitudes (10 kilometers), where transatlantic airplanes bene-fit from their presence when flying eastbound.

5. The contrast between ocean and landmasses near the equator results in differ-ential rising and ascending of air masses in a systematic manner. This is referred to as the "Walker circulation."

6. Most coupled models continue to simulate ENSO variability during the 21st century. However, slowly the long-term response to warming is a change in the Pacific sea surface temperature pattern resembling that of El Niño.

7. The A1B scenario, like the other scenarios in the A1 family of scenarios, assumes high economic growth, continued population growth until 2050, when global population peaks around 9 billion and then declines to about 7 billion by the end of the 21st century. The A1 family of scenarios also assumes convergence among regions, capacity building, and increased cultural and social interactions, with a substantial reduction in regional differences in per capita income. In terms of changes in energy technology, the A1B scenario assumes balance across all energy sources, meaning not relying too heavily on one particular energy source, on the assumption that similar improvement rates apply to all energy supply and end-use technologies. This places A1B in the middle range with respect to greenhouse gas emissions; it predicts carbon dioxide emissions increasing until around 2050 and then decreasing after that. For a more detailed description see Nakicenovic and others (2000).

8. For a more detailed description see Nakicenovic and others (2000).

9. Global temperature increase for the A1B scenario as depicted in IPCC 2007, figure SPM.5 and table TS.6, reproduced above in table A.1.

10. See the most recent versions of "State of the Climate," in the *Bulletin of the American Meteorological Society*; and Arguez 2007; Levinson 2005; Levinson and Lawrimore 2008; and Shein 2006 in the same journal.

# References

Arguez, A., ed. 2007. "State of the Climate in 2006." *Bulletin of the American Meteorological Society* 88: S1–S135.

Bales, R. C., D. M. Liverman, and B. J. Morehouse. 2004. "Integrated Assessment as a Step toward Reducing Climate Vulnerability in the Southwestern United States." *Bulletin of the American Meteorological Society* 85 (11): 1727.

Berbery, E. H., and E. A. Collini. 2000. "Springtime Precipitation and Water Vapor Flux Convergence over Southeastern South America." *Monthly Weather Review* 128: 1328–46.

Christensen, J. H., T. R. Carter, M. Rummukainen, and G. Amanatidis. 2007. "Evaluating the Performance and Utility of Regional Climate Models: The PRUDENCE Project." *Climatic Change* 81: supl. 1, 1–6, DOI:10.1007/s10584-006-9211-6.

Dettinger, M. D., D. S. Battista, G. J. McCabe, R. D. Garreaud, and C. M. Bitz. 2001. "Interhemispheric Effects of Interannual and Decadal ENSO–like Climate Variation on the Americas." In *Interhemispheric Climate Linkages*, ed. V. Markgraf. San Diego: Academic Press.

Doyle, M. E., and V. R. Barros. 2002. "Midsummer Low-Level Circulation and Precipitation in Subtropical South America and Related Sea Surface Temperature Anomalies in the South Atlantic." *Journal of Climate* 15: 3394–3410.

Emanuel, K. 2003. "Tropical Cyclones." *Annual Review of Earth and Planetary Science* 31: 75–104.

Frich, P., L. V. Alexander, P. Della-Marta, B. Gleason, M. Haylock, A. M. G. Klein Tank, and T. Peterson. 2002: "Observed Coherent Changes in Climate Extremes during the Second Half of the Twentieth Century." *Climate Research* 19: 193–212.

Hegerl, G. C., F. W. Zwiers, P. A. Stott, and V. V. Kharin. 2004. "Detectability of Anthropogenic Changes in Annual Temperature and Precipitation Extremes." *Journal of Climate* 17: 3683–700.

Garreaud, R. D. 2000. "A Multi-scale Analysis of the Summertime Precipitation over the Central Andes." *Monthly Weather Review* 127: 901–21.

Giorgi, F., and R. Francesco. 2000. "Evaluating Uncertainties in the Prediction of Regional Climate Change." *Geophysical Research Letters* 27: 1295–98.

Gray, W. M., 1984. "Atlantic Seasonal Hurricane Frequency. Part I: El Niño and 30 mb Quasi-biennial Oscillation Influences. " *Monthly Weather Review* 112: 1649–68.

IPCC (Intergovernmental Panel on Climate Change). 2001. *Special Report on Emission Scenarios.* Geneva: IPCC.

———. 2007. *Climate Change 2007: The Physical Science Basis.* Contribution of Working Group I to the Fourth Assessment Report of the IPCC. Geneva: IPCC.

Kidson, J. W. 1988. "Interannual Variations in the Southern Hemisphere Circulation." *Journal of Climate* 1: 1177–98.

Levinson, D. H., ed. 2005. "State of the Climate in 2004." *Bulletin of the American Meteorological Society* 86: S1–S86.

Levinson, D. H., and J. H. Lawrimore, eds. 2008. "State of the Climate in 2007." *Bulletin of the American Meteorological Society* 89: S1–S179.

Liebmann, B., G. N. Kiladis, J. A. Marengo, T. Ambrizzi, and J. D. Glick. 1999. "Submonthly Convective Variability over South America and the South Atlantic Convergence Zone." *Journal of Climate* 12: 1877–91.

Magaña, V., J. A. Amador, and S. Medina. 1999. "The Midsummer Drought over Mexico and Central America." *Journal of Climate* 12: 1577–88.

Mann, M. E., R. S. Bradley, and M. K. Hughes. 2000. "Long-Term Variability in the El Niño Southern Oscillation and Associated Teleconnections." In: *El Niño and the Southern Oscillation: Multiscale Variability and Its Impacts on Natural Ecosystems and Society*, ed. H. F. Diaz and V. Markgraf. 321–72. Cambridge, U.K.: Cambridge University Press.

Mearns, L. O., R. W. Arritt, G. Boer, D. Caya, P. Duffy, F. Giorgi, W. J. Gutowski, et al., 2005. "NARCCAP, North American Regional Climate Change Assessment Program, A Multiple AOGCM and RCM Climate Scenario Project over North America." *Preprints of the American Meteorological Society 16th Conference on Climate Variations and Change.* January 9–13, 2005. Paper J6.10, 235–38. Washington, DC: American Meteorological Society.

Meehl, G. A., T. F. Stocker, W. D. Collins, P. Friedlingstein, A. T. Gaye, J. M. Gregory, A. Kitoh, et al. 2007. "Global Climate Projections." In: *Climate Change 2007: The Physical Science Basis.* Contribution of Working Group I to the Fourth Assessment Report of the Intergovernmental Panel on Climate Change, ed. S. Solomon, D. Qin, M. Manning, Z. Chen, M. Marquis, K. B. Avery, M. Tignor, and H. L. Miller. Cambridge, U.K. and New York: Cambridge University Press.

Mo, K. C., and J. Nogués-Paegle. 2001. "The Pacific-South American Modes and Their Downstream Effects." *International Journal of Climatology* 21: 1211–29.

Nakicenovic, N., O. Davidson, G. Davis, A. Grübler, T. Kram, E. Lebre La Rovere, B. Metz, et al. 2000. *IPCC Special Report on Emissions Scenarios.* Cambridge, U.K.: Cambridge University Press.

Nogués-Paegle J., C. R. Mechoso, R. Fu, E. H. Berbery, W. C. Chao, T. C. Chen, K. H. Cook, et al. 2002. "Progress in Pan American CLIVAR Research: Understanding The South American Monsoon." *Meteorologica* 27(1, 2): 1–30.

Robertson, A. W., and C. R. Mechoso. 2000. "Interannual and Interdecadal Variability of the South Atlantic Convergence Zone." *Monthly Weather Review* 128: 2947–57.

Romero-Centeno, R., J. Zavala-Hidalgo, A. Gallegos, and J. J. O'Brien. 2003. "Isthmus of Tehuantepec Wind Climatology and ENSO Signal." *Journal of Climate* 16: 2628–39.

Ruosteenoja, K, T. R. Carter, K. Jylha, H. Tuomenvirta. 2003. "Future Climate in World Regions: An Intercomparison of Model-Based Projections for the New IPCC Emissions Scenarios." Helsinki: Finnish Environment Institute.

Shein, K. A., ed. 2006. "State of the Climate in 2005." *Bulletin of the American Meteorological Society* 87: S1–S102.

Silvestri, G. E., and C. S. Vera. 2003. "Antarctic Oscillation Signal on Precipitation Anomalies over Southeastern South America." *Geophysical Research Letters* 30 (21): 2115.

Tang, B. H., and J. D. Neelin. 2004. "ENSO Influence on Atlantic Hurricanes via Tropospheric Warming." *Geophysical Research Letters* 31: L24204, doi:10.1029/2004GL021072.

Taylor, M., and E. Alfero. 2005. "Climate of Central America and the Caribbean." In *The Encyclopedia of World Climatology*, ed. J. Oliver. Encyclopedia of Earth Sciences Series. Netherlands: Springer Press.

Tebaldi, C., K. Hayhoe, J. M. Arblaster, and G. E. Meehl. 2006. "Going to the Extremes: An Intercomparison of Model-Simulated Historical and Future Changes in Extreme Events." *Climate Change* 79: 185–211.

Thompson, D. W. J., and J. M. Wallace. 2000. "Annular Modes in the Extratropical Circulation. Part 1: Month-to-Month Variability." *Journal of Climate* 13: 1000–16.

Vera, C. S., and P. K. Vigliarolo. 2000. "A Diagnostic Study of Cold-Air Outbreaks over South America." *Monthly Weather Review* 128: 3–24.

Vernekar, A., B. Kirtman, and M. Fennessy. 2003. "Low-Level Jets and Their Effects on the South American Summer Climate as Simulated by the NCEP Eta Model." *Journal of Climate* 16: 297–311.

Verner, Dorte. 2010. *Reducing Poverty, Protecting Livelihoods and Building Assets in a Changing Climate: Social Implications of Climate Change in Latin America and the Caribbean*. Washington, DC: World Bank.

Zhang, X., Francis W. Zwiers, Gabriele C. Hegerl, F. Hugo Lambert, Nathan P. Gillett, Susan Solomon, Peter A. Stott, and Toru Nozawa. 2007. "Detection of Human Influence on Twentieth-Century Precipitation Trends." *Nature* 448: 461–64.

# Summaries of Likely Climate Change Impacts, by Country

## Jens Hesselbjerg Christensen

To understand observed climate change and variability, climate change impacts, and projections of future climate change at the level of individual countries requires that climate information—however imperfect it may be—be assessed at a local scale. Climate change and variability in South America, the American isthmus, and the Caribbean have for the larger part not been assessed at a country level, even less so in a coordinated fashion. This appendix is meant to serve as guidance for climate impact assessments at the scales where the changes may be perceived as important for individuals and society, so as to assist in the choice of appropriate regions to study in detail. However, the more details are sought about climate change, and the more precise information about "predicted" changes are requested, the more disappointed the researcher is likely to become, largely because of the complete lack of basic, reliable information. Thus far, climate research capabilities within Latin America and the Caribbean (LAC), in combination with expertise from outside the region, have not reached the same level as in the developed world. Limited trustworthy, scientifically scrutinized material has become available partly as a reaction to the claim in the 4th Assessment Report of the Intergovernmental Panel on Climate Change (IPCC 2007a; 2007b; 2007c) that basic information on climate and climate change is only available in a very limited amount for this region (more and more is emerging).

In the following, a condensed assessment of findings in observed trends in temperature and precipitation climates, along with notions of recent severe weather-related events, is provided for each of the countries in Latin America and the Caribbean. The assessment largely builds upon figures and tables available in the IPCC 4th Assessment Report. Most of the material is included toward the end of that document. In addition, information about recent events has been extracted from the National Oceanic and Atmospheric Administration, National Environmental Satellite, Data and Information Service, National Climatic Data Center, and American Meteorological Society State of the Climate reports 2001–2007 (*Bulletins of the American Meteorological Society*, or *BAMS* 2005; 2006; 2007; 2008).

The countries are listed alphabetical by region, starting with Mexico, followed by countries on the American isthmus, those in the Caribbean, and finally countries in South America. If going country by country, the reader may find some redundancies, which reflect the fact that for the smaller, neighboring countries, the differences between their experienced as well as projected climate changes are minor. Projected warming for each of the countries is left to the reader to interpret based on the maps, which are reproduced from the IPCC report in appendix A. It is assumed that issues related to precipitation and extreme weather occurrences are of most importance. For each country an ultra-short statement—an image—of the most pressing climate issues of the future is provided.

## Mexico

*Recent trends.* Overall temperatures have increased by 0.1–0.4°C per decade within the last 25 years (between 0.5° and more than 2.0°C since 1901). Notably, some regions have seen negative temperature changes. The geographical variation is large, and the long-term changes are most advanced in the northwest near Baja California, where most warming has occurred. Annual rainfall changes are well established throughout the country but also with strongly irregular patterns of positive trends versus negative trends. While the north shows trends of precipitation decreases, the rest of the country has experienced a weak precipitation increase during the last century. Sea level rise has reached 3 millimeters per year (mm/year) toward the Gulf of Mexico, while the increase is less at the Pacific coast.

**Figure B.1    Map of Mexico**

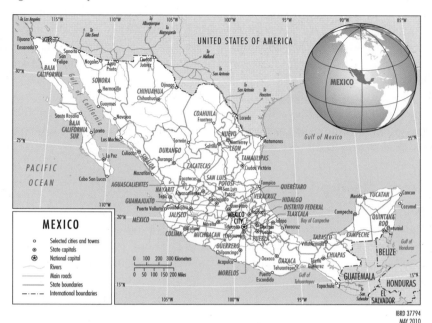

IBRD 37794
MAY 2010

*Source:* World Bank.

***Recent events.*** Intensifying El Niño occurrences and hurricanes: In 2005 Hurricane Wilma made several landfalls, mainly in the Yucatán peninsula. Also, Hurricanes Emily and Stan made landfall in 2005. Several hurricanes made landfall in 2006, and Hurricane Henriette made landfall in 2007.

General for the region: A positive tendency for intense precipitation and consecutive dry days; positive trend in the frequency of very heavy rains in central Mexico.

***Projection.*** See table 13.4 and figure 11.15 in IPCC Working Group II (WGII; IPCC 2007b) for mean temperature and mean precipitation. Mexico is located in a region dominated by atmospheric subsidence, and the signals for precipitation are relatively robust across general circulation models (GCMs) for the annual mean change, which seems very likely to be strengthened in a warmer world. Seasonal signals are also robust and point toward an overall reduction in June-July-August rainfall, possibly turning into an increase toward the interior north. In December-January-February the drying trend is very robust and largest in the northern part. The upward trend for precipitation frequency continues in the northern part and possibly along the east coast. Finally, the dry day frequency trend will continue upward and sea level rise (SLR) will compare with global increase.

Images: It is very likely that overall conditions will become drier, and thus risks of droughts will increase. There is also some certainty that heavy rains will increase in frequency and intensity. Finally, it is likely that tropical storms and hurricanes will become more intense.

## Central America

### Belize

**Recent trends.** Temperatures have increased by roughly from 0.1° to 0.2°C per decade over the last 25 years (about 0.5°C since 1901). Rainfall shows no consistent trend and SLR has increased by about 1.5 to 2 mm/year.

**Figure B.2    Map of Central America**

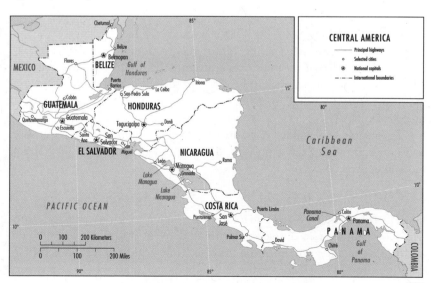

*Source:* World Bank.

*Recent events.* Intensifying El Niño occurrences and hurricanes: The 2005 record hurricane season in the Caribbean Basin produced several hurricanes with landings nearby that affected Belize. The last serious hurricane in Belize was Hurricane Iris in 2001.

General for the region: A positive tendency for intense precipitation and consecutive dry days. A positive trend in the frequency of very heavy rains has been observed, and an increase in the frequency of intense hurricanes has occurred.

*Projections.* See table 13.4 and figure 11.15 in WGII (IPCC 2007b) for mean temperature and mean precipitation. Belize is located in a region dominated by atmospheric subsidence, and the signals for precipitation are relatively robust across GCMs for the annual mean change, which seems very likely to be strengthened in a warmer world. It is likely that during the next century the risk of hurricanes will increase, and more-intense hurricanes will occur. Notably, it is not clear if there will be more hurricanes per se. Sea level rise will compare with global increase.

Images: Somewhat drier conditions and increased risk of tropical storm and hurricane incidents.

## Costa Rica
*Recent trends.* Over the last 25 years, temperature increases have been less than or about 0.1°C per decade (about 0.5°C since 1901). Rainfall, in general, shows a positive trend, while SLR is about 1.5 to 2 mm/year, with highest values occurring in the Atlantic coast.

*Recent events.* Intensifying El Niño occurrences and hurricanes: The 2005 record hurricane season in Caribbean Basin produced several hurricanes with landings nearby that affected Costa Rica, Hurricane Stan in particular.

Droughts: Recent years have seen an increase in periods with dry to droughtlike conditions.

General for the region: A positive tendency for intense precipitation and consecutive dry days. A positive trend in the frequency of very heavy rains has been observed, and an increase in the frequency of intense hurricanes has occurred.

*Projections.* See table 13.4 and figure 11.15 in WGII (IPCC 2007b) for mean temperature and mean precipitation. The precipitation signal is not very robust across the GCMs, which indicates that the country is near the boundary between increase and decrease in precipitation. The upward trend of precipitation frequency continues but not at a strong and significant signal. The frequency of dry days will increase, and over the next century the risk of intensifying hurricanes is likely to increase. Notably, it is not clear if there will be more hurricanes per se. SLR will compare with global increase.

Images: Overall conditions are likely to become drier, and the risk of more intense hurricanes will continue.

### El Salvador

**Recent trends.** Temperatures have increased less than or about 0.1°C per decade over the last 25 years (about 0.5°C since 1901), and rainfall shows a general positive trend. Yet reductions in rainfall are occurring inland. Sea level rise has occurred at about 1.5 mm/year.

**Recent events.** Intensifying El Niño occurrences and hurricanes: The 2005 record hurricane season in the Caribbean Basin affected the country severely when Hurricane Stan hit.

Droughts: In recent years periods with dry to droughtlike conditions have increased.

General for the region: A positive tendency toward intense precipitation and consecutive dry days. A positive trend in the frequency of very heavy rains has been observed, and an increase in the frequency of intense hurricanes has occurred.

**Projections.** See table 13.4 and figure 11.15 in WGII (IPCC 2007b) for mean temperature and mean precipitation. The signal is relatively robust across GCMs. The upward trend in the frequency of intense precipitation will continue but not at a strong and significant rate. The dry day frequency trend will continue upward, and during the next century the risk of intensifying hurricanes will increase. Notably, it is not clear if there will be more hurricanes per se. Finally, SLR will compare with global increase.

Images: Somewhat drier conditions, and continuing risk of hurricane incidents.

### Guatemala

**Recent trends.** Temperatures have increased less than or about 0.1°C per decade over the last 25 years (about 0.5°C since 1901). Rainfall shows no consistent trend, and SLR is about 1.5 to 2 mm/year, with the highest values on the Atlantic coast.

*Recent events.* Intensifying El Niño occurrences and hurricanes: The 2005 record hurricane season in the Caribbean Basin affected the country severely when Hurricane Stan hit.

Droughts: In recent years periods with dry to droughtlike conditions have increased.

General for the region: A positive tendency for intense precipitation and consecutive dry days. A positive trend in the frequency of very heavy rains has been observed, and an increase in the frequency of intense hurricanes has occurred.

*Projections.* See table 13.4 and figure 11.15 in WGII (IPCC 2007b) for mean temperature and mean precipitation. The signal is relatively robust across GCMs. The upward trend in the frequency of intense precipitation will continue but not at a strong and significant rate. The dry day frequency trend will continue upward, and during the next century the risk of intensifying hurricanes will increase. Notably, it is not clear if there will be more hurricanes per se. Finally, SLR will compare with global increase.

Images: Somewhat drier conditions and continuing risk of hurricane incidents.

## Honduras

*Recent trends.* Temperatures have increased less than or about 0.1°C per decade over the last 25 years (about 0.5°C since 1901). Rainfall in general shows a positive trend, but inland areas in the eastern part experience a reduction in rainfall. Sea level rise is about 1.5 to 2 mm/year, with the highest values occurring on the Atlantic coast.

*Recent events.* Intensifying El Niño occurrences and hurricanes: The 2005 record hurricane season in the Caribbean Basin affected the country little; last serious landing by Hurricane Felix in 2007.

Droughts: In recent years periods with dry to droughtlike conditions have increased.

General for the region: A positive tendency for intense precipitation and consecutive dry days. A positive trend in the frequency of very heavy rains has been observed, and an increase in the frequency of intense hurricanes has occurred.

*Projections.* See table 13.4 and figure 11.15 in WGII (IPCC 2007b) for mean temperature and mean precipitation. The signal is relatively robust across GCMs. The upward trend in the frequency of intense precipitation will continue but not at a strong and significant rate. The dry day frequency trend will continue upward, and during the next century the risk of intensifying hurricanes will increase. Notably, it is not clear if there will be more hurricanes per se. Finally, SLR will compare with global increases.

Images: Somewhat drier conditions and continuing risk of hurricane incidents.

### Nicaragua

**Recent trends.** Temperatures have increased at about 0.1°C per decade over the last 25 years (about 0.5°C since 1901). Rainfall in general shows a positive trend that is not completely consistent. Sea level rise is about 1.5 to 2 mm/year, with highest values on the Atlantic coast.

**Recent events.** Intensifying El Niño occurrences and hurricanes: The 2005 record hurricane season in the Caribbean Basin affected the country when Hurricanes Stan and Beta made landfall. The last serious landing was Hurricane Felix in 2007.

Droughts: In recent years periods with dry to droughtlike conditions have increased.

General for the region: A positive tendency for intense precipitation and consecutive dry days. A positive trend in the frequency of very heavy rains has been observed, and an increase in the frequency of intense hurricanes has occurred.

**Projections.** See table 13.4 and figure 11.15 in WGII (IPCC 2007b) for mean temperature and mean precipitation. The precipitation signal is not very robust across the GCMs, which indicates that the country is near the boundary between increase and decrease in precipitation. The upward trend of precipitation frequency continues but not at a strong and significant signal. The frequency of dry days will increase, and over the next century the risk of intensifying hurricanes is likely to increase. Notably, it is not clear if there will be more hurricanes per se. Finally, SLR will compare with global increase.

Images: Possibly overall drier conditions and continuing risk of hurricane incidents.

### Panama

**Recent trends.** Temperatures have increased at about 0.1°C per decade over the last 25 years (about 0.5°C since 1901). Rainfall in general shows a positive trend, but it is not consistent across the country. Sea level rise occurs at about 1 to 2 mm/year, with the highest values on the Atlantic coast.

*Recent events.* Intensifying El Niño occurrences and hurricanes: The 2005 record hurricane season in the Caribbean Basin affected the country little.

Precipitation: Has experienced below-normal precipitation during the wet season.

General for the region: A positive tendency for intense precipitation and consecutive dry days. A positive trend in the frequency of very heavy rains has been observed, and an increase in the frequency of intense hurricanes has occurred.

*Projections.* See table 13.4 and figure 11.15 in WGII (IPCC 2007b) for mean temperature and mean precipitation. The precipitation signal is not very robust across the GCMs, which indicates that the country is near the boundary between increase and decrease in precipitation. The upward trend of precipitation frequency continues but not at a strong and significant signal. The frequency of dry days will increase, and over the next century the risk of intensifying hurricanes is likely to increase. Notably, it is not clear if there will be more hurricanes per se. Finally, SLR will compare with global increase.

Images: Possibly overall drier conditions and continuing risk of hurricane incidents.

## The Caribbean

### Antigua and Barbuda

*Recent trends.* Temperatures have increased by roughly 0.1° to 0.2°C per decade over the last 25 years (about 1.0°C since 1901). Rainfall shows mostly an upward trend, while SLR is 2 to 3 mm/year.

*Recent events.* Intensifying hurricanes: The record hurricane season of 2005 in Caribbean Basin brought the island several landings; last serious landing by Hurricane Jose in 1999.

General for the region: A positive tendency for intense precipitation and consecutive dry days, as well as a positive trend in the frequency of very heavy rains can be expected. An increased number of intense hurricanes is likely.

**Figure B.3   Map of the Caribbean**

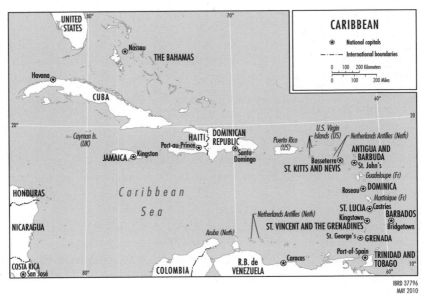

Source: World Bank.

***Projections.*** See table 13.4 and figure 11.15 in WGII (IPCC 2007b) for mean temperature and mean precipitation. The signal is relatively robust across GCMs. The trend for precipitation frequency is not clear. Yet the intensity of downpours may increase in connection with hurricanes. Dry day frequency trends are expected to continue upward. Over the next century, an increasing risk of more of the most intense hurricanes is expected. Notably, it is not clear if there will be more hurricanes per se. Sea level rise will compare with global increase.

Images: Somewhat drier conditions, and increased risk of tropical storm and hurricane incidents.

### Barbados

***Recent trends.*** Temperatures have increased roughly 0.1° to 0.2°C per decade over the last 25 years (about 1.0°C since 1901). Rainfall shows mostly an upward trend, while SLR is 2 to 3 mm/year.

***Recent events.*** Intensifying hurricanes: The record hurricane season of 2005 in the Caribbean Basin affected the island with several landings; last serious landing by Hurricane Dean in 2007.

General for the region: A positive tendency for intense precipitation and consecutive dry days, as well as a positive trend in the frequency of very heavy rains can be expected. An increased number of intense hurricanes is likely.

**Projections.** See table 13.4 and figure 11.15 in WGII (IPCC 2007b) for mean temperature and mean precipitation. The signal is relatively robust across GCMs. The trend for precipitation frequency is not clear. Yet in connection with hurricanes, intensity of downpours may increase. Dry day frequency trends are expected to continue upward. Over the next century, an increasing risk of more of the most intense hurricanes is expected. Notably, it is not clear if more hurricanes per se will appear. Sea level rise will compare with global increase.

Images: Somewhat drier conditions and increased risk of tropical storm and hurricane incidents.

### Cuba

**Recent trends.** Temperatures have increased by roughly 0.1° to 0.2°C per decade over the last 25 years (about 1.0°C since 1901). Rainfall shows mostly an upward trend, while SLR is 2 to 3 mm/year.

**Recent events.** Intensifying hurricanes: The record hurricane season of 2005 in Caribbean Basin affected the island with several landings; last serious landings were by Hurricanes Denis and Wilma in 2005.

General for the region: A positive tendency for intense precipitation and consecutive dry days, as well as a positive trend in the frequency of very heavy rains can be expected. An increased number of intense hurricanes is likely.

**Projections.** See table 13.4 and figure 11.15 in WGII (IPCC 2007b) for mean temperature and mean precipitation. The signal is relatively robust across GCMs. The trend for precipitation frequency is not clear. In connection with hurricanes, intensity of downpours may increase. Dry day frequency trends are expected to continue upward. Over the next century, an increasing risk for more of the most intense hurricanes is expected. Notably, it is not clear if there will be more hurricanes per se. Sea level rise will compare with global increase.

Images: Somewhat drier conditions and increased risk of tropical storm and hurricane incidents.

## Dominica

**Recent trends.** Temperatures have increased by roughly 0.1° to 0.2°C per decade over the last 25 years (about 1.0°C since 1901). Rainfall shows mostly an upward trend, while SLR is 2 to 3 mm/year.

**Recent events.** Intensifying hurricanes: Record hurricane season of 2005 in the Caribbean Basin affected the island with several landings; last serious landing was by Hurricane Dean in 2007.

General for the region: A positive tendency toward intense precipitation and consecutive dry days, as well as a positive trend in the frequency of very heavy rains can be expected. An increased number of intense hurricanes is likely.

**Projections.** See table 13.4 and figure 11.15 in WGII (IPCC 2007b) for mean temperature and mean precipitation. The signal is relatively robust across GCMs. The trend for precipitation frequency is not clear. In connection with hurricanes, intensity of downpours may increase. Dry day frequency trends are expected to continue upward. Over the next century, an increasing risk of more of the most intense hurricanes is expected. Notably, it is not clear if there will be more hurricanes per se. Sea level rise will compare with global increase.

Images: Somewhat drier conditions and increased risk of tropical storm and hurricane incidents.

## Dominican Republic

**Recent trends.** Temperatures have increased by roughly 0.1° to 0.2°C per decade over the last 25 years (about 1.0°C since 1901). Rainfall shows mostly an upward trend, while SLR is 2 to 3 mm/year.

**Recent events.** Intensifying hurricanes: Record hurricane season of 2005 in the Caribbean Basin affected the island with several landings; last serious landing was by Hurricane Dean in 2007.

General for the region: A positive tendency for intense precipitation and consecutive dry days, as well as a positive trend in the frequency of very heavy rains can be expected. An increased number of intense hurricanes is likely.

**Projections.** See table 13.4 and figure 11.15 in WGII (IPCC 2007b) for mean temperature and mean precipitation. The signal is relatively robust across GCMs. The trend for precipitation frequency is not clear.

In connection with hurricanes, intensity of downpours may increase. Dry day frequency trends are expected to continue upward. Over the next century, an increasing risk of more of the most intense hurricanes is expected. Notably, it is not clear if there will be more hurricanes per se. Sea level rise will compare with global increase.

Images: Somewhat drier conditions and increased risk of tropical storm and hurricane incidents.

### Grenada

**Recent trends.** Temperatures have increased by roughly 0.1° to 0.2°C per decade over the last 25 years (about 1.0°C since 1901). Rainfall shows mostly an upward trend, while SLR is 2 to 3 mm/year.

**Recent events.** Intensifying hurricanes: Record hurricane season of 2005 in the Caribbean Basin affected the island with several landings; last serious landing was by Hurricane Emily in 2005.

General for the region: A positive tendency for intense precipitation and consecutive dry days, as well as a positive trend in the frequency of very heavy rains can be expected. An increased number of intense hurricanes is likely.

**Projections.** See table 13.4 and figure 11.15 in WGII (IPCC 2007b) for mean temperature and mean precipitation. The signal is relatively robust across GCMs. The trend for precipitation frequency is not clear. In connection with hurricanes, intensity of downpours may increase. Dry day frequency trends are expected to continue upward. Over the next century, an increasing risk of more of the most intense hurricanes is expected. Notably, it is not clear if there will be more hurricanes per se. Sea level rise will compare with global increase.

Images: Somewhat drier conditions and increased risk of tropical storm and hurricane incidents.

### Haiti

**Recent trends.** Temperatures have increased by roughly 0.1° to 0.2°C per decade over the last 25 years (about 1.0°C since 1901). Rainfall shows mostly an upward trend, while SLR is 2 to 3 mm/year.

**Recent events.** Intensifying hurricanes: Record hurricane season of 2005 in the Caribbean Basin affected the island with several landings; last serious landing was by Hurricane Dean in 2007.

General for the region: A positive tendency for intense precipitation and consecutive dry days, as well as a positive trend in the frequency of very heavy rains can be expected. An increased number of intense hurricanes is likely.

**Projections.** See table 13.4 and figure 11.15 in WGII (IPCC 2007b) for mean temperature and mean precipitation. Signal is relatively robust across GCMs. The trend for precipitation frequency is not clear. But connection with hurricanes, intensity of downpours may increase. Dry day frequency trends are expected to continue upward. Over the next century, an increasing risk of more of the most intense hurricanes is expected. Notably, it is not clear if there will be more hurricanes per se. Sea level rise will compare with global increase.

    Images: Somewhat drier conditions and increased risk of tropical storm and hurricane incidents.

### Jamaica

**Recent trends.** Temperatures have increased by roughly 0.1° to 0.2°C per decade over the last 25 years (about 1.0°C since 1901). Rainfall shows mostly an upward trend, while SLR is 2 to 3 mm/year.

**Recent events.** Intensifying hurricanes: Record hurricane season of 2005 in the Caribbean Basin affected the island with several landings. Last serious landing was by Hurricane Dean in 2007.

    General for the region: A positive tendency for intense precipitation and consecutive dry days, as well as a positive trend in the frequency of very heavy rains can be expected. An increased number of intense hurricanes is likely.

**Projections.** See table 13.4 and figure 11.15 in WGII (IPCC 2007b) for mean temperature and mean precipitation. The signal is relatively robust across GCMs. The trend for precipitation frequency is not clear. In connection with hurricanes, intensity of downpours may increase. Dry day frequency trends are expected to continue upward. Over the next century, an increasing risk of more of the most intense hurricanes is expected. Notably, it is not clear if there will be more hurricanes per se. Sea level rise will compare with global increase.

    Images: Somewhat drier conditions and increased risk of tropical storm and hurricane incidents.

## St. Kitts and Nevis

***Recent trends.*** Temperatures have increased roughly 0.1° to 0.2°C per decade over the last 25 years (about 1.0°C since 1901). Rainfall shows mostly an upward trend, while SLR is 2 to 3 mm/year.

***Recent events.*** Intensifying hurricanes: Record hurricane season of 2005 in the Caribbean Basin affected the island with several landings. Last serious landing was by Hurricane George in 1998.

General for the region: A positive tendency for intense precipitation and consecutive dry days, as well as a positive trend in the frequency of very heavy rains can be expected. An increased number of intense hurricanes is likely.

***Projections.*** See table 13.4 and figure 11.15 in WGII (IPCC 2007b) for mean temperature and mean precipitation. The signal is relatively robust across GCMs. The trend for precipitation frequency is not clear. In connection with hurricanes, intensity level of downpours may increase. Dry day frequency trends are expected to continue upward. Over the next century, an increasing risk of more of the most intense hurricanes is expected. Notably, it is not clear if there will be more hurricanes per se. Sea level rise will compare with global increase.

Images: Somewhat drier conditions and increased risk of tropical storm and hurricane incidents.

## St. Lucia

***Recent trends.*** Temperatures have increased by roughly 0.1° to 0.2°C per decade over the last 25 years (about 1.0°C since 1901). Rainfall shows mostly an upward trend, while SLR is 2 to 3 mm/year.

***Recent events.*** Intensifying hurricanes: Record hurricane season of 2005 in the Caribbean Basin affected the island with several landings. Last serious landing was by Hurricane Dean in 2007.

General for the region: A positive tendency for intense precipitation and consecutive dry days, as well as a positive trend in the frequency of very heavy rains can be expected. An increased number of intense hurricanes is likely.

***Projections.*** See table 13.4 and figure 11.15 in WGII (IPCC 2007b) for mean temperature and mean precipitation. The signal is relatively robust across GCMs. The trend for precipitation frequency is not clear. In connection with hurricanes, intensity of downpours may increase. Dry day frequency trends are expected to continue upward. Over the next century, an increasing risk of more of the most intense hurricanes is expected. Notably, it is not clear if there will be more hurricanes per se. Sea level rise will compare with global increase.

Images: Somewhat drier conditions and increased risk of tropical storm and hurricane incidents.

### St. Vincent and the Grenadines

*Recent trends.* Temperatures have increased roughly 0.1° to 0.2°C per decade over the last 25 years (about 1.0°C since 1901). Rainfall shows mostly an upward trend, while SLR is 2 to 3 mm/year.

*Recent events.* Intensifying hurricanes: Record hurricane season of 2005 in the Caribbean Basin affected the island with several landings. Last serious landing was by Hurricane Dean in 2007.

General for the region: A positive tendency for intense precipitation and consecutive dry days, as well as a positive trend in the frequency of very heavy rains can be expected. An increased number of intense hurricanes is likely.

*Projections.* See table 13.4 and figure 11.15 in WGII (IPCC 2007b) for mean temperature and mean precipitation. The signal is relatively robust across GCMs. The trend for precipitation frequency is not clear. In connection with hurricanes, intensity level of downpours may increase. Dry day frequency trends are expected to continue upward. During the next century, an increasing risk for more of the most intense hurricanes is expected. Notably, it is not clear if there will be more hurricanes per se. Sea level rise will compare with global increase.

Images: Somewhat drier conditions and increased risk of tropical storm and hurricane incidents.

## South America

### Argentina

*Recent trends.* A temperature increase of 0.1°C per decade has been observed north of Buenos Aires, but no change is seen to the south during the past 25 years (0.5°C warming since 1901, except near Buenos Aires, where temperatures have increased by roughly 1.5°C). Annual

**Figure B.4    Map of South America**

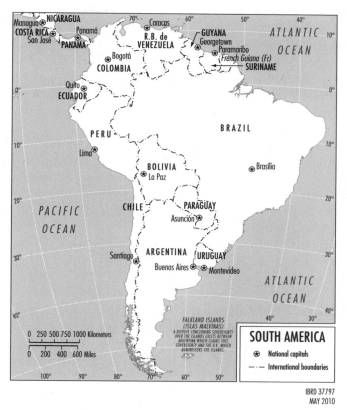

IBRD 37797
MAY 2010

*Source:* World Bank.

rainfall has generally increased, especially in the central and northeastern parts, while a decrease has been observed in the southwestern parts. Sea level rise at Buenos Aires is 1.7 mm/year but generally varies between 2 and 3 mm/year.

***Recent events.*** Increased El Niño occurrences.

Droughts: Droughts occurred in the Chaco region in 2004. The Pampas flooded in 2000 and 2002, and unprecedented and destructive hailstorms occurred in Buenos Aires in 2006.

Strong Zonda: Extreme dry-air events occurred in 2006.

General for the region: A positive tendency for intense precipitation and consecutive dry days has occurred. In addition, the region has experienced a positive trend in the frequency of very heavy rains. Glaciers are retreating drastically.

*Projections.* See table 13.4 and figure 11.15 in WGII (IPCC 2007b) for mean temperature and mean precipitation changes. The precipitation signal is quite robust across GCMs. While the central part is likely to become drier, wetter or unchanging conditions may occur in the north and in the south. The upward trend in precipitation frequency continues but not at a strong and significant signal. The dry day frequency trend generally continues upward, except in the far south. Over the next century, the risk for more of the most intense extratropical cyclones and associated storm surges will increase. Sea level rise will compare with global increase. However, greater sea level rise is estimated to occur near the mouth of River Plata. Finally, glaciers will decrease in size and volume.

Images: Wetter in the north, drier central, and increased risk of drought and heavy rains.

## Bolivia

*Recent trends.* Temperatures have increased at about 0.1° to 0.2°C per decade over the last 25 years (around 0.5°C since 1901, but information only sparsely available). Rainfall shows a varying trend, which is consistent across the Andes. In particular, reductions in rainfall are seen toward the west.

*Recent events.* Increased El Niño occurrences.

Droughts: The country experienced a severe drought throughout 2004–05.

Precipitation: The country experienced a destructive hailstorm in 2002, as well as heavy rains leading to severe flooding in 2006.

General for the region: A positive tendency for intense precipitation and consecutive dry days has occurred. In addition, the region has experienced a positive trend in the frequency of very heavy rains. Finally, glaciers are retreating drastically.

*Projections.* See table 13.4 and figure 11.15 in WGII (IPCC 2007b) for mean temperature and mean precipitation changes. Note that the precipitation trend will be mainly positive and the upward trend in precipitation intensity will continue. The dry day frequency trend is also expected to continue upward.

Images: Glaciers will retreat and in some places disappear completely. Extreme rain events with the possibility for flash floods are likely to occur more often.

## Brazil

**Recent trends.** Temperatures have increased by 0.1° to 0.4°C per decade within the last 25 years (between 0.5° and more than 2.0°C since 1901). The geographical variation within the country is large, and long-term changes are expected to occur mostly in the southeast, where the greatest warming has occurred so far. Annual rainfall changes are not well established throughout the country. In general, however, rainfall has decreased in the northern part of Amazonia, whereas both positive and negative changes are seen in southern Amazonia. Toward the southeast, precipitation has been rising significantly. Sea level rise has reached 4 mm/year in several ports, around 3 mm/year around the river mouth of Amazonas, 2 to 3 mm/year on the southeast coast, and around 2 mm/year elsewhere.

**Recent events.** Increased El Niño occurrences.

Hurricanes: In 2004 Brazil was hit by Hurricane Catarina, which was the first hurricane in the South Atlantic. The hurricane severely affected southern Brazil and eastern Amazonia.

Droughts: A severe drought occurred in central and southwestern Amazonia during 2004–06, also affecting Rio Grande do Sul.

Temperatures: Extreme heat waves were experienced in 2006.

General for the region: A positive tendency for intense precipitation and consecutive dry days has occurred. In addition, the region has experienced a positive trend in the frequency of very heavy rains. Finally, glaciers are retreating drastically.

**Projections.** See table 13.4 and figure 11.15 in WGII (IPCC 2007b) for mean temperature and mean precipitation. The precipitation signal is not very robust across GCMs for the annual mean change, indicating that the region is near the boundary between increase and decrease in precipitation. Seasonal signals are more robust and point toward an overall reduction in June-July-August rainfall—possibly turning into an increase toward the Andes countries. The upward trend in precipitation frequency continues, possibly with the exception of areas along the east coast. The dry day frequency trend continues upward, and SLR will compare with global increase.

Images: Possible overall drier conditions and increasing risk for drought and heavy rains.

## Chile

**Recent trends.** Modest temperature increases of 0.1°C per decade have occurred in the north, while no change, or even a 0.1°C cooling per decade, has occurred toward the south over the last 25 years (often less than 0.5°C of either warming or cooling, so no significant trend since 1901). Annual rainfall has generally decreased, by as much as by 50 percent in the central part of Chile. Sea level rise is on the order of 1 to 2 mm/year, with higher levels in the southern part.

**Recent events.** Precipitation: Heavy snowfalls occurred in the Andes, and heavy rains occurred in central Chile, during 2005, resulting in landslides.

General for the region: A positive tendency for intense precipitation and consecutive dry days has occurred. In addition, the region has experienced a positive trend in the frequency of very heavy rains. Glaciers are retreating drastically.

**Projections.** See table 13.4 and figure 11.15 in WGII (IPCC 2007b) for mean temperature and mean precipitation. The precipitation signal is quite robust across GCMs, emphasizing drying everywhere except the southernmost part. Overall the picture is constant throughout the year. The upward trend for precipitation frequency does not appear significant—in fact a reduction seems more likely. The dry day frequency trend continues upward. Additionally, over the next century the risk for more of the most intense cyclones will increase. Sea level rise will compare with global increases.

Images: Overall drier conditions and enhanced risk for droughts. Glaciers are disappearing or being reduced considerably in volume and area coverage.

## Colombia

**Recent trends.** Temperatures have increased by 0.1° to 0.2°C per decade over the last 25 years (0.5° to 1.0°C since 1901, but information is only sparsely available). Annual rainfall shows no clear trend throughout the country, though a negative trend has generally been observed in northern Amazonia. Sea level rise is roughly 2 to 3 mm/year toward the Atlantic and around 1 mm/year in the Pacific.

*Recent events.* Increased El Niño occurrences.

Precipitation: The country experienced severe heavy rains in 2005 and 2006, with associated flooding.

General for the region: A positive tendency for intense precipitation and consecutive dry days has occurred. In addition, the region has experienced a positive trend in the frequency of very heavy rains. Glaciers are retreating drastically.

*Projections.* See table 13.4 and figure 11.15 in WGII (IPCC 2007b) for mean temperature and mean precipitation. Note that the precipitation trend will be mainly positive in the Andes, while somewhat negative in Amazonia. The signal is relatively robust across GCMs. The country is located near the dividing line; hence the significance of the exact position of the divide between wetter and drier conditions is speculative. The precipitation intensity trend and the dry day frequency trend continue upward, but the trends are not robust everywhere. In fact, the Pacific part of the country may realize a reduction in dry days. Sea level rise will compare with global increases.

Images: Wetter conditions in the west but with large interannual variability further east; increased chance for fast floods; disappearing glaciers.

## Ecuador

*Recent trends.* Temperatures have increased by 0.1°C per decade over the last 25 years (between 0.5° and 1°C since 1901, but coverage is poor). Annual rainfall has generally been increasing, and SLR has occurred at about 1 mm/year.

*Recent events.* Increased El Niño occurrences.

Precipitation: Heavy rains occurred in 2003 and again in 2006, both times with associated flooding.

General for the region: A positive tendency for intense precipitation and consecutive dry days has occurred. In addition, the region has experienced a positive trend in the frequency of very heavy rains. Glaciers are retreating drastically.

*Projections.* See table 13.4 and figure 11.15 in WGII (IPCC 2007b) for mean temperature and mean precipitation. The signal is relatively robust across GCMs. A clear tendency for wetter conditions is showing year-round. The upward trend in precipitation intensity continues, while the trend in dry day frequency is turning downward. Hence the country may realize a reduction in dry days. Sea level rise will compare with global increases.

Images: Wetter conditions, increased chance for fast floods, disappearing glaciers.

### French Guiana

**Recent trends.** Modest temperature increases of 0.1°C per decade have been observed within the last 25 years (about 0.5°C since 1901). Annual rainfall changes are not well established, and SLR occurs at about 2 mm/year.

**Recent events.** Increased El Niño occurrences.

Precipitation: Several events with heavy rains have occurred in recent years.

General for the region: A positive tendency for intense precipitation and consecutive dry days has occurred. In addition, the region has experienced a positive trend in the frequency of very heavy rains. Glaciers are retreating drastically.

**Projections.** See table 13.4 and figure 11.15 in WGII (IPCC 2007b) for mean temperature and mean precipitation. Note that the precipitation trend will be mainly negative, but the signal is not very robust across GCMs, particularly toward the south. The country is located near the dividing line between wetter and drier conditions for some parts of the year. Hence, the significance of the exact position of the divide between wetter and drier conditions is speculative. Precipitation intensity will show a downward trend, while the dry day frequency trend continues upward. Sea level rise will compare with global increases.

Images: Drier conditions but with some interannual variability.

### Guyana

**Recent trends.** Modest temperature increases have occurred at about 0.1° to 0.2°C per decade in the last 25 years (about 0.5° to 1.0°C since 1901). Annual rainfall changes are not well established, and sea level rise is occurring at roughly 2 mm/year.

**Recent events.** Increased El Niño occurrences.

Precipitation: Heavy rains occurred in 2005, resulting in severe flooding in the capital, Georgetown.

General for the region: A positive tendency for intense precipitation and consecutive dry days has occurred. In addition, the region has experienced a positive trend in the frequency of very heavy rains. Finally, glaciers are retreating drastically.

***Projections.*** See table 13.4 and figure 11.15 in WGII (IPCC 2007b) for mean temperature and mean precipitation. Note that the precipitation trend will be mainly negative, but the signal is not very robust across GCMs, particularly toward the south. The country is located near the dividing line between wetter and drier conditions for some parts of the year. Hence, the significance of the exact position of the divide between wetter and drier conditions is speculative. Precipitation intensity will show a downward trend, while the trend in dry day frequency continues upward. Sea level rise will compare with global increases.

Images: Drier conditions but with some interannual variability.

## Paraguay

***Recent trends.*** Temperatures have increased by 0.2°C per decade over the last 25 years (about 1°C since 1901), and annual rainfall has significantly increased as well.

***Recent events.*** Increased El Niño occurrences.

Droughts: The country was exposed to a drought in 2004–06, though it was less severe than the one in Brazil.

General for the region: A positive tendency for intense precipitation and consecutive dry days has occurred. In addition, the region has experienced a positive trend in the frequency of very heavy rains. Finally, glaciers are retreating drastically.

***Projections.*** See table 13.4 and figure 11.15 in WGII (IPCC 2007b) for mean temperature and mean precipitation. Note that the precipitation trend is close to zero, although the signal is not very robust across GCMs. That is expected, however, since small positive and negative values would tend to dominate the disagreement. The country is located near several dividing lines between wetter and drier conditions. Hence, the significance of the exact position of the divide between wetter and drier conditions is speculative. Precipitation intensity shows an upward trend, and the trend of dry day frequency continues upward as well.

Images: Relatively unchanged precipitation regimes but with increasing risk for extreme rain and drought.

## Peru

***Recent trends.*** Temperatures have generally increased by 0.1°C per decade over the last 25 years (between 0.5 and 1°C since 1901, but coverage is poor). Annual rainfall has generally been decreasing, except toward the northwest, where an increase is seen. Sea level rise is 1 to 2 mm/year, with highest values in the southern part.

***Recent events.*** Increased El Niño occurrences.

Precipitation: Anomalously wet in 2003.

General for the region: A positive tendency for intense precipitation and consecutive dry days has occurred. In addition, the region has experienced a positive trend in the frequency of very heavy rains. Glaciers are retreating drastically.

***Projections.*** See table 13.4 and figure 11.15 in WGII (IPCC 2007b) for mean temperature and mean precipitation. Note that the precipitation trend will be mainly positive throughout the country, though with some variation showing a negative trend in the southern part. The signal is relatively robust across GCMs. The country is located near the dividing line; hence the significance of the exact position of the divide between wetter and drier conditions is speculative. The precipitation intensity trend, as well as the trend of dry day frequency, continues upward but is not robust everywhere. In fact the northern part of the country may realize a reduction in dry days. Sea level rise will compare with global increases.

Images: Generally wetter conditions but with large interannual variability farther south; increased risk of fast floods.

## Suriname

***Recent trends.*** A modest temperature increase of 0.1°C per decade in the last 25 years (about 0.5°C since 1901) has been observed. Annual rainfall changes are not well established, and sea level rise has occurred at about 2 mm/year.

***Recent events.*** Increased El Niño occurrences.

Precipitation: Heavy rains and flooding occurred in 2006.

General for the region: A positive tendency for intense precipitation and consecutive dry days has occurred. In addition, the region has experienced a positive trend in the frequency of very heavy rains. Glaciers are retreating drastically.

*Projections.* See table 13.4 and figure 11.15 in WGII (IPCC 2007b) for mean temperature and mean precipitation. Note that the precipitation trend will be mainly negative, but the signal is not very robust across GCMs, in particular toward the south. The country is located near the dividing line between wetter and drier conditions for some parts of the year. Hence the significance of the exact position of the divide between wetter and drier conditions is speculative. Precipitation intensity will show a downward trend, while the trend in dry day frequency continues upward. Sea level rise will compare with global increases.

Images: Drier conditions but with some interannual variability.

### Trinidad and Tobago

*Recent trends.* Modest temperature increases of about 0.1°C per decade have been observed in the last 25 years (about 1.0°C since 1901). Annual rainfall changes show an upward trend in coastal South America, but little information is available from the islands themselves. Sea level rise is close to 2 mm/year.

*Recent events.* Increased El Niño occurrences.

Precipitation: Some heavy rain events have occurred.

General for the region: A positive tendency for intense precipitation and consecutive dry days has occurred. In addition, the region has experienced a positive trend in the frequency of very heavy rains. Glaciers are retreating drastically.

*Projections.* See table 13.4 and figure 11.15 in WGII (IPCC 2007b) for mean temperature and mean precipitation. Note that the precipitation trend will be mainly negative, but the signal is not very robust across GCMs, particularly toward the south. The country is located near the dividing line between wetter and drier conditions for some parts of the year. Hence the significance of the exact position of the divide between wetter and drier conditions is speculative. Precipitation intensity will show a downward trend, while the trend of dry day frequency continues upward. Sea level rise will compare with global increase.

Images: Drier conditions but with some interannual variability.

### Uruguay

*Recent trends.* Overall a modest temperature increase of 0.1°C per decade has been observed during the past 25 years (1.5°C since 1901).

Annual rainfall has significantly increased and sea level rise has occurred at up to 3 mm/year.

**Recent events.** Hurricanes (cyclones): An extratropical cyclone occurred in 2005, with strong winds and a storm surge.

Droughts: Dry to droughtlike conditions were observed in the northern part of the country in 2006.

General for the region: A positive tendency for intense precipitation and consecutive dry days has occurred. In addition, the region has experienced a positive trend in the frequency of very heavy rains. Finally, glaciers are retreating drastically.

**Projections.** See table 13.4 and figure 11.15 in WGII (IPCC 2007b) for mean temperature and mean precipitation. Note that the precipitation trend will be mainly positive throughout the year, but the signal is not always robust across GCMs. The country is located near the dividing line between wetter and drier conditions for some parts of the year. Hence the significance of the exact position of the divide between wetter and drier conditions is speculative. Precipitation intensity and dry day frequency will both show an upward trend. Sea level rise will be higher than global increases.

Images: Wetter conditions and increased risk for fast floods and droughts.

### Venezuela, R. B. De

**Recent trends.** Temperature increases at a rate of 0.1° to 0.2°C per decade have been observed over the last 25 years (0.5° to 1.0°C since 1901, but information is only sparsely available). Annual rainfall shows an increasing trend near the coast, while a negative trend is generally observed in northern Amazonia. Sea level rise occurs at about 1.5 to 2 mm/year.

**Recent events.** Increased El Niño occurrences.

Precipitation: Heavy rains occurred in 2005 on the central coast and in the Andes, with associated flooding. Very dry to droughtlike conditions occurred in the southern part (Amazonia) during 2003–04.

General for the region: A positive tendency for intense precipitation and consecutive dry days has occurred. In addition, the region has experienced a positive trend in the frequency of very heavy rains. Glaciers are retreating drastically.

**Figure B.5    Multimodel Projections of Sea Level Change**
*(in meters)*

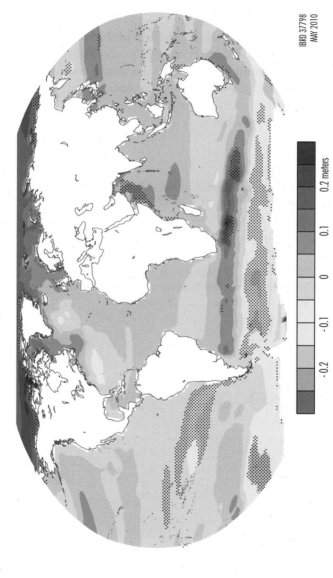

IBRD 37798
MAY 2010

-0.2    -0.1    0    0.1    0.2 meters

*Source:* Adapted from IPCC 2007b.

*Note:* Local sea level change (meters) due to ocean density and circulation change relative to the global average (positive values indicate greater local sea level change than global) during the 21st century, calculated as the difference between averages for 2080–99 and those for 1980–99, as an ensemble mean over 16 AOGCMs forced with the Special Report on Emissions Scenarios A1B scenario. Stippling denotes regions where the magnitude of the multimodel ensemble mean divided by the multimodel standard deviation exceeds 1.0.

**Projections.** See table 13.4 and figure 11.15 in WGII (IPCC 2007b) for mean temperature and mean precipitation. Note that the precipitation trend will be mainly negative near the coast but positive toward the Andes. Yet the signal is not very robust across GCMs. The country is located near the dividing line between wetter and drier conditions for some parts of the year. Hence the significance of the exact position of the divide between wetter and drier conditions is speculative. Precipitation intensity will show a downward trend in the north and increase toward the Andes. The trend in dry day frequency continues upward. Sea level rise will compare with global increases.

Images: Drier conditions but with large geographical variations.

## References

IPCC (Intergovernmental Panel on Climate Change). 2007a. *Climate Change 2007: The Physical Science Basis.* Contribution of Working Group I to the Fourth Assessment Report of the IPCC. Geneva: IPCC.

———. 2007b. *Climate Change 2007: Impacts, Adaptation, and Vulnerability.* Contribution of Working Group II to the Fourth Assessment Report of the IPCC. Geneva: IPCC.

———. 2007c. *Synthesis Report: An Assessment of the Intergovernmental Panel on Climate Change.* Valencia, Spain, 12–17 November.

# Methodology and Data

## Lykke Andersen, Inger Brisson, Claus Pörtner, and Dorte Verner

This appendix first describes the sustainable livelihoods framework as used in this study, and then describes the data and quantitative methods that were used to explore whether there are correlations between climate change and variability and life expectancy or child mortality (in chapter 6), poverty and inequality (in chapter 7), and migration (in chapter 8).

### Sustainable Livelihoods Framework

The framework, adapted from U.K. Department for International Development's sustainable livelihoods framework (SLF; DFID 2004), represents the main factors that affect people's livelihoods (figure C.1).

In focusing on people's access to assets, the SLF reflects the multidimensionality of poverty. The SLF is a tool for assessing the vulnerability of the poor and their capacity to cope in the presence of shocks and adapt to changing trends—which is highly important seen from a climate change perspective. Vulnerability, and their added vulnerability due to climate change and climatic variability, can be assessed for different population groups, together with their ability to adapt to climate change within their specific environmental context. To address people's vulnerability, the SLF focuses on five aspects to assess livelihood outcomes. This book has augmented the SLF with one more asset, cultural capital.

**Figure C.1    Sustainable Livelihoods Framework**

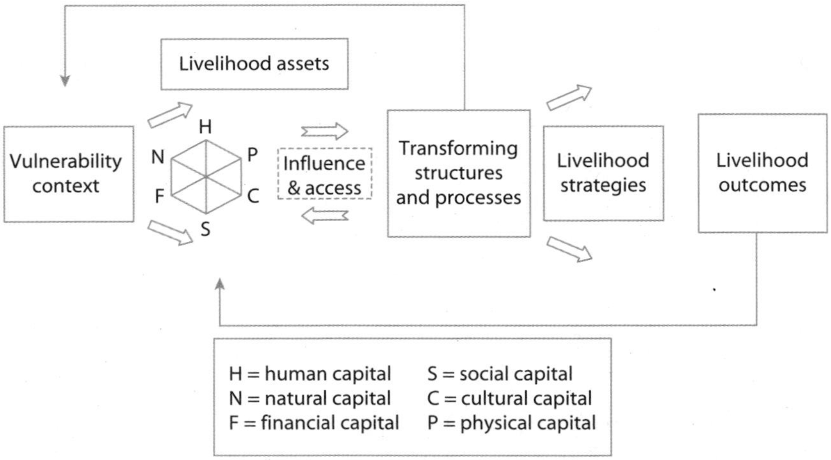

H = human capital    S = social capital
N = natural capital    C = cultural capital
F = financial capital    P = physical capital

*Source:* Augmented from DFID 2004.

## Vulnerability Context

The "vulnerability context" refers to the external environment in which people live. People's livelihoods and the wider availability of assets are largely affected by external trends, shocks, seasonality, and climatic variability. For the purpose of this book, the trends considered are gradual climate change and global warming, which may produce a wide range of shocks (flooding, hurricanes, tornadoes, droughts), which in turn may result in negative outcomes, such as economic and health conditions that may lead to conflict, migration, and so on. These shocks will be indicative of the current vulnerability of the populations examined. Climate change may also affect the seasonality of climatic resources and increase climatic variability, also affecting vulnerability.

## Livelihood Assets

In this next step, the SLF considers the assets that individuals, households, or communities have available. The assets considered include physical, financial, human, social, cultural, and natural capital (table C.1).[1]

It is important to note that a single asset can generate multiple benefits. For example, if a household has secure access to fertile land (natural capital), it may also be well endowed with financial capital because it can use the land not only directly for productive activities but also as collateral (DFID 2004). Hence the SLF can aid in gauging the availability of

Table C.1    Definition of Livelihood Assets

| Capital | Assets |
|---------|--------|
| Physical | The stock or plant. Equipment, infrastructure, and other productive resources owned by individuals, the business sector, or the country itself. |
| Financial | The financial resources available to people (savings, supplies of credit). |
| Human | Investments in education, health, and the nutrition of individuals. Labor is linked to investments in human capital; health status determines people's capacity to work, and skill and education determine the returns from their labor. |
| Social | An intangible asset, defined as the rules, norms, obligations, reciprocity, and trust embedded in social relations, social structures, and societies' institutional arrangement. It is embedded at the microinstitutional level (communities and households) as well as in the rules and regulations governing formalized institutions in the marketplace, political system, and civil society. |
| Cultural | The knowledge, experience, and connections people have had throughout their lives, which enable them to succeed better than someone from a less-experienced background. |
| Natural | The stock of environmentally provided assets such as soil, atmosphere, forests, minerals, water, and wetlands. In rural communities land is a critical productive asset for the poor; in urban areas, land for shelter is also a critical productive asset. |

*Source:* Augmented from DFID 2004.

assets and determine how these interact with, or affect, one another, and it can map which populations may be more affected by shocks than others. It is understood that the more assets a person has available, the less vulnerable he or she is.

DFID's original SLF includes five types of livelihood assets, namely, physical, natural, financial, social, and human capital. For this book, a sixth type of livelihood asset, cultural capital, has been included. Field research has revealed that the cultural dimension of livelihood strategies and social institutions is particularly important for understanding and describing the impacts of climate change and climatic variability experienced by indigenous peoples.

"Cultural capital" can be defined as the knowledge, experience, and connections that people have had throughout their lives, which enable them to succeed better than someone without such a background. Cultural capital is a sociological concept that has gained widespread popularity since it was first articulated by Bourdieu.[2] Cultural capital acts as a social relation within a system of exchange, and the term is extended to all the goods—material and symbolic—that present themselves as rare and worthy of being sought after in a particular social formation. Thus, cultural capital acts as a social relation within a system of exchange, which includes the accumulated cultural knowledge that confers power and status. In other words, cultural capital is created, through social relations, from cultural knowledge that in turn generates power and status. Cultural capital, in this book's use of the concept, has much in common with cultural practices, in the sense of adding cohesion, structure, stability, and a degree of conservatism to societies. So whereas "social capital" refers to the network of people and interactions, "cultural capital" refers to the network of shared ideas and beliefs. The deviation of Bourdieu's theory of capital from the dominant, exclusively economic conception, allows an understanding of the other factors that have an important role in society, especially social and cultural factors; it permits a complete and open analysis of the issues and contexts and makes it possible to generate solutions specific to the problems that arise in each case.

### Transforming Structures and Processes

The SLF directs attention to the structures and processes at play in the community. The transforming structures and processes within the SLF are the institutions, organizations, policies, and legislation that shape livelihoods. They operate at all levels, from the household, community, and municipality levels to the national and international levels, and in all spheres, from the most private to the most public. They effectively determine

- access (to various types of assets, to livelihood strategies, and to decision-making bodies and sources of influence);
- the terms of exchange between different types of assets; and
- returns (economic and otherwise) on any given livelihood strategy.

In addition, structures and processes directly influence whether people are able to achieve a feeling of inclusion and well-being. Because culture is included, they also account for other "unexplained" differences in the way things are done in different societies. Much of this book focuses on these structures and processes, how they amplify vulnerabilities, or how they can be harnessed to enhance adaptive capacity and resilience.

### Livelihood Strategies

The framework provides insight into how vulnerability and assets, as well as structures and processes, influence livelihood strategies and how these may be improved. The expansion of choice and value is important because it provides people with opportunities for self-determination and the flexibility to adapt over time. It is most likely to be achieved by working to improve poor people's access to assets—the building blocks for livelihood strategies—and to make the structures and processes that transform these into livelihood outcomes more responsive to their needs.

### Livelihood Outcomes

Finally, the framework examines how the livelihood strategies, given the other factors, result in different livelihood outcomes and how the livelihood outcomes feed back into available assets, creating either a virtuous or vicious circle. Specifically, the framework helps to pinpoint entry points for how to raise income, increase well-being, reduce vulnerability, improve food security, and achieve more sustainable use of resources.

The study described in this book used the SLF to analyze

- how climate change and climatic variability affect the vulnerability context through their effects on trends, shocks, and seasonality;
- how climate change affects the available assets;
- the nature of the adaptive capacity, in terms of transforming structures and processes, and how existing transforming structures and processes may obstruct adaptation; and
- how the livelihoods of individuals and communities can be improved through adaptation, by influencing assets, and transforming structures and processes and thus building resilience.

Livelihood assets and the transforming structures and processes together determine the livelihood strategies available to people, while their chosen strategy, given the vulnerability context and so on, determines their livelihood outcome.

For the disadvantaged and socially excluded, the livelihood outcomes are dismal, a situation that will only be compounded by the effects of climate change. To correct this, their access to livelihood assets must be changed through changes in transforming structures and processes to improve resilience. This is the entry point for policy measures and governance transformation.

## Quantitative Analyses

### Weather-Related Disasters and Health Outcomes

Chapter 6, on the health impacts of climate change and variability, includes analyses of the relationship between weather-related disasters and health outcomes in Guatemala. The data used are described in table C.2.

**Table C.2    Data Used in Analysis of Relationships between Weather-Related Disasters and Health Outcomes**

|        | Name | Source | Year | Area | Sample |
|--------|------|--------|------|------|--------|
| Health | Demographic and Health Survey | www.measuredhs.com | 1987 | Guatemala | Rural, aged 3–59 months |
|        | Demographic and Health Survey | www.measuredhs.com | 1995 | Guatemala | Rural, aged 3–59 months |
|        | Demographic and Health Survey | www.measuredhs.com | 1998 | Guatemala | Rural, aged 3–59 months |
| Shocks | Desastres Naturales y Zonas de Riesgo en Guatemala | UNICEF (2000) | 1880–1997 | Guatemala | All |

The hazards examined are the following: strong winds, flooding, heavy rain, hurricanes, frost and freezing, and droughts. Exposure to hazards is measured as the number of shocks that occurred during the six months preceding the month of the interview. In addition, the hazard variables are interacted with a subset of the other explanatory variables, to allow for the fact that the effects of hazards depend partly on the characteristics of the individual children and households.

Child health is measured by two anthropometric measures—height for age and weight for height—and three symptoms of illness—diarrhea, fever, and cough during the two weeks prior to the survey. As is standard in most studies of child health, the anthropometric variables are converted into Z-scores, which is the deviation of a child's value from the median value of the reference population, divided by the standard deviation of the reference population (WHO 2006). In general, children in the sample do very poorly, with a mean Z-score of –2.45 for height for age, although girls do better than boys.

Estimation is done using ordinary least squares regression. To capture differences among areas that may directly or indirectly affect child health, department dummies are included, together with a fourth-order polynomial for the altitude of the municipality.[3] Dummies for the survey rounds are included, to allow for differences in economic development over time, which might affect child health. One potential issue here is that it is not possible to calculate reliable measures of risk for most of the hazards, since the time-series available are not long enough. This may bias the results (Pörtner 2008). In that case, the level of risk will be included in the error term, and hence create a correlation between the error term and the shock variable. Which direction the bias will take depends on how risk affects child health directly. If households respond to higher risk by having more children, we might expect that effect to be negative for the anthropometric measures because of closer spacing of children and a more binding resource constraint. Obviously, there is a positive correlation between risk and the number of shocks that occur in a department, which means that the bias, in this case, will be downward.

The household- and individual-level explanatory variables used are the sex of the child; the age of the child in months at the time of the survey; the age of the mother at the time of the child's birth; her education and literacy levels; the father's education level; and the ethnicity of the child. For both parental age and education levels, the squares of the variables are also included. Unfortunately, the data do not include direct information on land ownership, and so a dummy variable is included as a proxy: this takes the value of 1 if either the mother or the father responds that they work on their own or family land, and zero otherwise.

### The Effects of Climate Change on Life Expectancy/Child Mortality and on Poverty and Inequality

To assess how climate change is likely to affect a population, it is necessary to understand how climate is currently affecting them and how climate is changing.

A simple way to evaluate how climate affects human development is to compare human development across regions that have different climates. This has been done by Horowitz (2006), who used a cross section of 156 countries to estimate the relationship between temperature and income level. The overall relationship found is very strongly negative, with a 1.1°C increase in global temperatures implying a 13 percent drop in income.[4] This is very dramatic, but the relationship is thought to be mostly historical and thus not very relevant for projecting the effects of future climate change. Horowitz proposes using local-level climate and development data for large, heterogeneous countries as a way of controlling the long-term, historical effects of climate and thus obtaining instead a short-run relationship (decades instead of centuries), which would be more relevant for evaluating the likely impacts of current climate change.

Following that suggestion, the researchers used municipal-level data on climate, income, and life expectancy (or child mortality, where life expectancy data are not available) for five countries in Latin America and the Caribbean (Brazil, Mexico, Chile, Peru, and Bolivia) to estimate the likely relationships between climate, on one hand, and life expectancy/child mortality (chapter 6) and poverty and inequality (chapter 7) on the other. Both average annual temperature and its square are included in the regression model, and so is rainfall squared. The regressions control for other factors that may affect the level of development but are likely to be insensitive to climate change in the short run. The key control variables are education level and urbanization level.

Apart from income level as a measure of development, life expectancy is also used. The life expectancy regression takes the same form as the income regressions, except that the natural logarithm is not applied to the dependent variable. All regressions are weighted ordinary least squares regressions, where the weights consist of the population size in each municipality.

Once estimated, the country-specific relationships between climate and income and climate and life expectancy are used to gauge the likely effects of climate change for each municipality in the countries under investigation. The effects of the climate change that has taken place during the last 50 years, as documented by local temperature records, are analyzed, as are the effects of expected future climate change over the next 50 years, as projected by IPCC's Atmosphere-Ocean General Circulation Models. Evaluating the effects for each municipality enables us to assess whether climate change is contributing to increased inequality and poverty, by testing whether initially poorer municipalities are more adversely affected by climate change than are initially richer municipalities.

### Data Used in the Life Expectancy and Poverty and Inequality Analyses

*For the time-series analysis of climate,* to detect long-term trends in subregional climate, the database Monthly Climatic Data for the World, collected by the U.S. National Climatic Data Center (NCDC), was used. That project started in May 1948, with 100 stations spread across the world, including 40 in Latin America and the Caribbean. Since then, many more stations have been included in the database, and 104 stations from the six case study countries have contributed data sufficiently regularly to permit trends for the 1948–2008 period to be estimated. *For the municipality-level cross-section analysis,* see table C.3.

**Table C.3  Key Variables in the Municipal-Level Databases for Bolivia, Brazil, Chile, Mexico, Nicaragua, and Peru**

| Variable | Definition | Unit | Source |
|---|---|---|---|
| Total population | The number of inhabitants in the municipality. | – | Latest population census. |
| Urbanization rate | The share of the population living in urban areas, according to the official national classification. | – | Latest population census. |
| Years of education | Average number of years of education of the population aged 15 or above. | Years | Latest population census. |
| Life expectancy | Life expectancy at birth for each municipality. | Years | Calculated by UNDP[a] for its Human Development Index. |
| Per capita income or per capita expenditure | Average household income or expenditure per capita in each municipality. | PPP[b]-adjusted US$/month | Calculated by UNDP[a] for its Human Development Index. |
| Latitude | Latitude of the main city in the municipality. | Degrees south of equator | Google Earth, GeoMaker. |
| Longitude | Longitude of the main city in the *comuna*. | Degrees west of the Prime Meridian | Google Earth, GeoMaker. |
| Elevation | Elevation of the main city in the *comuna*. | Kilometers above sea level | Google Earth, GeoMaker. |
| Normal average annual temperature | The average annual temperature in the main city of the municipality as measured over a reference period (typically 1961–90 but sometimes longer). | °C | National meteorological offices and www.worldclimate.com. |
| Normal annual rainfall | The average annual rainfall in the main city of the municipality as measured over a reference period (usually 1961–90 but sometimes longer). | Meters | National meteorological offices and www.worldclimate.com. |

*Source:* Authors.

a. UNDP = United Nations Development Program.

b. PPP = Purchasing power parity.

## Regressions for the Impact of Climate Change on Life Expectancy (or Child Mortality)

The level of health under a situation of no climate change can be written as:

$$health_{i,NCC} = \hat{\beta}_1 \cdot temp_{i,NCC} + \hat{\beta}_2 \cdot temp^2_{i,NCC} + \hat{\beta}_3 \cdot rain_{i,NCC}$$

$$+ \hat{\beta}_4 \cdot rain^2_{i,NCC} + \sum_{j=1}^{k} \hat{\alpha}_j X_{j,i} + \hat{\varepsilon}_i$$

where the index $i$ refers to municipality $i$; *temp* and *rain* are the temperature and rainfall variables; the $\hat{\beta}$s are the estimated coefficients on the temperature and rainfall variables; the $X_j$s are the remaining $j$ explanatory variables including the constant term; the $\hat{\alpha}_i$s are the coefficient to these variables; and $\hat{\varepsilon}_i$ are the estimated error terms for each municipality.

Equivalently, the level of health under the assumption of climate change can be written as follows:

$$\hat{health}_{i,CC} = \hat{\beta}_1 \cdot temp_{i,CC} + \hat{\beta}_2 \cdot temp^2_{i,CC}$$

$$+ \hat{\beta}_3 \cdot rain_{i,CC} + \hat{\beta}_4 \cdot rain^2_{i,CC} + \sum_{j=1}^{k} \hat{\alpha}_j X_{j,i}$$

where the only differences are the temperature and rainfall variables. The control variables are held constant, so as to isolate the effects of climate change and variability.

The difference in life expectancy that can be directly attributed to climate change and variability can be found as the difference between the two scenarios:

$$\Delta_{CC} \hat{health}_i = \hat{health}_{i,CC} - \hat{health}_{i,NCC}$$

$$= \hat{\beta}_1 \cdot (temp_{i,CC} - temp_{i,NCC}) + \hat{\beta}_2 \cdot (temp^2_{i,CC} - temp^2_{i,NCC})$$

$$+ \hat{\beta}_3 \cdot (rain_{i,CC} - rain_{i,NCC}) + \hat{\beta}_4 \cdot (rain^2_{i,CC} - rain^2_{i,NCC})$$

## Regressions for the Impact of Climate Change on Poverty and Inequality

The regression for analyzing the short-run implications of climate change takes the following form:

$$\ln y_i = \alpha + \beta_1 \cdot temp_i + \beta_2 \cdot temp^2_i + \beta_3 \cdot rain_i$$

$$+ \beta_4 \cdot rain^2_i + \beta_5 \cdot edu_i + \beta_6 \cdot urb_i + \beta_7 \cdot urb^2_i + \varepsilon_i$$

where $y_i$ is a measure of the income level in municipality $i$, $temp_i$ and $rain_i$ are normal average annual temperature and normal accumulated annual rainfall in municipality $i$, $edu_i$ is a measure of the education level (average years of schooling, or level of literacy, depending on which variables were available), $urb_i$ is the urbanization rate of the municipality, and $\varepsilon_i$ is the error term for municipality $i$.

### Simulation of Impacts of Climate Change on Income in LAC Countries

To estimate the changes in income that can be attributed to future climate change, we compare two scenarios: No Climate Change (NCC) and Climate Change (CC) in each municipality. The NCC level of income for municipality $i$ can be written as follows:

$$\ln(Y_{i,NCC}) = \hat{\alpha} + \hat{\beta}_1 \cdot temp_{i,NCC} + \hat{\beta}_2 \cdot temp^2_{i,NCC} + \hat{\beta}_3 \cdot rain_{i,NCC} + \hat{\beta}_4 \cdot rain^2_{i,NCC}$$
$$+ \hat{\beta}_5 \cdot edu_i + \hat{\beta}_6 \cdot urb_i + \hat{\beta}_7 \cdot urb^2_i + \hat{\varepsilon}_i \qquad ,$$

where the $temp_{i,NCC}$ is the average temperature in 2058 in municipality $i$ in the absence of climate change, the $\hat{\beta}$s are the estimated coefficients from Table C.3 above, and $\hat{\varepsilon}_i$ are the estimated error terms for each municipality. Equivalently, the CC level of income can be written as follows:

$$\ln(\hat{Y}_{i,CC}) = \hat{\alpha} + \hat{\beta}_1 \cdot temp_{i,CC} + \hat{\beta}_2 \cdot temp^2_{i,CC} + \hat{\beta}_3 \cdot rain_{i,CC} + \hat{\beta}_4 \cdot rain^2_{i,CC}$$
$$+ \hat{\beta}_5 \cdot edu_i + \hat{\beta}_6 \cdot urb_i + \hat{\beta}_7 \cdot urb^2_i \qquad ,$$

where the only differences are the temperature and rainfall levels. All other variables are held constant in the simulation. The ratio of Climate Change Income to No Climate Change Income can then be written as follows:

$$\Delta_{CC}\hat{Y}_i = \frac{\hat{Y}_{i,CC}}{\hat{Y}_{i,NCC}}$$
$$= \frac{\exp\left\{\hat{\beta}_1 \cdot temp_{i,CC} + \hat{\beta}_2 \cdot temp^2_{i,CC} + \hat{\beta}_3 \cdot rain_{i,CC} + \hat{\beta}_4 \cdot rain^2_{i,CC}\right\}}{\exp\left\{\hat{\beta}_1 \cdot temp_{i,NCC} + \hat{\beta}_2 \cdot temp^2_{i,NCC} + \hat{\beta}_3 \cdot rain_{i,NCC} + \hat{\beta}_4 \cdot rain^2_{i,NCC}\right\}}$$

After estimating this ratio for each municipality, it is easy to calculate the percentage change in income that can be attributed to climate change.

## Climate Change and Migration

The effects of climate change on internal and international migration are quantified using municipal-level data from Bolivia and Mexico, respectively. The data used to estimate the migration models in both cases come from the latest population census (2000 in the case of Mexico and 2001 in the case of Bolivia), which asked everybody about their current place of residence, as well as their residence 5 years earlier, among other things. Those data were used to estimate a complete municipality-to-municipality migration matrix in the case of Bolivia, and a Mexico–United States migration intensity index in the case of Mexico.

To analyze internal migration in Bolivia, the 314 municipalities that existed at the time of the 2001 census were used. A complete, municipality-to-municipality migration matrix therefore includes $314 \times 313 = 98,282$ entries. For each entry the migration rate, $m_{ij}$, is calculated as follows:

$$m_{ij} = \frac{M_{ij}}{pob_i},$$

where $M_{ij}$ is the number of migrants moving from municipality $i$ to municipality $j$ within a given time period (the 5 years before the 2001 census), and $pob_i$ is the population in municipality $i$ at the beginning of the period.

The migration rate can be decomposed in two parts:

$$m_{ij} = \left(1 - \frac{M_{ii}}{pob_i}\right) \cdot \left(\frac{M_{ij}}{\sum_{k=1,k\neq i}^{n} M_{ik}}\right)$$

where the first part is the proportion of persons who decide to migrate from municipality $i$ and the second part is the probability that they choose destination $j$, given that they have decided to migrate. This specification reflects the two step process of migration: first people decide whether to migrate or not, and then they decide where to go.

Given that, three different models can be estimated to analyze the determinants of migration:

Model 1 (out-migration):

$$1 - \frac{M_{ij}}{pob_i} = \alpha + \beta'(X_i)$$

Model 2 (destination choice):

$$\frac{M_{ij}}{\sum\limits_{k=1,k\neq i}^{n} M_{ik}} = \alpha + \beta'(d_{ij}, X_j).$$

Model 3 (migration rate):

$$m_{ij} = \alpha + \beta'(d_{ij}, X_i, X_j)$$

Model 1 establishes the factors that cause a municipality to retain or expel its population. All the possible explanatory variables are characteristics of the municipality itself, including climate variables as well as economic variables.

Model 2 establishes the factors that make municipalities more or less attractive as a migration destination. All the possible explanatory variables are characteristics of the receiving municipality, plus the physical distance between the sending and the receiving municipality.

Finally, Model 3 establishes the determinants of migration rates from one specific municipality to another, using the distance between the two municipalities $(d_{ij})$, characteristics of the sending region $(X_i)$, and characteristics of the receiving region $(X_j)$.

Included as possible explanatory variables are temperature, temperature squared, rainfall, and rainfall squared, as well as a set of control variables. In the case of Mexico, only the out-migration model is estimated, as the destination is always the United States.

## Notes

1. The shape of the asset hexagon (see figure C.1) can be used to show graphically the variation in people's access to assets. The center point of the hexagon, where the lines meet, represents zero access to assets. The greater the distance from the center, the greater is the access to any given asset. Thus, on this basis, in principle, differently shaped hexagons can be drawn for different communities or social groups within communities.

2. Pierre Bourdieu and Jean-Claude Passeron first used the term in *Cultural Reproduction and Social Reproduction* (Bourdieu 1973). In that work, Bourdieu attempted to explain differences in educational outcomes in France during the 1960s. The concept has since been elaborated and developed in terms of

other types of capital in *The Forms of Capital* (Bourdieu 1983, 1986) and in terms of higher education in, for instance, in *The State Nobility* (Bourdieu, de Saint Martin, and Clough 1996).

3. There are 22 departments in Guatemala, with a total of 331 municipalities.

4. Horowitz looked at a 2°F increase in temperature, which is equivalent to a 1.1°C increase.

## References

Bourdieu, Pierre. 1973. "Cultural Reproduction and Social Reproduction." In *Knowledge, Education, and Cultural Change*, ed. Richard Brown. London: Willmer Brothers Ltd.

————. 1983, 1986. "The Forms of Capital." Trans. by Richard Nice. In *Handbook of Theory and Research for the Sociology of Education*, ed. J. G. Richardson. 241–58. New York: Greenwood Press.

Bourdieu, Pierre, Monique de Saint Martin, and Laurette C. Clough. 1996. *The State Nobility*. Stanford, CA: Stanford University Press.

DFID (Department for International Development, U.K.). 2004. *Climate Change in Latin America*. Key Fact Sheet No. 12. Key Fact Sheets on Climate Change and Poverty. http://www.dfid.gov.uk/pubs/files/climatechange/keysheetsindex.asp.

Horowitz, J. K. 2006. "The Income-Temperature Relationship in a Cross-Section of Countries and Its Implications for Global Warming." Department of Agricultural and Resource Economics, University of Maryland, Submitted manuscript, July. http://faculty.arec.umd.edu/jhorowitz/Income-Temp-i.pdf.

Pörtner, Claus C. 2008. "Natural Hazards and Child Health." Department of Economics, University of Washington. http://faculty.washington.edu/cportner/nathaz.pdf.

UNICEF. 2000. "Desastres Naturales Y Zonas De Riesgo En Guatemala." Unpublished paper, UNICEF.

WHO (World Health Organization). 2006. "WHO Child Growth Standards." World Health Organization, Geneva. http://www.who.int/childgrowth/standards/en/.

# Index

*Boxes, figures, notes, and tables are indicated by b, f, n, or t following the page number.*